Lecture Notes in Artificial Intelligence   12322

Subseries of Lecture Notes in Computer Science

More information about this series at http://www.springer.com/series/1244

Jesse Davis · Karim Tabia (Eds.)

Scalable Uncertainty Management

14th International Conference, SUM 2020
Bozen-Bolzano, Italy, September 23–25, 2020
Proceedings

 Springer

Editors
Jesse Davis (iD)
KU Leuven
Heverlee, Belgium

Karim Tabia (iD)
Artois University
Lens, France

ISSN 0302-9743 ISSN 1611-3349 (electronic)
Lecture Notes in Artificial Intelligence
ISBN 978-3-030-58448-1 ISBN 978-3-030-58449-8 (eBook)
https://doi.org/10.1007/978-3-030-58449-8

LNCS Sublibrary: SL7 – Artificial Intelligence

This Springer imprint is published by the registered company Springer Nature Switzerland AG
The registered company address is: Gewerbestrasse 11, 6330 Cham, Switzerland

Preface

This volume contains papers from the 14th International Conference on Scalable Uncertainty Management (SUM 2020). Established in 2007, the SUM conferences are annual events which aim to gather researchers with a common interest in managing and analyzing imperfect information from a wide range of fields, such as Artificial Intelligence and Machine Learning, Databases, Information Retrieval and Data Mining, the Semantic Web, and Risk Analysis. It aims to foster collaboration and cross-fertilization of ideas from these different communities.

SUM 2020 was initially planned to be held in Bolzano, Italy, during September 23–25, 2020. Moreover, it was supposed to take place in the context of the Bolzano Summer of Knowledge, which aimed to bring together researchers from multiple different disciplines such as Philosophy, Knowledge Representation, Logic, Conceptual Modeling and Ontology Engineering, Biology, Medicine, Cognitive Science, and Neuroscience. The idea was to have several weeks of conferences, workshops, and summer schools related to these areas, all with an emphasis on exploring themes around knowledge. Unfortunately, the COVID-19 pandemic forced the postponement of this event. Therefore, SUM 2020 was changed to a fully virtual conference.

Prior to the conference, SUM 2020 solicited three types of paper submissions. Long papers could report on original research, or provide a survey that synthesizes some current research trends. Short papers could be about promising work in progress, system descriptions, or position papers on controversial issues. Finally, extended abstracts could report on recently published work in a relevant journal or top-tier conference. A special feature of SUM 2020 was the addition of a PhD track for papers where the first author was a PhD student. We received 30 submissions, all of which were reviewed by at least three members of the Technical Program Committee. On the basis of these reviews, 12 submissions were accepted as long papers and 9 as short papers.

The conference also included two invited talks. The first invited speaker was Gabriella Pasi from the University of Milano-Bicocca, Italy. She gave a talk on "Assessing Information Credibility in the Social Web." The second invited speaker was V. S. Subrahmanian from Dartmouth College, USA, and his talk was on "Deception, Deterrence and Security."

Additionally, there were five invited tutorials. Vaishak Belle from the University of Edinburgh and the Alan Turing Institute, UK, gave a tutorial entitled "Symbolic Logic meets Machine Learning: A Brief Survey in Infinite Domains." Leopoldo Bertossi from Adolfo Ibáñez University, Chile, gave a tutorial entitled "Score-Based Explanations in Data Management and Machine Learning." Davide Ciucci from the University of Milano-Bicocca, Italy, gave a tutorial on "Rough sets." Frédéric Pichon from Artois University, France, spoke about "Information fusion using belief functions: source quality and conflict." Finally, Steven Schockaert from Cardiff University, UK, spoke about "Representing Knowledge using Vector Space Embeddings." The tutorial

authors also had a chance to submit a 14-pages paper, that was reviewed by the
Program Committee co-chairs, to be included in these proceedings. Vaishak Belle and
Leopoldo Bertossi have such a paper.

There are a number of people we would like to thank for their support in preparing
this conference. Our appreciation is particularly warranted this year, due to the additional stresses, uncertainties, and complications posed by the worldwide COVID-19
pandemic. Firstly, we would like to thank Rafael Peñaloza who was initially in charge
of local arrangements and maintaining the conference's web presence. He then pivoted
towards helping coordinate the online component of the conference. Secondly, we
would like to thank the SUM Steering Committee, which was chaired by Henri Prade.
They provided invaluable advice along the way by proposing potential tutorial speakers
and helping us navigate the transition from a physical to virtual conference. Thirdly, we
would like to thank the members of the Technical Program Committee for providing
high-quality and timely reviews. We would like also to thank all authors who submitted
papers to the conference. Finally, we are very grateful to Springer for sponsoring the
Best Student Paper Award as well as for the ongoing support of its staff in publishing
this volume.

July 2020   Jesse Davis
                                                                  Karim Tabia

Organization

General Chair

Rafael Peñaloza   University of Milano-Bicocca, Italy

General Chair of the Bolzano Summer of Knowledge

Oliver Kutz   Free University of Bozen-Bolzano, Italy

Steering Committee

Salem Benferhat   Artois University, France
Didier Dubois   IRIT-CNRS, France
Lluis Godo   Artificial Intelligence Research Institute, IIIA-CSIC, Spain
Eyke Hüllermeier   Universität Paderborn, Germany
Anthony Hunter   University College London, UK
Henri Prade   IRIT-CNRS, France
Steven Schockaert   Cardiff University, UK
V. S. Subrahmanian   Dartmouth College, USA

Program Committee Chairs

Jesse Davis   KU Leuven, Belgium
Karim Tabia   Artois University, France

Program Committee

Hadj Ali Allel   LIAS, ENSMA, France
Alessandro Antonucci   IDSIA, Switzerland
Jessa Bekker   KU Leuven, Belgium
Nahla Ben Amor   Institut Supérieur de Gestion de Tunis, Tunisia
Salem Benferhat   Artois University, France
Leopoldo Bertossi   Adolfo Ibáñez University, Data Observatory Foundation, IMFD Santiago, Chile
Fernando Bobillo   University of Zaragoza, Spain
Imen Boukhris   Université de Tunis, ISG Tunis, Tunisia
Davide Ciucci   University of Milano-Bicocca, Italy
Thierry Denoeux   Universite de Technologie de Compiègne, France
Sébastien Destercke   CNRS, UMR Heudiasyc, France
Zied Elouedi   Institut Supérieur de Gestion de Tunis, Tunisia
Rainer Gemulla   Universität Mannheim, Germany

Lluis Godo	Artificial Intelligence Research Institute, IIIA-CSIC, Spain
John Grant	Towson University, USA
Manuel Gómez-Olmedo	University de Granada, Spain
Arjen Hommersom	Open University of the Netherlands, The Netherlands
Angelika Kimmig	Cardiff University, UK
Eric Lefèvre	Artois University, France
Philippe Leray	Nantes University, France
Sebastian Link	The University of Auckland, New Zealand
Thomas Lukasiewicz	University of Oxford, UK
Silviu Maniu	Université Paris-Sud, France
Serafin Moral	University de Granada, Spain
Francesco Parisi	DIMES, University of Calabria, Italy
Nico Potyka	Universität Osnabrueck, IKW, Germany
Henri Prade	IRIT-CNRS, France
Andrea Pugliese	University of Calabria, Italy
Benjamin Quost	Université de Technologie de Compiègne, France
Steven Schockaert	Cardiff University, UK
Umberto Straccia	ISTI-CNR, Italy
Andrea Tettamanzi	University of Nice Sophia Antipolis, France
Matthias Thimm	Universität Koblenz-Landau, Germany
Pietro Totis	KU Leuven, Belgium
Barbara Vantaggi	Sapienza University of Rome, Italy
Maurice van Keulen	University of Twente, The Netherlands

Additional Reviewers

Francesca Mangili
Tjitze Rienstra

Abstracts of Invited Talks

Assessing Information Credibility in the Social Web

Gabriella Pasi

Università degli Studi di Milano Bicocca, Italy
gabriella.pasi@unimib.it

Abstract. In the context of the Social Web, where a large amount of User Generated Content is diffused through Social Media without any form of trusted external control, the risk of running into misinformation is not negligible. For this reason, the issue of assessing the credibility of "potential" information is of increasing interest and importance. In the last few years several approaches have been proposed to automatically assess the credibility of UCG in Social Media. Most are data-driven approaches, based on machine learning techniques, but recently model-driven approaches are also being investigated, in particular, approaches relying on the Multi Criteria Decision Making paradigm. In this talk an overview of the approaches aimed at tackling this problem are addressed, with particular emphasis on model driven approaches; their application to specific problems will also be addressed.

Deception, Deterrence and Security

V. S. Subrahmanian

Department of Computer Science, Institute for Security, Technology, and Society
Dartmouth College, Hanover, NH 03755
vs@dartmouth.edu

Abstract. Deception is at the heart of many security issues. For instance, phishing and spear-phishing attacks use deception. So do man in the middle attacks in which, for instance, a fake cell tower deceives individual mobile devices to connect to them. However, deception can also be used for "good" in order to inject uncertainty and inflict costs on a malicious adversary. In this talk, I will go over 2 major case studies involving deception for good which have a deterrent effect on a malicious adversary. In the first, I will discuss how selective disclosure of probabilistic logic-based behavioral models can help shape the actions of terrorist groups, making their behavior more predictable (for us) and hence more defendable. In a second application, this time in cybersecurity, I will show methods and a prototype system to inflict costs on an adversary who steals valuable intellectual property by populating a network with automatically generated fake documents that masquerade as intellectual property.

Contents

Tutorial Papers

Symbolic Logic Meets Machine Learning:
A Brief Survey in Infinite Domains

Vaishak Belle[1,2(⊠)]

[1] University of Edinburgh, Edinburgh, UK
[2] Alan Turing Institute, London, UK
vaishak@ed.ac.uk

Abstract. The tension between deduction and induction is perhaps the most fundamental issue in areas such as philosophy, cognition and artificial intelligence (AI). The deduction camp concerns itself with questions about the expressiveness of formal languages for capturing knowledge about the world, together with proof systems for reasoning from such knowledge bases. The learning camp attempts to generalize from examples about partial descriptions about the world. In AI, historically, these camps have loosely divided the development of the field, but advances in cross-over areas such as statistical relational learning, neuro-symbolic systems, and high-level control have illustrated that the dichotomy is not very constructive, and perhaps even ill-formed.

In this article, we survey work that provides further evidence for the connections between logic and learning. Our narrative is structured in terms of three strands: logic versus learning, machine learning for logic, and logic for machine learning, but naturally, there is considerable overlap. We place an emphasis on the following "sore" point: there is a common misconception that logic is for discrete properties, whereas probability theory and machine learning, more generally, is for continuous properties. We report on results that challenge this view on the limitations of logic, and expose the role that logic can play for learning in infinite domains.

1 Introduction

The tension between *deduction* and *induction* is perhaps the most fundamental issue in areas such as philosophy, cognition and artificial intelligence (AI). The deduction camp concerns itself with questions about the expressiveness of formal languages for capturing knowledge about the world, together with proof systems for reasoning from such knowledge bases. The learning camp attempts to generalize from examples about partial descriptions about the world. In AI, historically, these camps have loosely divided the development of the field, but advances in cross-over areas such as *statistical relational learning* [38,83], *neuro-symbolic systems* [28,37,60], and *high-level control* [50,59] have illustrated that the dichotomy is not very constructive, and perhaps even ill-formed. Indeed, logic emphasizes high-level reasoning, and encourages structuring the world in terms of objects, properties, and relations. In contrast, much of the inductive machinery

The author was supported by a Royal Society University Research Fellowship. He is grateful to Ionela G. Mocanu, Paulius Dilkas and Kwabena Nuamah for their feedback.

© Springer Nature Switzerland AG 2020
J. Davis and K. Tabia (Eds.): SUM 2020, LNAI 12322, pp. 3–16, 2020.
https://doi.org/10.1007/978-3-030-58449-8_1

assume random variables to be independent and identically distributed, which can be problematic when attempting to exploit symmetries and causal dependencies between groups of objects. But the threads connecting logic and learning go deeper, far beyond the apparent flexibility that logic offers for modeling relations and hierarchies in noisy domains. At a conceptual level, for example, although there is much debate about what precisely commonsense knowledge might look like, it is widely acknowledged that concepts such as time, space, abstraction and causality are essential [68, 98]. In that regard, (classical, or perhaps non-classical) logic can provide the formal machinery to reason about such concepts in a rigorous way. At a pragmatic level, despite the success of methods such as deep learning, it is now increasingly recognized that owing to a number of reasons, including model re-use, transferability, causal understanding, relational abstraction, explainability and data efficiency, those methods need to be further augmented with logical, symbolic and/or programmatic artifacts [17, 35, 97]. Finally, for building intelligent agents, it is recognized that low-level, data-intensive, reactive computations needs to be tightly integrated with high-level, deliberative computations [50, 59, 67], the latter possibly also engaging in hypothetical and counterfactual reasoning. Here, a parallel is often drawn to Kahneman's so-called *System 1* versus *System 2* processing in human cognition [51], in the sense that experiential and reactive processing (learned behavior) needs to be coupled with cogitative processing (reasoning, deliberation and introspection) for sophisticated machine intelligence.

The purpose of this article is not to resolve this debate, but rather provide further evidence for the connections between logic and learning. In particular, our narrative is inspired by a recent symposium on logic and learning [13], where the landscape was structured in terms of three strands:

1. *Logic vs. Machine Learning, including the study of problems that can be solved using either logic-based techniques or via machine learning, ...;*
2. *Machine Learning for Logic, including the learning of logical artifacts, such as formulas, logic programs, ...; and*
3. *Logic for Machine Learning, including the role of logics in delineating the boundary between tractable and intractable learning problems, ..., and the use of logic as a declarative framework for expressing machine learning constructs.*

In this article, we particularly focus on the following "sore" point: there is a common misconception that logic is for discrete properties, whereas probability theory and machine learning, more generally, is for continuous properties. It is true that logical formulas are discrete structures, but they can very easily also express properties about countably infinite or even uncountably many objects. Consequently, in this article we survey some recent results that tackle the integration of logic and learning in infinite domains. In particular, in the context of the above three strands, we report on the following developments. On (1), we discuss approaches for logic-based probabilistic inference in continuous domains. On (2), we cover approaches for learning logic programs in continuous domains, as well as learning formulas that represent countably infinite sets of objects. Finally, on (3), we discuss attempts to use logic as a declarative framework for common tasks in machine learning over discrete and continuous features, as well as using logic as a meta-theory to consider notions such as the *abstraction* of a probabilistic model.

We remark that this survey is undoubtedly a biased view, as the area of research is large, but we do attempt to briefly cover the major threads. Readers are encouraged to refer to discussions in [13,38,83], among others, to get a sense of the breadth of the area.

2 Logic vs. Machine Learning

To appreciate the role and impact of logic-based solvers for machine learning systems, it is perhaps useful to consider the core computational problem underlying (probabilistic) machine learning: the problem of inference, including evaluating the partition function (or conditional probabilities) of a probabilistic graphical model such as a Bayesian network.

When leveraging Bayesian networks for machine learning tasks [56], the networks are often learned using local search to maximize a likelihood or a Bayesian quantity. For example, given data $\mathcal{D}$ and the current guess for the network $\mathcal{N}$, we might estimate the "goodness" of the guess by means of a score: $score(\mathcal{N}, \mathcal{D}) \propto \log \Pr(\mathcal{D} \mid \mathcal{N}) - size(\mathcal{N})$. That is, we want to maximize the fit of the data wrt the current guess, but we would like to penalize the model complexity, to avoid overfitting. Then, we would opt for a second guess $\mathcal{N}'$ only if $score(\mathcal{N}', \mathcal{D}) > score(\mathcal{N}, \mathcal{D})$. Needless to say, even with a reasonable local search procedure, the most significant computational effort here is that of probabilistic inference.

Reasoning in such networks becomes especially challenging with logical syntax. The prevalence of large-scale social networks, machine reading domains, and other types of relational knowledge bases has led to numerous formalisms that borrow the syntax of predicate logic for probabilistic modeling [78,81,85,93]. This has led to a large family of solvers for the *weighted model counting* (WMC) problem [20,39]. The idea is this: given a Bayesian network, a relational Bayesian network, a factor graph, or a probabilistic program [84], one considers an encoding of the formalism as a *weighted propositional theory*, consisting of a propositional theory Δ and a weight function w that maps atoms in Δ to $\mathbb{R}^+$. Recall that SAT is the problem of finding an assignment to such a Δ, whereas #SAT counts the number of assignments for Δ. WMC extends #SAT by computing the sum of the weights of all assignments: that is, given a set of models $\mathcal{M}(\Delta) = \{M \mid M \models \Delta\}$, we evaluate the quantity $W(\Delta) = \sum_{M \in \mathcal{M}(\Delta)} w(M)$ where $w(M)$ is factorized in terms of the atoms true at M. To obtain the conditional probability of a query q against evidence e (wrt the theory Δ), we define $\Pr(q \mid e) = W(\Delta \wedge q \wedge e)/W(\Delta \wedge e)$.

The popularity of WMC can be explained as follows. Its formulation elegantly decouples the logical or symbolic representation from the numeric representation, which is encapsulated in the weight function. When building solvers, this allows us to reason about logical equivalence and reuse SAT solving technology (such as constraint propagation and clause learning). WMC also makes it more natural to reason about deterministic, hard constraints in a probabilistic context [20]. Both exact solvers, based on knowledge compilation [23], as well as approximate solvers [19] have emerged in the recent years, as have lifted techniques [95] that exploit the relational syntax during inference (but in a finite domain setting). For ideas on generating such representations randomly to assess scalability and compare inference algorithms, see [29], for example.

On the point of modelling finite vs infinite properties, note that owing to the underlying propositional language, the formulation is limited to discrete random variables. A similar observation can be made for SAT, which for the longest time could only be applied in discrete domains. This changed with the increasing popularity of *satisfiability modulo theories* (SMT) [4], which enable us to, for example, reason about the satisfiability of linear constraints over the rationals. Extending earlier insights on piecewise-polynomial weight functions [88,89], the formulation of *weighted model integration* (WMI) was proposed in [12]. WMI extends WMC by leveraging the idea that SMT theories can represent mixtures of Boolean and continuous variables: for example, a formula such as $p \wedge (x > 5)$ denotes the logical conjunction of a Boolean variable p and a real-valued variable x taking values greater than 5. For every assignment to the Boolean and continuous variables, the WMI problem defines a weight. The total WMI is computed by integrating these weights over the domain of solutions to Δ, which is a mixed discrete-continuous (or simply *hybrid*) space. Consider, for example, the special case when Δ has no Boolean variables, and the weight of every model is 1. Then, the WMI simplifies to computing the volume of the polytope encoded in Δ. When we additionally allow for Boolean variables in Δ, this special case becomes the hybrid version of #SAT, known as #SMT [21]. Since that proposal, numerous advances have been made on building efficient WMI solvers (e.g., [69,74,99]) including the development of compilation targets [53,54,100].

Note that WMI proposes an extension of WMC for uncountably infinite (i.e., continuous) domains. What about countably infinite domains? The latter type is particularly useful for reasoning in (general) first-order settings, where we may say that a property such as $\forall x, y, z(parent(x, y) \wedge parent(y, z) \supset grandparent(x, z))$ applies to every possible x, y and z. Of course, in the absence of the finite domain assumption, reasoning in the first-order setting suffers from undecidability properties, and so various strategies have emerged for reasoning about an *open universe* [87]. One popular approach is to perform *forward reasoning*, where samples needed for probability estimation are obtained from the facts and declarations in the probabilistic model [45,87]. Each such sample corresponds to a possible world. But there may be (countably or uncountably) infinitely many worlds, and so exact inference is usually sacrificed. A second approach is to restrict the model wrt the query and evidence atoms and define estimation from the resulting finite sub-model [41,70,90], which may also be substantiated with exact inference in special cases [6,7].

Given the successes of logic-based solvers for inference and probability estimation, one might wonder whether such solvers would also be applicable to learning tasks in models with relational features and hard, deterministic constraints? These, in addition to other topics, are considered in the next section.

3   Machine Learning for Logic

At least since the time of Socrates, inductive reasoning has been a core issue for the logical worldview, as we need a mechanism for obtaining axiomatic knowledge. In that regard, the learning of logical and symbolic artifacts is an important issue in AI, and computer science more generally [43]. There is a considerable body of work on learning propositional and relational formulas, and in context of probabilistic information, learning weighted formulas [13,26,75,83]. Approaches can be broadly lumped together as follows.

1. *Entailment-based scoring:* Given a logical language $\mathcal{L}$, background knowledge $\mathcal{B} \subset \mathcal{L}$, examples $\mathcal{D}$ (usually a set of $\mathcal{L}$-atoms), find a hypothesis $\mathcal{H} \in \overline{\mathcal{H}}, \mathcal{H} \subset \mathcal{L}$ such that $\mathcal{B} \cup \mathcal{H}$ entail the instances in $\mathcal{D}$. Here, the set $\overline{\mathcal{H}}$ places restrictions of the syntax of $\mathcal{H}$ so as to control model complexity and generalization. (For example, $\mathcal{H} = \mathcal{D}$ is a trivial hypothesis that satisfies the entailment stipulation.)
2. *Likelihood-based scoring:* Given $\mathcal{L}$ and $\mathcal{D}$ as defined above, find $\mathcal{H} \subset \mathcal{L}$ such that $score(\mathcal{H}, \mathcal{D}) > score(\mathcal{H}', \mathcal{D})$ for every $\mathcal{H}' \neq \mathcal{H}$. As discussed before, we might define $score(\mathcal{H}, \mathcal{D}) \propto \log \Pr(\mathcal{D} \mid \mathcal{H}) - size(\mathcal{H})$. Here, like $\overline{\mathcal{H}}$ above, $size(\mathcal{H})$ attempts to the control model complexity and generalization.

Many recipes based on these schemes are possible. For example, we may use entailment-based inductive synthesis for an initial estimate of the hypothesis, and then resort to Bayesian scoring models [85]. The synthesis step might invoke neural machinery [35]. We might not require that the hypothesis entails every example in $\mathcal{D}$ but only the largest consistent subset, which is sensible when we expect the examples to be noisy [26]. We might compile $\mathcal{B}$ to an efficient data structure, and perform likelihood-based scoring on that structure [63], and so $\mathcal{B}$ could be seen as deterministic domain-specific constraints. Finally, we might stipulate the conditions under which a "correct" hypothesis may be inferred wrt unknown ground truth, only a subset of which is provided in $\mathcal{D}$. This is perhaps best represented by the (probably approximately correct) PAC-semantics that captures the quality possessed by the output of learning algorithm whilst costing for the number of examples that need to be observed [22,94]. (But other formulations are also possible, e.g., [42].)

This discussion pertained to finite domains. What about continuous spaces? By means of arithmetic fragments and formulations like WMI, it should be clear that it now becomes possible to extend the above schemes to learn continuous properties. For example, one could learn linear expressions from data [55]. For an account that also tries to evaluate a hypothesis that is correct wrt unknown ground truth, see [72]. If the overall objective is to obtain a distribution of the data, other possibilities present themselves. In [77], for example, real-valued data points are first lumped together to obtain atomic continuous random variables. From these, relational formulas are constructed so as to yield hybrid probabilistic programs. The learning is based on likelihood scoring. In [91], the real-valued data points are first intervalized, and polynomials are learned for those intervals based on likelihood scoring. These weighted atoms are then used for learning clauses by entailment judgements [26].

Such ideas can also be extended to data structures inspired by knowledge compilation, often referred to as *circuits* [20,82]. Knowledge compilation [25] arose as a way to represent logical theories in a manner where certain kinds of computations (e.g., checking satisfiability) is significantly more effective, often polynomial in the size of the circuit. In the context of probabilistic inference, the idea was to then position probability estimation to also be computable in time polynomial in the size of the circuit [20,82]. Consequently, (say) by means of likelihood-based scoring, the learning of circuits is particularly attractive because once learned, the bottleneck of inference is alleviated [63,66]. In [15,73], along the lines of the work above on learning logical formulas in continuous domains, it is shown that the learning of circuits can also be coupled with WMI.

What about countably infinite domains? In most pragmatic instances of learning logical artifacts, the difference between the uncountable and countably infinite setting is this: in the former, we see finitely many real-valued samples as being drawn from an (unknown) interval, and we could inspect these samples to crudely infer a lower and upper bound. In the latter, based on finitely many relational atoms, we would need to infer a universally quantified clause, such as $\forall x, y, z(parent(x, y) \land parent(y, z) \supset grandparent(x, z))$. If we are after a hypothesis that is simply guaranteed to be consistent wrt the observed examples, then standard rule induction strategies would suffice [75], and we could interpret the rules as quantifying over a countably infinite domain. But this is somewhat unsatisfactory, as there is no distinction between the rules learned in the standard finite setting and its supposed applicability to the infinite setting. What is really needed is an analysis of what rule learning would mean wrt the infinitely many examples that have *not* been observed. This was recently considered via the PAC-semantics in [10], by appealing to ideas on reasoning with open universes discussed earlier [6].

Before concluding this section, it is worth noting that although the above discussion is primarily related to the learning of logical artifacts, it can equivalently be seen as a class of machine learning methods that leverage symbolic domain knowledge [30]. Indeed, logic-based probabilistic inference over deterministic constraints, and entailment-based induction augmented with background knowledge are instances of such a class. Analogously, the automated construction of relational and statistical knowledge bases [18,79] by combining background knowledge with extracted tuples (obtained, for example, by applying natural language processing techniques to large textual data) is another instance of such a class.

In the next section, we will consider yet another way in which logical and symbolic artifacts can influence learning: we will see how such artifacts are useful to enable tractability, correctness, modularity and compositionality.

4   Logic for Machine Learning

There are two obvious ways in which a logical framework can provide insights on machine learning theory. First, consider that computational tractability is of central concern when applying logic in computer science, knowledge representation, database theory and search [62,65,71]. Thus, the natural question to wonder is whether these ideas would carry over to probabilistic machine learning. On the one hand, probabilistic extensions to tractable knowledge representation frameworks could be considered [57]. But on the other, as discussed previously, ideas from knowledge compilation, and the use of circuits, in particular, are proving very effective for designing tractable paradigms for machine learning. While there has always been an interest in capturing tractable distributions by means of low tree-width models [2], knowledge compilation has provided a way to also represent high tree-width models and enable exact inference for a range of queries [63,82]. See [24] for a comprehensive view on the use of knowledge compilation for machine learning.

The other obvious way logic can provide insights on machine learning theory is by offering a formal apparatus to reason about *context*. Machine learning problems are often positioned as atomic tasks, such as a classification task where regions of images need to be labeled as cats or dogs. However, even in that limited context, we imagine the resulting classification system as being deployed as part of a larger system, which includes various modules that communicate or interface with the classification system. We imagine an implicit accountability to the labelling task in that the detected object is either a cat or a dog, but not both. If there is information available that all the entities surrounding the object of interest have been labelled as lions, we would want to accord a high probability to the object being a cat, possibly a wild cat. There is a very low chance of the object being a dog, then. If this is part of a vision system on a robot, we should ensure that the robot never tramples on the object, regardless of whether it is a type of cat or a dog. To inspect such patterns, and provide meta-theory for machine learning, it can be shown that symbolic, programmatic and logical artifacts are enormously useful. We will specifically consider correctness, modularity and compositionality to explore the claim.

On the topic of correctness, the classical framework in computer science is *verification*: can we provide a formal specification of what is desired, and can the system be checked against that specification? In a machine learning context, we might ask whether the system, during or after training, satisfies a specification. The specification here might mean constraints about the physical laws of the domain, or notions of perturbation in the input space while ensuring that the labels do not change, or insisting that the prediction does not label an object as being both a cat and a dog, or otherwise ensuring that outcomes are not subject to, say, gender bias. Although there is a broad body of work on such issues, touching more generally on *trust* [86], we discuss approaches closer to the thrust of this article. For example, [49] show that a trained neural network can be verified by means of an SMT encoding of the network. In recent work, [96] show that the loss function of deep learning systems can be adjusted to logical constraints by insisting that the distribution on the predictions is proportional to the weighted model count of those constraints. In [63], prior (logical) constraints are compiled to a circuit to be used for probability estimation. In [80], circuits are shown to be amenable to training against probabilistic and causal prior constraints, including assertions about fairness, for example.

In [32, 67], a somewhat different approach to respecting domain constraints is taken: the low-level prediction is obtained as usual from a machine learning module, which is then interfaced with a probabilistic relational language and its symbolic engine. That is, the reasoning is positioned to be tackled directly by the symbolic engine. In a sense, such approaches cut across the three strands: the symbolic engine uses weighted model counting, the formulas in the language could be obtained by (say) entailment based scoring, and the resulting language supports modularity and compositionality (discussed below).

While there is not much to be said about the distinction between finite vs infinite wrt correctness, many of these ideas are likely amenable to extensions to an infinite setting in the ways discussed in the previous sections (e.g., considering constraints of a continuous or a countably infinite nature).

On the topic of modularity, recall that the general idea is to reduce, simplify or otherwise abstract a (probabilistic) computation as an atomic entity, which is then to be referenced in another, possibly more complex, entity. In standard programming languages, this might mean the compartmentalization and interrelation of computational entities. For machine learning, approaches such as probabilistic programming [27,40] support probabilistic primitives in the language, with the intention of making learning modules re-usable and modular. It can be shown, for example, that the computational semantics of some of these languages reduce to WMC [36,48]. Thus, in the infinite case, a corresponding reduction to WMI follows [1,31,91].

A second dimension to modularity is the notion of *abstraction*. Here, we seek to model, reason and explain the behavior of systems in a more tractable search space, by omitting irrelevant details. The idea is widely used in natural and social sciences. Think of understanding the political dynamics of elections by studying micro level phenomena (say, voter grievances in counties) versus macro level events (e.g., television advertisements, gerrymandering). In particular, in computer science, it is often understood as the process of mapping one representation onto a simpler representation by suppressing irrelevant information. In fact, integrating low-level behavior with high-level reasoning, exploiting relational representations to reduce the number of inference computations, and many other search space reduction techniques can all loosely be seen as instances of abstraction [8].

While there has been significant work on abstraction in deterministic systems [3], for machine learning, however, a probabilistic variant is clearly needed. In [47], an account of abstraction for loop-free propositional probabilistic programs is provided, where certain parts of the program (possibly involving continuous properties) can be reduced to a Bernoulli random variable. For example, suppose every occurrence of the continuous random variable x, drawn uniformly on the interval $[0,1]$, in a program is either of the form $x \leq 7$ or of the form $x > 7$. Then, we could use a discrete random variable b with a 0.7 probability of being true to capture $x \leq 7$; and analogously, $\neg b$ to capture $x > 7$. The resulting program is likely to be simpler. In [8], an account of abstraction for probabilistic relational models is considered, where the notion of abstraction also extends to deterministic constraints and complex formulas. For example, a single probabilistic variable in the abstracted model could denote a complex logical formula in the original model. Moreover, the logical properties that enable verifying and inducing abstractions are also considered, and it is shown how WMC is sufficient for the computability of these properties (also see [48]).

Incidentally, abstraction brings to light a reduction between finite vs infinite: it is shown in [8] that the modelling of piecewise densities as weighted propositions, which is leveraged in WMI [12,31], is a simple case of the more general account. Therefore, it is worthwhile to investigate whether this or other accounts of abstraction could emerge as general-purpose tools that allow us to inspect the conditions under which infinitary statements reduce to finite computations.

A broader point here is the role abstraction might play in generating explanations [44]. For example, a user's understanding of the domain is likely to be different from the low-level data that a machine learning system interfaces with [92], and so, abstractions can capture these two levels in a formal way.

Finally, we turn to the topic of compositionality, which, of course, is closely related to modularity in that we want to distinct modules to come together to form a complex composition. Not surprisingly, this is of great concern in AI, as it is widely acknowledged that most AI systems will involve heterogeneous components, some of which may involve learning from data, and others reasoning, search and symbol manipulation [68]. In continuation with the above discussion, probabilistic programming is one such endeavor that purports to tackle this challenge by allowing modular components to be composed over programming and/or logical connectives [5,11,16,27,32,40,46,67,76,85]. (See [34,64,71] for ideas in deterministic systems.) However, probabilistic programming only composes probabilistic computations, but does not offer an obvious means to capture other types of search-based computations, such as SAT, and integer and convex programming.

Recall that the computational semantics of probabilistic programs reduces to WMC [36,48]. Following works such as [14,33], an interesting observation made in [52] is that by appealing to a sum of products computation over different semiring structures, we can realize a large number of tasks such as satisfiability, unweighted model counting, sensitivity analysis, gradient computations, in addition to WMC. It was then shown in [9] that the idea could be generalized further for infinite domains: by defining a measure on first-order models, WMI and convex optimization can also be captured. As the underlying language is a logical one, composition can already be defined using logical connectives. But an additional, more involved, notion of composition is also proposed, where a sum of products over different semirings can be concatenated. To reiterate, the general idea behind these proposals [9,33,52] is to arrive at a principled paradigm that allows us to interface learned modules with other types of search and optimization computations for the compositional building of AI systems. See also [58] for analogous discussions, but where a different type of coupling for the underlying computations is suggested. Overall, we observed that a formal apparatus (symbolic, programmatic and logical artifacts) help us define such compositional constructions by providing a meta-theory.

5 Conclusions

In this article, we surveyed work that provides further evidence for the connections between logic and learning. Our narrative was structured in terms of three strands: logic versus learning, machine learning for logic, and logic for machine learning, but naturally, there was considerable overlap.

We covered a large body of work on what these connections look like, including, for example, pragmatic concerns such as the use of hard, domain-specific constraints and background knowledge, all of which considerably eases the requirement that all of the agent's knowledge should be derived from observations alone. (See discussions in [61] on the limitations of learned behavior, for example.) Where applicable, we placed an emphasis on how extensions to infinite domains are possible. In the very least, logical artifacts can help in constraining, simplifying and/or composing machine learning entities, and in providing a principled way to study the underlying representational and computational issues.

In general, this type of work could help us move beyond the narrow focus of the current learning literature so as to deal with time, space, abstraction, causality, quantified generalizations, relational abstractions, unknown domains, unforeseen examples, among other things, in a principled fashion. In fact, what is being advocated is the tackling of problems that symbolic logic and machine learning might struggle to address individually. One could even think of the need for a recursive combination of strands 2 and 3: purely reactive components interact with purely cogitative elements, but then those reactive components are learned against domain constraints, and the cogitative elements are induced from data, and so on. More broadly, making progress towards a formal realization of *System 1* versus *System 2* processing might also contribute to our understanding of human intelligence, or at least capture human-like intelligence in automated systems.

References

1. Albarghouthi, A., D'Antoni, L., Drews, S., Nori, A.V.: Quantifying program bias. CoRR, abs/1702.05437 (2017)
2. Bach, F.R., Jordan, M.I.: Thin junction trees. In: Advances in Neural Information Processing Systems, pp. 569–576 (2002)
3. Banihashemi, B., De Giacomo, G., Lespérance, Y.: Abstraction in situation calculus action theories. In: AAAI, pp. 1048–1055 (2017)
4. Barrett, C., Sebastiani, R., Seshia, S.A., Tinelli, C.: Satisfiability modulo theories. In: Handbook of Satisfiability, chap. 26, pp. 825–885. IOS Press (2009)
5. Belle, V.: Logic meets probability: towards explainable AI systems for uncertain worlds. In: IJCAI (2017)
6. Belle, V.: Open-universe weighted model counting. In: AAAI, pp. 3701–3708 (2017)
7. Belle, V.: Weighted model counting with function symbols. In: UAI (2017)
8. Belle, V.: Abstracting probabilistic models: relations, constraints and beyond. Knowl.-Based Syst. **199**, 105976 (2020). https://www.sciencedirect.com/science/article/abs/pii/S0950705120302914
9. Belle, V., De Raedt, L.: Semiring programming: a declarative framework for generalized sum product problems. In: AAAI Workshop: Statistical Relational Artificial Intelligence (2020)
10. Belle, V., Juba, B.: Implicitly learning to reason in first-order logic. In: Advances in Neural Information Processing Systems, pp. 3376–3386 (2019)
11. Belle, V., Levesque, H.J.: Allegro: belief-based programming in stochastic dynamical domains. In: IJCAI (2015)
12. Belle, V., Passerini, A., Van den Broeck, G.: Probabilistic inference in hybrid domains by weighted model integration. In: IJCAI, pp. 2770–2776 (2015)
13. Benedikt, M., Kersting, K., Kolaitis, P.G., Neider, D.: Logic and learning (dagstuhl seminar 19361). Schloss Dagstuhl-Leibniz-Zentrum fuer Informatik (2020)
14. Bistarelli, S., Montanari, U., Rossi, F.: Semiring-based constraint logic programming: syntax and semantics. TOPLAS **23**(1), 1–29 (2001)
15. Bueff, A., Speichert, S., Belle, V.: Tractable querying and learning in hybrid domains via sum-product networks. In: KR Workshop on Hybrid Reasoning (2018)
16. Bundy, A., Nuamah, K., Lucas, C.: Automated reasoning in the age of the internet. In: Fleuriot, J., Wang, D., Calmet, J. (eds.) AISC 2018. LNCS (LNAI), vol. 11110, pp. 3–18. Springer, Cham (2018). https://doi.org/10.1007/978-3-319-99957-9_1

17. Bunel, R., Hausknecht, M., Devlin, J., Singh, R., Kohli, P.: Leveraging grammar and reinforcement learning for neural program synthesis. arXiv preprint arXiv:1805.04276 (2018)
18. Carlson, A., Betteridge, J., Kisiel, B., Settles, B., Hruschka Jr., E.R., Mitchell, T.M.: Toward an architecture for never-ending language learning. In: AAAI, pp. 1306–1313 (2010)
19. Chakraborty, S., Fremont, D.J., Meel, K.S., Seshia, S.A., Vardi, M.Y.: Distribution-aware sampling and weighted model counting for SAT. In: AAAI, pp. 1722–1730 (2014)
20. Chavira, M., Darwiche, A.: On probabilistic inference by weighted model counting. Artific. Intell. **172**(6–7), 772–799 (2008)
21. Chistikov, D., Dimitrova, R., Majumdar, R.: Approximate counting in SMT and value estimation for probabilistic programs. TACAS **9035**, 320–334 (2015)
22. Cohen, W.W.: PAC-learning nondeterminate clauses. In: AAAI, pp. 676–681 (1994)
23. Darwiche, A.: New advances in compiling CNF to decomposable negation normal form. In: ECAI, pp. 328–332 (2004)
24. Darwiche, A.: Three modern roles for logic in AI. In: Proceedings of the 39th ACM SIGMOD-SIGACT-SIGAI Symposium on Principles of Database Systems, pp. 229–243 (2020)
25. Darwiche, A., Marquis, P.: A knowledge compilation map. J. Artif. Intell. Res. **17**, 229–264 (2002)
26. De Raedt, L., Dries, A., Thon, I., Van den Broeck, G., Verbeke, M.: Inducing probabilistic relational rules from probabilistic examples. In: Twenty-Fourth International Joint Conference on Artificial Intelligence (2015)
27. De Raedt, L., Kimmig, A.: Probabilistic (logic) programming concepts. Mach. Learn. **100**(1), 5–47 (2015)
28. De Raedt, L., Manhaeve, R., Dumancic, S., Demeester, T., Kimmig, A.: Neuro-symbolic= neural+ logical+ probabilistic. In: NeSy 2019@ IJCAI, The 14th International Workshop on Neural-Symbolic Learning and Reasoning, pp. 1–4 (2019)
29. Dilkas, P., Belle, V.: Generating random logic programs using constraint programming. CoRR, abs/2006.01889 (2020)
30. Domingos, P.: The Master Algorithm: How the Quest for the Ultimate Learning Machine Will Remake Our World. Basic Books (2015)
31. Dos Martires, P.Z., Dries, A., De Raedt, L.: Exact and approximate weighted model integration with probability density functions using knowledge compilation. In: Proceedings of the AAAI Conference on Artificial Intelligence, vol. 33, pp. 7825–7833 (2019)
32. Dries, A., Kimmig, A., Davis, J., Belle, V., De Raedt, L.: Solving probability problems in natural language. In: IJCAI (2017)
33. Eisner, J., Filardo, N.W.: Dyna: extending datalog for modern AI. In: de Moor, O., Gottlob, G., Furche, T., Sellers, A. (eds.) Datalog 2.0 2010. LNCS, vol. 6702, pp. 181–220. Springer, Heidelberg (2011). https://doi.org/10.1007/978-3-642-24206-9_11
34. Ensan, A., Ternovska, E.: Modular systems with preferences. In: IJCAI, pp. 2940–2947 (2015)
35. Evans, R., Grefenstette, E.: Learning explanatory rules from noisy data. J. Artif. Intell. Res. **61**, 1–64 (2018)
36. Fierens, D., Van den Broeck, G., Thon, I., Gutmann, B., De Raedt, L.: Inference in probabilistic logic programs using weighted CNF's. In: UAI, pp. 211–220 (2011)
37. d'Avila Garcez, A., Gori, M., Lamb, L.C., Serafini, L., Spranger, M., Tran, S.N.: Neural-symbolic computing: an effective methodology for principled integration of machine learning and reasoning. arXiv preprint arXiv:1905.06088 (2019)
38. Getoor, L., Taskar, B. (eds.): An Introduction to Statistical Relational Learning. MIT Press, Cambridge (2007)
39. Gomes, C.P., Sabharwal, A., Selman, B.: Model counting. In: Handbook of Satisfiability. IOS Press (2009)

40. Goodman, N.D., Mansinghka, V.K., Roy, D.M., Bonawitz, K., Tenenbaum, J.B.: Church: a language for generative models. In: Proceedings of UAI, pp. 220–229 (2008)
41. Grohe, M., Lindner, P.: Probabilistic databases with an infinite open-world assumption. In: Proceedings of the 38th ACM SIGMOD-SIGACT-SIGAI Symposium on Principles of Database Systems, pp. 17–31 (2019)
42. Grohe, M., Ritzert, M.: Learning first-order definable concepts over structures of small degree. In: 2017 32nd Annual ACM/IEEE Symposium on Logic in Computer Science (LICS), pp. 1–12. IEEE (2017)
43. Gulwani, S.: Dimensions in program synthesis. In: PPDP, pp. 13–24. ACM (2010)
44. Gunning, D.: Explainable artificial intelligence (XAI). Technical report, DARPA/I20 (2016)
45. Gutmann, B., Thon, I., Kimmig, A., Bruynooghe, M., De Raedt, L.: The magic of logical inference in probabilistic programming. Theor. Pract. Logic Program. **11**(4–5), 663–680 (2011)
46. Halpern, J.Y.: Reasoning about Uncertainty. MIT Press (2003)
47. Holtzen, S., Millstein, T.: and G. Van den Broeck. Probabilistic program abstractions, In UAI (2017)
48. Holtzen, S., Van den Broeck, G., Millstein, T.: Dice: compiling discrete probabilistic programs for scalable inference. arXiv preprint arXiv:2005.09089 (2020)
49. Huang, X., Kwiatkowska, M., Wang, S., Wu, M.: Safety verification of deep neural networks. In: Majumdar, R., Kunčak, V. (eds.) CAV 2017. LNCS, vol. 10426, pp. 3–29. Springer, Cham (2017). https://doi.org/10.1007/978-3-319-63387-9_1
50. Kaelbling, L.P., Lozano-Pérez, T.: Integrated task and motion planning in belief space. I. J. Robotic Res. **32**(9–10), 1194–1227 (2013)
51. Kahneman, D.: Thinking, Fast and Slow. Macmillan (2011)
52. Kimmig, A., Van den Broeck, G., De Raedt, L.: Algebraic model counting. J. Appl. Log. **22**, 46–62 (2017)
53. Kolb, S., Mladenov, M., Sanner, S., Belle, V., Kersting, K.: Efficient symbolic integration for probabilistic inference. In: IJCAI (2018)
54. Kolb, S., et al.: The PYWMI framework and toolbox for probabilistic inference using weighted model integration (2019). https://www.ijcai.org/proceedings/2019/
55. Kolb, S., Teso, S., Passerini, A., De Raedt, L.: Learning SMT (LRA) constraints using SMT solvers. In: IJCAI, pp. 2333–2340 (2018)
56. Koller, D., Friedman, N.: Probabilistic Graphical Models - Principles and Techniques. MIT Press (2009)
57. Koller, D., Levy, A., Pfeffer, A.: P-classic: a tractable probablistic description logic. In: Proceedings of the AAAI/IAAI, pp. 390–397 (1997)
58. Kordjamshidi, P., Roth, D., Kersting, K.: Systems AI: a declarative learning based programming perspective. In: IJCAI, pp. 5464–5471 (2018)
59. Lakemeyer, G., Levesque, H.J.: Cognitive robotics. In: Handbook of Knowledge Representation, pp. 869–886. Elsevier (2007)
60. Lamb, L., Garcez, A., Gori, M., Prates, M., Avelar, P., Vardi, M.: Graph neural networks meet neural-symbolic computing: a survey and perspective. arXiv preprint arXiv:2003.00330 (2020)
61. Levesque, H.J.: Common Sense, the Turing Test, and the Quest for Real AI. MIT Press (2017)
62. Levesque, H.J., Brachman, R.J.: Expressiveness and tractability in knowledge representation and reasoning. Comput. Intell. **3**, 78–93 (1987)
63. Liang, Y., Bekker, J., Van den Broeck, G.: Learning the structure of probabilistic sentential decision diagrams. In: Proceedings of the 33rd Conference on Uncertainty in Artificial Intelligence (UAI) (2017)

64. Lierler, Y., Truszczynski, M.: An abstract view on modularity in knowledge representation. In: AAAI, pp. 1532–1538 (2015)
65. Liu, Y., Levesque, H.: Tractable reasoning with incomplete first-order knowledge in dynamic systems with context-dependent actions. In: Proceedings of the IJCAI, pp. 522–527 (2005)
66. Lowd, D., Domingos, P.: Learning arithmetic circuits. In: Proceedings of the 24th Conference in Uncertainty in Artificial Intelligence (UAI), pp. 383–392 (2008)
67. Manhaeve, R., Dumancic, S., Kimmig, A., Demeester, T., De Raedt, L.: Deepproblog: neural probabilistic logic programming. In: Advances in Neural Information Processing Systems, pp. 3749–3759 (2018)
68. Marcus, G., Davis, E.: Rebooting AI: Building Artificial Intelligence We Can Trust. Pantheon (2019)
69. Merrell, D., Albarghouthi, A., D'Antoni, L.: Weighted model integration with orthogonal transformations. In: Proceedings of the Twenty-Sixth International Joint Conference on Artificial Intelligence (2017)
70. Milch, B., Marthi, B., Sontag, D., Russell, S.J., Ong, D.L., Kolobov, A.: Approximate inference for infinite contingent Bayesian networks. In: AISTATS, pp. 238–245 (2005)
71. Mitchell, D.G., Ternovska, E.: A framework for representing and solving NP search problems. In: AAAI, pp. 430–435 (2005)
72. Mocanu, I.G., Belle, V., Juba, B.: Polynomial-time implicit learnability in SMT. In: ECAI (2020)
73. Molina, A., Vergari, A., Di Mauro, N., Natarajan, S., Esposito, F., Kersting, K.: Mixed sum-product networks: a deep architecture for hybrid domains. In: Thirty-Second AAAI Conference on Artificial Intelligence (2018)
74. Morettin, P., Passerini, A., Sebastiani, R.: Advanced SMT techniques for weighted model integration. Artif. Intell. **275**, 1–27 (2019)
75. Muggleton, S., De Raedt, L.: Inductive logic programming: theory and methods. J. Logic Program. **19**, 629–679 (1994)
76. Nitti, D., Belle, V., De Laet, T., De Raedt, L.: Planning in hybrid relational mdps. Mach. Learn. **106**(12), 1905–1932 (2017)
77. Nitti, D., Ravkic, I., Davis, J., Raedt, L.D.: Learning the structure of dynamic hybrid relational models. In: Proceedings of the Twenty-second European Conference on Artificial Intelligence, pp. 1283–1290. IOS Press (2016)
78. Niu, F., Ré, C., Doan, A., Shavlik, J.: Tuffy: scaling up statistical inference in markov logic networks using an rdbms. Proc. VLDB Endowment **4**(6), 373–384 (2011)
79. Niu, F., Zhang, C., Ré, C., Shavlik, J.W.: Deepdive: web-scale knowledge-base construction using statistical learning and inference. VLDS **12**, 25–28 (2012)
80. Papantonis, I., Belle, V.: On constraint definability in tractable probabilistic models. arXiv preprint arXiv:2001.11349 (2020)
81. Poole, D.: First-order probabilistic inference. In: Proceedings of the IJCAI, pp. 985–991 (2003)
82. Poon, H., Domingos, P.: Sum-product networks: a new deep architecture. In: UAI, pp. 337–346 (2011)
83. Raedt, L.D., Kersting, K., Natarajan, S., Poole, D.: Statistical relational artificial intelligence: logic, probability, and computation. Synth. Lect. Artif. Intell. Mach. Learn. **10**(2), 1–189 (2016)
84. Renkens, J., et al.: ProbLog2: from probabilistic programming to statistical relational learning. In: Roy, D., Mansinghka, V., Goodman, N. (eds.) Proceedings of the NIPS Probabilistic Programming Workshop, December 2012. Accepted
85. Richardson, M., Domingos, P.: Markov logic networks. Mach. Learn. **62**(1), 107–136 (2006)

86. Rudin, C., Ustun, B.: Optimized scoring systems: toward trust in machine learning for healthcare and criminal justice. Interfaces **48**(5), 449–466 (2018)
87. Russell, S.J.: Unifying logic and probability. Commun. ACM **58**(7), 88–97 (2015)
88. Sanner, S., Abbasnejad, E.: Symbolic variable elimination for discrete and continuous graphical models. In: AAAI (2012)
89. Shenoy, P., West, J.: Inference in hybrid Bayesian networks using mixtures of polynomials. Int. J. Approximate Reasoning **52**(5), 641–657 (2011)
90. Singla, P., Domingos, P.M.: Markov logic in infinite domains. In: UAI, pp. 368–375 (2007)
91. Speichert, S., Belle, V.: Learning probabilistic logic programs in continuous domains. In: ILP (2019)
92. Sreedharan, S., Srivastava, S., Kambhampati, S.: Hierarchical expertise level modeling for user specific contrastive explanations. In: IJCAI, pp. 4829–4836 (2018)
93. Suciu, D., Olteanu, D., Ré, C., Koch, C.: Probabilistic databases. Synth. Lect. Data Manage. **3**(2), 1–180 (2011)
94. Valiant, L.G.: Robust logics. Artif. Intell. **117**(2), 231–253 (2000)
95. Van den Broeck, G.: Lifted Inference and Learning in Statistical Relational Models. Ph.D. thesis. KU Leuven (2013)
96. Xu, J., Zhang, Z., Friedman, T., Liang, Y., Van den Broeck, G.: A semantic loss function for deep learning with symbolic knowledge. In: International Conference on Machine Learning, pp. 5502–5511 (2018)
97. Xu, K., Li, J., Zhang, M., Du, S.S., Kawarabayashi, K.-I., Jegelka, S.: What can neural networks reason about? arXiv preprint arXiv:1905.13211 (2019)
98. Zellers, R., Bisk, Y., Schwartz, R., Choi, Y.: Swag: a large-scale adversarial dataset for grounded commonsense inference. arXiv preprint arXiv:1808.05326 (2018)
99. Zeng, Z., Van den Broeck, G.: Efficient search-based weighted model integration. arXiv preprint arXiv:1903.05334 (2019)
100. Zuidberg Dos Martires, P., Dries, A., De Raedt, L.: Knowledge compilation with continuous random variables and its application in hybrid probabilistic logic programming. arXiv preprint arXiv:1807.00614 (2018)

Score-Based Explanations in Data Management and Machine Learning

Leopoldo Bertossi[(⊠)]

Universidad Adolfo Ibáñez, Data Observatory Foundation, IMFD, Santiago, Chile
leopoldo.bertossi@uai.cl

Abstract. We describe some approaches to explanations for observed outcomes in data management and machine learning. They are based on the assignment of numerical scores to predefined and potentially relevant inputs. More specifically, we consider explanations for query answers in databases, and for results from classification models. The described approaches are mostly of a causal and counterfactual nature. We argue for the need to bring domain and semantic knowledge into score computations; and suggest some ways to do this.

1 Introduction

In data management and machine learning one wants *explanations* for certain results. For example, for query results from databases, and for outcomes of classification models. Explanations, that may come in different forms, have been the subject of philosophical enquires for a long time, but, closer to home, they appear under different forms in model-based diagnosis and in causality as developed in artificial intelligence. In the last few years, explanations that are based on *numerical scores* assigned to elements of a model that may contribute to an outcome have become popular. These scores attempt to capture the degree of their contribution to an outcome, e.g. answering questions like these: What is the contribution of this tuple to the answer to this query? What is the contribution of this feature value of an entity to the displayed classification of the latter?

Let us consider, as an example, a financial institution that uses a learned classifier, e.g. a decision tree, to determine if clients should be granted loans or not, returning labels 0 or 1, resp. A particular client, an entity $\mathbf{e}$, applies for a loan, the classifier returns $M(\mathbf{e}) = 1$, i.e. the loan is rejected. The client requests an explanation. A common approach consists in giving scores to the feature values in $\mathbf{e}$, to quantify their relevance in relation to the classification outcome. The higher the score of a feature value, the more explanatory is that value. For example, the fact that the client has value "5 years" for feature *Age* could have the highest score.

Motivated, at least to a large extent, by the trend towards *explainable AI* [22], different explanation scores have been proposed in the literature. Among them, in data management, the *responsibility score* as found in actual causality [10,13] has been used to quantify the strength of a tuple as a cause for a query

© Springer Nature Switzerland AG 2020
J. Davis and K. Tabia (Eds.): SUM 2020, LNAI 12322, pp. 17–31, 2020.
https://doi.org/10.1007/978-3-030-58449-8_2

result [2, 19]. The Shapley-value, as found in coalition game theory, has been used for the same purpose [15]. In machine learning, in relation to results of classification models, the Shapley-value has been used to assign scores to feature values. In the form of the SHAP-score, it has become quite popular and influential [17, 18]. A responsibility-based score, RESP, was introduced in [5] to assign numbers to feature values of entities under classification. It is based on the notions of counterfactual intervention and causal responsibility.

Some scores used in machine learning appeal to the components of the mathematical model behind the classifier. There can be all kinds of explicit models, and some are easier to understand or interpret or use for this purpose. For example, the FICO-score proposed in [9], for the FICO dataset about loan requests, depends on the internal outputs and displayed coefficients of two nested logistic regression models. Decision trees [21], random forests [7], rule-based classifiers, etc., could be seen as relatively easy to understand and use for explanations on the basis of their components.

Other scores can be applied with *black-box* models, in that they use, in principle, only the input/output relation that represents the classifier, without having access to the internals details of the model. In this category we could find classifiers based on complex neural networks, or XGBoost [16]. They are opaque enough to be treated as black-box models. The SHAP-score and the RESP-score can be applied to this category. In [5], the SHAP-score, the RESP-score and the FICO-score are compared. In general, the computation of the first two is intractable.

The SHAP-score and the RESP-score can be applied with open-box models. In this case, an interesting question is whether having access to the mathematical model may make their computation tractable, at least for some classes of classifiers.

As suggested above, scores can be assigned to tuples in databases, to measure their contribution to a query answer, or to the violation of an integrity constraint. The responsibility score has been applied for this purpose [2, 19], and is based on causality in databases [19]. Also the Shapley-value has been used for this task [15].

In this article we survey some of the approaches to score-based explanations we just mentioned above, in databases and in classification in machine learning. This is not intended to be an exhaustive survey of these areas, but it is heavily influenced by our latest research. Next, we discuss the relevance of bringing *domain and semantic knowledge* into these score computations. We also show some first ideas and techniques on how this knowledge can be accommodated in the picture. To introduce the concepts and techniques we will use mostly examples, trying to convey the main intuitions and issues.

This paper is structured as follows. In Sect. 2 we concentrate on causal explanations in databases. In Sect. 3, we describe the use of the Shapley-value to provide explanation scores in databases. In Sect. 3, we describe score-based explanations for classification results. In Sect. 5, we show how semantic knowledge can be brought into the score computations. We conclude with some final remarks in Sect. 6.

2 Explanations in Databases

In data management we need to understand and compute *why* certain results are obtained or not, e.g. query answers, violations of semantic conditions, etc.; and we expect a database system to provide *explanations*.

2.1 Causal Responsibility

Here, we will consider *causality-based explanations* [19,20], which we will illustrate by means of an example.

Example 1. Consider the database D, and the Boolean conjunctive query (BCQ)

R	A	B
	a	b
	c	d
	b	b

S	A
	a
	c
	b

$Q: \exists x \exists y (S(x) \land R(x,y) \land S(y))$.

It holds: $D \models Q$, i.e. the query is true in D.

We ask about the causes for Q to be true: A tuple $\tau \in D$ is *counterfactual cause* for Q (being true in D) if $D \models Q$ and $D \smallsetminus \{\tau\} \not\models Q$.

In this example, $S(b)$ is counterfactual cause for Q: If $S(b)$ is removed from D, Q is no longer true.

Removing a single tuple may not be enough to invalidate the query. Accordingly, a tuple $\tau \in D$ is an *actual cause* for Q if there is a *contingency set* $\Gamma \subseteq D$, such that τ is a counterfactual cause for Q in $D \smallsetminus \Gamma$.

In this example, $R(a,b)$ is an actual cause for Q with contingency set $\{R(b,b)\}$: If $R(a,b)$ is removed from D, Q is still true, but further removing $R(b,b)$ makes Q false.   □

Notice that every counterfactual cause is also an actual cause, with empty contingent set. Actual but non-counterfactual causes need company to invalidate a query result. Now we ask how strong are these tuples as causes? For this we appeal to the *responsibility* of an actual cause τ for Q [19], defined by:

$$\rho_D(\tau) := \frac{1}{|\Gamma| + 1},$$

with $|\Gamma|$ = size of a smallest contingency set for τ, and 0, otherwise.

Example 2. (Example 1 cont.) The responsibility of $R(a,b)$ is $\frac{1}{2} = \frac{1}{1+1}$ (its several smallest contingency sets have all size 1).

$R(b,b)$ and $S(a)$ are also actual causes with responsibility $\frac{1}{2}$; and $S(b)$ is actual (counterfactual) cause with responsibility $1 = \frac{1}{1+0}$.   □

High responsibility tuples provide more interesting explanations. Causes in this case are tuples that come with their responsibilities as "scores". Actually, all tuples can be seen as actual causes and only the non-zero scores matter. Causality and responsibility in databases can be extended to the attribute-value level [2,4].

There is a connection between database causality and *repairs* of databases w.r.t. integrity constraints (ICs) [1], and also connections to *consistency-based diagnosis* and *abductive diagnosis*. These connections have led to new complexity and algorithmic results for causality and responsibility [2,3]. Actually, the latter turns out to be intractable. In [3], causality under ICs was introduced and investigated. This allows to bring semantic and domain knowledge into causality in databases.

Model-based diagnosis is an older area of knowledge representation where explanations are main characters. In general, the diagnosis analysis is performed on a logic-based model, and certain elements of the model are identified as explanations. Causality-based explanations are somehow more recent. In this case, still a model is used, which is, in general, a more complex than a database with a query. In the case of databases, actually there is an underlying logical model, the *lineage or provenance* of the query [8] that we will illustrate in Sect. 2.2, but it is still a relatively simple model.

The idea behind *actual causality* is the (potential) execution of *counterfactual interventions* on a *structural logico-probabilistic model* [13], with the purpose of answering hypothetical or counterfactual questions of the form: *What would happen if we change ...?*. It turns out that counterfactual interventions can also be used to define different forms of score-based explanations, in the same spirit of causal responsibility in databases (c.f. Sect. 4.2). Score-based explanations can also be defined in the absence of a model, and without counterfactual interventions (or at least with them much less explicit).

2.2   The Causal-Effect Score

Sometimes responsibility does not provide intuitive or expected results, which led to the consideration of an alternative score, the *causal-effect score*. We show the issues and this score by means of an example.

Example 3.   Consider the database E that represents the graph below, and the Boolean Datalog query Π that is true in E if there is a path from a to b. Here, $E \cup \Pi \models yes$.

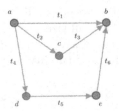

E	X	Y
t_1	a	b
t_2	a	c
t_3	c	b
t_4	a	d
t_5	d	e
t_6	e	b

$$yes \leftarrow P(a,b)$$
$$P(x,y) \leftarrow E(x,y)$$
$$P(x,y) \leftarrow P(x,z), E(z,y)$$

All tuples are actual causes since every tuple appears in a path from a to b. Also, all the tuples have the same causal responsibility, $\frac{1}{3}$, which may be counterintuitive, considering that t_1 provides a direct path from a to b.   □

In [27], the notion *causal effect* was introduced. It is based on three main ideas, namely, the transformation, for auxiliary purposes, of the database into a probabilistic database, interventions on the lineage of the query, and the use of expected values for the query. This is all shown in the next example.

Example 4. Consider the database D below, and a BCQ.

R	A	B
	a	b
	a	c
	c	b

S	B
	b
	c

Q : $\exists x \exists y (R(x,y) \wedge S(y))$, which is true in D.

The lineage of the query instantiated on D is given by the propositional formula:

$$\Phi_Q(D) = (X_{R(a,b)} \wedge X_{S(b)}) \vee (X_{R(a,c)} \wedge X_{S(c)}) \vee (X_{R(c,b)} \wedge X_{S(b)}),   (1)$$

where X_τ is a propositional variable that is true iff $\tau \in D$. Here, $\Phi_Q(D)$ takes value 1 in D.

Now, for illustration, we want to quantify the contribution of tuple $S(b)$ to the query answer. For this purpose, we assign probabilities, uniformly and independently, to the tuples in D, obtaining a a *probabilistic database* D^p [29]. Potential tuples outside D get probability 0.

R^p	A	B	prob
	a	b	$\frac{1}{2}$
	a	c	$\frac{1}{2}$
	c	b	$\frac{1}{2}$

S^p	B	prob
	b	$\frac{1}{2}$
	c	$\frac{1}{2}$

The X_τ's become independent, identically distributed Boolean random variables; and Q becomes a Boolean random variable. Accordingly, we can ask about the probability that Q takes the truth value 1 (or 0) when an *intervention* is performed on D.

Interventions are of the form $do(X = x)$, meaning making X take value x, with $x \in \{0,1\}$, in the *structural model*, in this case, the lineage. That is, we ask, for $\{y, x\} \subseteq \{0,1\}$, about the conditional probability $P(Q = y \mid do(X_\tau = x))$, i.e. conditioned to making X_τ false or true.

For example, with $do(X_{S(b)} = 0)$ and $do(X_{S(b)} = 1)$, the lineage in (1) becomes, resp., and abusing the notation a bit:

$$\Phi_Q(D|do(X_{S(b)} = 0) := (X_{R(a,c)} \wedge X_{S(c)}).$$
$$\Phi_Q(D|do(X_{S(b)} = 1) := X_{R(a,b)} \vee (X_{R(a,c)} \wedge X_{S(c)}) \vee X_{R(c,b)}.$$

On the basis of these lineages and D^p, when $X_{S(b)}$ is made false, the probability that the instantiated lineage becomes true in D^p is:

$$P(Q = 1 \mid do(X_{S(b)} = 0)) = P(X_{R(a,c)} = 1) \times P(X_{S(c)} = 1) = \frac{1}{4}.$$

Similarly, when $X_{S(b)}$ is made true, the probability of the lineage becoming true in D^p is:

$$P(Q = 1 \mid do(X_{S(b)} = 1)) = P(X_{R(a,b)} \vee (X_{R(a,c)} \wedge X_{S(c)}) \vee X_{R(c,b)} = 1) = \frac{13}{16}.$$

The *causal effect* of a tuple τ is defined by:

$$\mathcal{CE}^{D,Q}(\tau) := \mathbb{E}(Q \mid do(X_\tau = 1)) - \mathbb{E}(Q \mid do(X_\tau = 0)).$$

In particular, using the probabilities computed so far:

$$\mathbb{E}(Q \mid do(X_{S(b)} = 0)) = P(Q = 1 \mid do(X_{S(b)} = 0)) = \frac{1}{4},$$

$$\mathbb{E}(Q \mid do(X_{S(b)} = 1)) = P(Q = 1 \mid do(X_{S(b)} = 1)) = \frac{13}{16}.$$

Then, the causal effect for the tuple $S(b)$ is: $\mathcal{CE}^{D,Q}(S(b)) = \frac{13}{16} - \frac{1}{4} = \frac{9}{16} > 0$, showing that the tuple is relevant for the query result, with a relevance score provided by the causal effect, of $\frac{9}{16}$. $\square$

Let us now retake the initial example of this section.

Example 5. (Example 3 cont.) The Datalog query, here as a union of BCQs, has the lineage: $\Phi_Q(D) = X_{t_1} \vee (X_{t_2} \wedge X_{t_3}) \vee (X_{t_4} \wedge X_{t_5} \wedge X_{t_6})$. It holds:

$$\mathcal{CE}^{D,Q}(t_1) = 0.65625,$$
$$\mathcal{CE}^{D,Q}(t_2) = \mathcal{CE}^{D,Q}(t_3) = 0.21875,$$
$$\mathcal{CE}^{D,Q}(t_4) = \mathcal{CE}^{D,Q}(t_5) = \mathcal{CE}^{D,Q}(t_6) = 0.09375.$$

The causal effects are different for different tuples, and the scores are much more intuitive than the responsibility scores. $\square$

The definition of the causal-effect score may look rather *ad hoc* and arbitrary. We will revisit it in Sect. 3.2, where we will have yet another score for applications in databases. Actually, trying to take a new approach to measuring the contribution of a database tuple to a query answer, one can think of applying the *Shapley-value*, which is firmly established in game theory, and also used in several other areas.

The main idea is that *several tuples together* are necessary to violate an IC or produce a query result, much like players in a coalition game. Some may contribute more than others to the *wealth distribution function* (or simply, game function), which in this case becomes the query result, namely 1 or 0 if the query is Boolean, or a number if the query is an aggregation. The Shapley-value of a tuple can be used to assign a score to its contribution. This was done in [15], and will be retaken in Sect. 3.2. But first things first.

3   The Shapley-Value in Databases

3.1   The Shapley-Value

The Shapley value was proposed in game theory by Lloyd Shapley in 1953 [28], to quantify the contribution of a player to a coalition game where players share a wealth function.[1] It has been applied in many disciplines. In particular, it has been investigated in computer science under *algorithmic game theory* [23], and it has been applied to many and different computational problems. The computation of the Shapley-value is, in general, intractable. In many scenarios where it is applied its computation turns out to be #P-hard [11,12].

In particular, the Shapley value has been used in knowledge representation, to measure the degree of inconsistency of a propositional knowledge base [14]; in data management to measure the contribution of a tuple to a query answer [15] (c.f. Sect. 3.2); and in machine learning to provide explanations for the outcomes of classification models on the basis of numerical scores assigned to the participating feature values [18] (c.f. Sect. 4.1).

Consider a set of players D, and a game function, $\mathcal{G} : \mathcal{P}(D) \to \mathbb{R}$, where $\mathcal{P}(D)$ the power set of D. The Shapley-value of player p in D es defined by:

$$Shapley(D, \mathcal{G}, p) := \sum_{S \subseteq D \setminus \{p\}} \frac{|S|!(|D| - |S| - 1)!}{|D|!} (\mathcal{G}(S \cup \{p\}) - \mathcal{G}(S)). \quad (2)$$

Notice that here, $|S|!(|D| - |S| - 1)!$ is the number of permutations of D with all players in S coming first, then p, and then all the others. That is, this quantity is the expected contribution of player p under all possible additions of p to a partial random sequence of players followed by a random sequence of the rests of the players. Notice the counterfactual flavor, in that there is a comparison between what happens having p vs. not having it. The Shapley-value is the only function that satisfy certain natural properties in relation to games. So, it is a result of a categorical set of axioms or conditions.

3.2   Shapley for Query Answering

Back to query answering in databases, the players are tuples in the database D. We also have a Boolean query $\mathcal{Q}$, which becomes a game function, as follows: For $S \subseteq D$,

$$\mathcal{Q}(S) = \begin{cases} 1 & \text{if } S \models \mathcal{Q} \\ 0 & \text{if } S \not\models \mathcal{Q} \end{cases}$$

With this game elements we can define a specific Shapley-value for a database tuple τ:

$$Shapley(D, \mathcal{Q}, \tau) := \sum_{S \subseteq D \setminus \{\tau\}} \frac{|S|!(|D| - |S| - 1)!}{|D|!} (\mathcal{Q}(S \cup \{\tau\}) - \mathcal{Q}(S)).$$

---

[1] The original paper and related ones on the Shapley value can be found in the book edited by Alvin Roth [26]. Shapley and Roth shared the Nobel Prize in Economic Sciences 2012.

If the query is *monotone*, i.e. its set of answers never shrinks when new tuples are added to the database, which is the case of conjunctive queries (CQs), among others, the difference $\mathcal{Q}(S \cup \{\tau\}) - \mathcal{Q}(S)$ is always 1 or 0, and the average in the definition of the Shapley-value returns a value between 0 and 1. This value quantifies the contribution of tuple τ to the query result. It was introduced and investigated in [15], for BCQs and some aggregate queries defined over CQs. We report on some of the findings in the rest of this section. The analysis has been extended to queries with negated atoms in CQs [24].

A main result obtained in [15] is in relation to the complexity of computing this Shapley score. It is the following *Dichotomy Theorem*: For $\mathcal{Q}$ a BCQ without self-joins, if $\mathcal{Q}$ is *hierarchical*, then $Shapley(D, \mathcal{Q}, \tau)$ can be computed in polynomial-time (in the size of D); otherwise, the problem is $FP^{\#P}$-complete.

Here, $\mathcal{Q}$ is hierarchical if for every two existential variables x and y, it holds: (a) $Atoms(x) \subseteq Atoms(y)$, or $Atoms(y) \subseteq Atoms(x)$, or $Atoms(x) \cap Atoms(y) = \emptyset$. For example, $\mathcal{Q}: \exists x \exists y \exists z (R(x,y) \wedge S(x,z))$, for which $Atoms(x) = \{R(x,y), S(x,z)\}$, $Atoms(y) = \{R(x,y)\}$, $Atoms(z) = \{S(x,z)\}$, is hierarchical. However, $\mathcal{Q}^{nh}: \exists x \exists y (R(x) \wedge S(x,y) \wedge T(y))$, for which $Atoms(x) = \{R(x), S(x,y)\}$, $Atoms(y) = \{S(x,y), T(y)\}$, is not hierarchical.

These are the same criteria for (in)tractability that apply to BCQs over probabilistic databases [29]. However, the same proofs do not (seem to) apply. The intractability result uses query $\mathcal{Q}^{nh}$ above, and a reduction from counting independent sets in a bipartite graph.

The dichotomy results can be extended to summation over CQs, with the same conditions and cases. This is because the Shapley-value, as an expectation, is linear. Hardness extends to aggregates max, min, and avg over non-hierarchical queries.

For the hard cases, there is an *Approximation Result:* For every fixed BCQ $\mathcal{Q}$ (or summation over a CQ), there is a multiplicative fully-polynomial randomized approximation scheme (FPRAS), A, with

$$P(\tau \in D \mid \frac{Shapley(D, \mathcal{Q}, \tau)}{1 + \epsilon} \leq A(\tau, \epsilon, \delta) \leq (1 + \epsilon) Shapley(D, \mathcal{Q}, \tau)\}) \geq 1 - \delta.$$

A related and popular score, in coalition games and other areas, is the *Bahnzhaf Power Index*, which is similar to the Shapley-value, but the order of players is ignored, by considering subsets of players rather than permutations thereof:

$$Banzhaf(D, \mathcal{Q}, \tau) := \frac{1}{2^{|D|-1}} \cdot \sum_{S \subseteq (D \setminus \{\tau\})} (\mathcal{Q}(S \cup \{\tau\}) - \mathcal{Q}(S)).$$

The Bahnzhaf-index is also difficult to compute; provably #P-hard in general. The results in [15] carry over to this index when applied to query answering in databases.

In [15] it was proved that the causal-effect score of Sect. 2.2 coincides with the Banzhaf-index, which gives to the former a more fundamental or historical justification.

4   Score-Based Explanations for Classification

Let us consider a classifier, C, that receives a representation of a entity, $\mathbf{e}$, as a record of feature values, and outputs a label, $L(\mathbf{e})$, corresponding to the possible decision alternatives. We could see C as a black-box, in the sense that only by direct interaction with it, we have access to its input/output relation. We may not have access to the mathematical classification model inside C.

$$L(\mathbf{e}) \longleftarrow \qquad \longleftarrow \mathbf{e}$$

To simplify the presentation we will assume that the entities and the classifier are binary, that is, in the representation $\mathbf{e} = \langle x_1, \ldots, x_n \rangle$ of an entity, the feature values are binary (0 or 1), corresponding to propositional features (false or true, resp.). The label is always 0 or 1. For example, we could have a client of a financial institution requesting a loan, but the classifier, on the basis of his/her feature values, assigns the label 1, for rejection. An explanation is requested by the client. Of course, the same situation may occur if we have an explicit classification model, e.g. a classification tree or a logistic regression model, in which cases, we might be in a better position to given an explanation, because we can inspect the internals of the model [25]. However, we will put ourselves in the "worst scenario" in which we do not have access to the internal model.

An approach to explanations that has become popular, specially in the absence of the model, assigns numerical *scores*, trying to answer the question about which of the feature values x_i of $\mathbf{e}$ contribute the most to the received label.

Score-based methodologies are sometimes based on counterfactual interventions: What would happen with the label if we change this value, leaving the others fixed? Or the other way around: What if we leave this value fixed, and change the others? The resulting labels can be aggregated, leading to a score for the feature value under inspection.

In the next two sections we briefly introduce two scores. Both can be applied with open-box or black-box models.

4.1   The SHAP-Score

We will consider until further announcement the *uniform probability space*. Actually, since we consider only binary feature values taking values 0 or 1, this is the uniform distribution on $E = \{0,1\}^n$, assigning probability $P^u(\mathbf{e}) = \frac{1}{2^n}$ to $\mathbf{e} \in E$. One could consider appealing to other, different distributions.

In the context of classification, the Shapley-value has taken the form of the SHAP-score [17], which we briefly introduce. Given the binary classifier, C, on binary entities, it becomes crucial to identify a suitable game function. In this case, it will be expressed in terms of expected values (not unlike the causal-effect score), which requires an underlying probability space on the population of entities. For the latter we use, as just said, the *uniform distribution over* $\{0,1\}^n$.

Given a set of features $\mathcal{F} = \{F_1, \dots, F_n\}$, and an entity $\mathbf{e}$ whose label is to be explained, the set of players D in the game is $\mathcal{F}(\mathbf{e}) := \{F(\mathbf{e}) \mid F \in \mathcal{F}\}$, i.e. the set of feature values of $\mathbf{e}$. Equivalently, if $\mathbf{e} = \langle x_1, \dots, x_n \rangle$, then $x_i = F_i(\mathbf{e})$. We assume these values have implicit feature identifiers, so that duplicates do not collapse, i.e. $|\mathcal{F}(\mathbf{e})| = n$. The game function is defined as follows. For $S \subseteq \mathcal{F}(\mathbf{e})$,

$$\mathcal{G}_\mathbf{e}(S) := \mathbb{E}(L(\mathbf{e}') \mid \mathbf{e}'_S = \mathbf{e}_S),$$

where $\mathbf{e}_S$: is the projection of $\mathbf{e}$ on S. That is, the expected value of the label for entities $\mathbf{e}'$ when their feature values are fixed and equal to those in in S for $\mathbf{e}$. Other than that, the feature values of $\mathbf{e}'$ may independently vary over $\{0,1\}$.

Now, one can instantiate the general expression for the Shapley-value in (2), using this game function, as $Shapley(\mathcal{F}(\mathbf{e}), \mathcal{G}_\mathbf{e}, F(\mathbf{e}))$, obtaining, for a particular feature value $F(\mathbf{e})$:

$$\mathsf{SHAP}(\mathcal{F}(\mathbf{e}), \mathcal{G}_\mathbf{e}, F(\mathbf{e})) := \sum_{S \subseteq \mathcal{F}(\mathbf{e}) \setminus \{F(\mathbf{e})\}} \frac{|S|!(n - |S| - 1)!}{n!} \times$$
$$(\mathbb{E}(L(\mathbf{e}'|\mathbf{e}'_{S \cup \{F(\mathbf{e})\}} = \mathbf{e}_{S \cup \{F(\mathbf{e})\}}) - \mathbb{E}(L(\mathbf{e}')|\mathbf{e}'_S = \mathbf{e}_S)).$$

Here, the label L acts as a Bernoulli random variable that takes values through the classifier. We can see that the SHAP-score is a weighted average of differences of expected values of the labels [17].

4.2 The RESP-Score

In the same setting of Sect. 4.1, let us consider the following score introduced in [5]. For $F \in \mathcal{F}$, and an entity $\mathbf{e}$ for which we have obtained label 1, the "negative" outcome one would like to see explained:

$$\mathsf{COUNTER}(\mathbf{e}, F) := L(\mathbf{e}) - \mathbb{E}(L(\mathbf{e}') \mid \mathbf{e}'_{\mathcal{F} \setminus \{F\}} = \mathbf{e}_{\mathcal{F} \setminus \{F\}}). \tag{3}$$

This score measures the expected difference between the label for $\mathbf{e}$ and those for entities that coincide in feature values everywhere with $\mathbf{e}$ but on feature F. Notice the essential counterfactual nature of this score, which is reflected in all the possible hypothetical changes of features values in $\mathbf{e}$.

The $\mathsf{COUNTER}$-score can be applied in same scenarios as SHAP, it is easier to compute, and gives reasonable and intuitive results, and also behaves well in experimental comparisons with other scores [5]. As with the SHAP-score, one could consider different underlying probability distributions (c.f. [5] for a discussion). Again, so as for SHAP, there is no need to access the internals of the classification model.

One problem with $\mathsf{COUNTER}$ is that changing a single value, no matter how, may not switch the original label, in which case no explanations are obtained. In order to address this problem, we can bring in *contingency sets* of feature values, which leads to the RESP-score introduced in [5]. We just give the idea and a simplified version of it by means of an example.

Example 6. In the picture below, the black box is the classifier. Entities have three feature values. The table on the right-hand side shows all the possible entities with their labels. We want to explain the label 1 obtained by entity e_1.

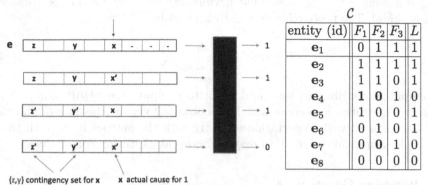

entity (id)	F_1	F_2	F_3	L
e_1	0	1	1	1
e_2	1	1	1	1
e_3	1	1	0	1
e_4	1	0	1	0
e_5	1	0	0	1
e_6	0	1	0	1
e_7	0	0	1	0
e_8	0	0	0	0

{z,y} contingency set for x   x actual cause for 1

Through counterfactual interventions we change feature values in e_1, trying to change the label to 0. This process is described in the figure above, on the left-hand side, where we are attempting to quantify the contribution of value $\mathbf{x} = F(\mathbf{e}_1)$. Let us assume that by changing $\mathbf{x}$ into any $\mathbf{x}'$, we keep obtaining label 1. So, we leave $\mathbf{x}$ as it is, and consider changing other original values, $\mathbf{y}$ and $\mathbf{z}$, into $\mathbf{y}'$ and $\mathbf{z}'$, still getting 1. However, if we now, in addition, change $\mathbf{x}$ into $\mathbf{x}'$, we get label 0. Then, in the spirit of actual causality, as seen in Sect. 2.1, we can say that the feature value $\mathbf{x}$ is an actual cause for the original label 1, with $\mathbf{y}$ and $\mathbf{z}$ forming a contingency set for $\mathbf{x}$; in this case, of size 2.

On this basis, we can define [6]: (a) $\mathbf{x}$ is a *counterfactual explanation* for $L(\mathbf{e}) = 1$ if $L(\mathbf{e}\frac{\mathbf{x}}{\mathbf{x}'}) = 0$, for some $\mathbf{x}' \in Dom(F)$ (the domain of feature F). (b) $\mathbf{x}$ is an *actual explanation* for $L(\mathbf{e}) = 1$ if there is a set of values $\mathbf{Y}$ in $\mathbf{e}$, with $\mathbf{x} \notin \mathbf{Y}$, and new values $\mathbf{Y}' \cup \{\mathbf{x}'\}$, such that $L(\mathbf{e}\frac{\mathbf{Y}}{\mathbf{Y}'}) = 1$ and $L(\mathbf{e}\frac{\mathbf{xY}}{\mathbf{x'Y'}}) = 0$. Here, as usual, $\frac{\mathbf{x}}{\mathbf{x}'}$, denotes the replacement of value $\mathbf{x}$ by $\mathbf{x}'$, and so on.

Contingency sets may come in sizes from 0 to $n - 1$ for feature values in records of length n. Accordingly, we can define for the actual cause $\mathbf{x}$: If $\mathbf{Y}$ is a minimum size contingency set for $\mathbf{x}$, $\mathsf{RESP}(\mathbf{x}) := \frac{1}{1+|\mathbf{Y}|}$; and as 0 when $\mathbf{x}$ is not an actual cause. This score can be formulated in terms of expected values, generalizing expression (3) through the introduction of contingency sets [5].

Coming back to the entities in the figure above, due to e_7, $F_2(\mathbf{e}_1)$ is counterfactual explanation; with $\mathsf{RESP}(F_2(\mathbf{e}_1)) = 1$. Due to e_4, $F_1(\mathbf{e}_1)$ is actual explanation; with $\{F_2(\mathbf{e}_1)\}$ as contingency set, and $\mathsf{RESP}(F_1(\mathbf{e}_1)) = \frac{1}{2}$. □

5   Bringing-In Domain Knowledge

The uniform space gives equal probability to each entity in the underlying population. One can argue that this is not realistic, in that certain combinations of feature values may be more likely than others; or that certain correlations among them exist. One can consider assigning or modifying probabilities in the hope of capturing correlations and logical relationships between feature values.

5.1   Empirical Distributions

An alternative consists in using an *empirical distribution* as a proxy. In this case we have a sample $S \subseteq E$ (we could have repetitions, but we do not consider this case here). The probability of $\mathbf{e} \in E$, is given by:

$$P_S(\mathbf{e}) := \begin{cases} \frac{1}{|S|} & \text{if } \mathbf{e} \in S \\ 0 & \text{if } \mathbf{e} \notin S \end{cases} \tag{4}$$

The empirical distribution was used in [5] to compute the SHAP-score. More precisely, the entities in S come with labels obtained via the classifier $\mathcal{C}$; and the score is computed with expectations directly with the entities in S, with their labels. The empirical distribution may be better at capturing correlations.

5.2   Imposing Constraints

One can introduce *constraints* that prohibit certain combinations of values, in the spirit of *denial constraints* in databases, but in this case admitting positive and negative atoms. For example, we may not want the combination of *"The age is not greater than 20"* and *"Gets an account overdraft above $50M"* to hold simultaneously.

These constraints, which are satisfied or violated by a single entity at a time, are of the form:

$$\chi : \neg(\bigwedge_{i \in S} F_i \wedge \bigwedge_{j \in S'} \bar{F}_j), \tag{5}$$

where $S \cup S' \subseteq F$, $S \cap S' = \emptyset$, and $F_i, \bar{F}_j$ mean that features F_i, F_j take values 1 and 0, resp. In the example, it would be of the form $\neg(\overline{Age} \wedge OverDr50M)$. The events, i.e. subsets of E, associated to the violation of χ should get zero probability.

A way to accommodate a constraint, χ, consists in defining an event associated to it:

$$A(\chi) = \{\mathbf{e} \in E \mid \mathbf{e} \models \chi\},$$

where $\mathbf{e} \models \chi$ has the obvious meaning of satisfaction of χ by entity $\mathbf{e}$.

Given the uniform probability space $\langle E, P^u \rangle$, we can redefine the probability in order to enforce χ. For $A \subseteq E$,

$$P_\chi^u(A) := P^u(A|A(\chi)) = \frac{P^u(A \cap A(\chi))}{P^u(A(\chi))}. \tag{6}$$

Since χ is logically consistent (it is satisfied by some entities in E), the conditional distribution is well-defined. Notice that the probability of χ's violation set, i.e. of $E \smallsetminus A(\chi)$, is now:

$$P_\chi^u(E \smallsetminus A(\chi)) = \frac{P^u(\emptyset)}{P^u(A(\chi))} = 0.$$

This definition can be extended to finite sets, Θ, of constraints, as long as it is consistent (i.e. satisfiable in E), by using $\wedge\Theta$, the conjunction of the constraints in Θ: $P_\Theta^u(A) := P_{\wedge\Theta}^u(A)$.

Of course, one could go beyond constraints of the form (5), applying the same ideas, and consider any propositional formula that is intended to be evaluated on a single entity at a time, as opposed to considering combinations of feature values for different entities.

The resulting modified distributions that accommodate constraints could be used in the computation of any of the scores expressed in terms of expected values (or in probabilistic terms, in general).

6 Final Remarks

Explainable AI (XAI) is an effervescent area of research. Its relevance can only grow considering that legislation around explainability, transparency and fairness of AI/ML systems is being produced and enforced. There are different approaches and methodologies in relation to explanations, and causality, counterfactuals and scores have a relevant role to play.

Much research is still needed on the use of contextual, semantic and domain knowledge. Some approaches may be more appropriate, e.g. declarative ones [6].

Still fundamental research is needed on *what is a good explanation*, and in particular, on what are the desired properties of an explanation score. After all, the original, general Shapley-value emerged from a list of *desiderata* in relation to coalition games. Although the Shapley value is being used in XAI, in particular in its SHAP incarnation, there could be a different and specific set of desired properties of explanation scores that could lead to a still undiscovered explanation-score.

Acknowledgments. L. Bertossi is a member of the Academic Network of RelationalAI Inc., where his interest in explanations in ML started.

References

1. Bertossi, L.: Database Repairing and Consistent Query Answering. Synthesis Lectures in Data Management. Morgan & Claypool (2011)
2. Bertossi, L., Salimi, B.: From causes for database queries to repairs and model-based diagnosis and back. Theor. Comput. Syst. **61**(1), 191–232 (2017)
3. Bertossi, L., Salimi, B.: Causes for query answers from databases: datalog abduction, view-updates, and integrity constraints. Int. J. Approx. Reason. **90**, 226–252 (2017)
4. Bertossi, L.: Characterizing and computing causes for query answers in databases from database repairs and repair programs. In: Ferrarotti, F., Woltran, S. (eds.) FoIKS 2018. LNCS, vol. 10833, pp. 55–76. Springer, Cham (2018). https://doi.org/10.1007/978-3-319-90050-6_4

5. Bertossi, L., Li, J., Schleich, M., Suciu, D., Vagena, Z.: Causality-based explanation of classification outcomes. In: Proceedings of the Fourth Workshop on Data Management for End-To-End Machine Learning, DEEM@SIGMOD 2020, pp. 6:1–6:10 (2020)
6. Bertossi, L.: An ASP-based approach to counterfactual explanations for classification. To appear in Proceedings of the RuleML-RR 2020 (2020). Arxiv:2004.13237
7. Breiman, L., Friedman, J., Stone, C.J., Olshen, R.A.: Classification and Regression Trees. CRC Press, Boca Raton (1984)
8. Buneman, P., Khanna, S., Wang-Chiew, T.: Why and where: a characterization of data provenance. In: Van den Bussche, J., Vianu, V. (eds.) ICDT 2001. LNCS, vol. 1973, pp. 316–330. Springer, Heidelberg (2001). https://doi.org/10.1007/3-540-44503-X_20
9. Chen, C., Lin, K., Rudin, C., Shaposhnik, Y., Wang, S., Wang, T.: An interpretable model with globally consistent explanations for credit risk. CoRR, abs/1811.12615 (2018)
10. Chockler, H., Halpern, J.: Responsibility and blame: a structural-model approach. J. Artif. Intell. Res. **22**, 93–115 (2004)
11. Deng, X., Papadimitriou, C.: On the complexity of cooperative solution concepts. Math. Oper. Res. **19**(2), 257–266 (1994)
12. Faigle, U., Kern, W.: The Shapley value for cooperative games under precedence constraints. Int. J. Game Theor. **21**, 249–266 (1992)
13. Halpern, J., Pearl, J.: Causes and explanations: a structural-model approach. Part I: causes. Br. J. Philos. Sci. **56**(4), 843–887 (2005)
14. Hunter, A., Konieczny, S.: On the measure of conflicts: Shapley inconsistency values. Artif. Intell. **174**(14), 1007–1026 (2010)
15. Livshits, E., Bertossi, L., Kimelfeld, B., Sebag, M.: The Shapley value of tuples in query answering. In: 23rd International Conference on Database Theory, ICDT 2020, Copenhagen, Denmark, March 30–April 2, 2020, vol. 155, pp. 20:1–20:19 (2020)
16. Lucic, A., Haned, H., de Rijke, M.: Explaining predictions from tree-based boosting ensembles. CoRR, abs/1907.02582 (2019)
17. Lundberg, S., et al.: From local explanations to global understanding with explainable AI for trees. Nat. Mach. Intell. **2**(1), 2522–5839 (2020)
18. Lundberg, S., Lee, S.: A unified approach to interpreting model predictions. In: Advances in Neural Information Processing Systems 30: Annual Conference on Neural Information Processing Systems 2017, Long Beach, CA, USA, 4–9 December 2017, pp. 4765–4774 (2017)
19. Meliou, A., Gatterbauer, W., Moore, K.F., Suciu, D.: The complexity of causality and responsibility for query answers and non-answers. Proc. VLDB, 34–41 (2010)
20. Meliou, A., Gatterbauer, W., Halpern, J.Y., Koch, C., Moore, K.F., Suciu, D.: Causality in databases. IEEE Data Eng. Bull. **33**(3), 59–67 (2010)
21. Mitchell, T.M.: Machine Learning. McGraw Hill Series in Computer Science. McGraw-Hill (1997)
22. Molnar, C.: Interpretable Machine Learning: A Guide for Making Black Box Models Explainable (2020). https://christophm.github.io/interpretable-ml-book
23. Halpern, D., Shah, N.: Fair division with subsidy. In: Fotakis, D., Markakis, E. (eds.) SAGT 2019. LNCS, vol. 11801, pp. 374–389. Springer, Cham (2019). https://doi.org/10.1007/978-3-030-30473-7_25
24. Reshef, A., Kimelfeld, B., Livshits, E.: The impact of negation on the complexity of the Shapley value in conjunctive queries. In: Proceedings of the PODS 2020, pp. 285–297 (2020)

25. Rudin, C.: Stop explaining black box machine learning models for high stakes decisions and use interpretable models instead. Nat. Mach. Intell. **1**, 206–215 (2019). Also arXiv:1811.10154 (2018)
26. Roth, A.E. (ed.) The Shapley Value: Essays in Honor of Lloyd S. Shapley. Cambridge University Press (1988)
27. Salimi, B., Bertossi, L., Suciu, D., Van den Broeck, G.: Quantifying causal effects on query answering in databases. In: Proceedings of the 8th USENIX Workshop on the Theory and Practice of Provenance (TaPP) (2016)
28. Shapley, L.S.: A value for n-person games. Contrib. Theor. Games **2**(28), 307–317 (1953)
29. Suciu, D., Olteanu, D., Re, C., Koch, C.: Probabilistic Databases. Synthesis Lectures on Data Management. Morgan & Claypool (2011)

Regular Papers

From Possibilistic Rule-Based Systems to Machine Learning - A Discussion Paper

Didier Dubois and Henri Prade[✉]

IRIT, CNRS & Univ. Paul Sabatier, 118 route de Narbonne,
31062 Toulouse Cedex 9, France
{dubois,prade}@irit.fr

Abstract. This paper is a plea for developing possibilistic learning methods that would be consistent with if-then rule-based reasoning. The paper first recall the possibility theory-based handling of cascading sets of parallel if-then rules. This is illustrated by an example describing a classification problem. It is shown that the approach is both close to a possibilistic logic handling of the problem and can also be put under the form of a max-min-based matrix calculus describing a function underlying a structure somewhat similar to a max-min neural network. The second part of the paper discusses how possibility distributions can be obtained from precise or imprecise statistical data, and then surveys the few existing works on learning in a possibilistic setting. A final discussion emphasizes the interest of handling learning and reasoning in a consistent way.

1 Introduction

Nowadays, many people tend to oppose new artificial intelligence (AI) based on data and machine learning with so-called "old fashioned AI", oriented towards reasoning, exploiting knowledge and having its roots in expert systems. Indeed, at first glance, they look quite different. The former is fond of numerical methods, black box models, and often relies on neural net approaches and/or statistical views. The latter privileges logic-based modeling, often referred as 'symbolic AI', and is more suitable for explanations. A closer analysis reveals that there are several kinds of meeting points between machine learning (ML) on the one hand, and knowledge representation and reasoning (KRR) on the other hand [8]. In this paper, we consider a problem that is at the crossroads of ML and KRR: the classification of items. Such a problem can be envisaged as the learning of a function that maps sets of feature values into classes, taking advantage of a set of examples. Alternatively, a set of if-then rules may also describe a classification process.

There are a number of examples where logical and functional views co-exist. Quite early the authors have pointed out that a set of rules could to do a job similar to a multiple criteria aggregation function [19]. Conversely, an aggregation function may need to be "decomposed" into rules for explanation purposes,

© Springer Nature Switzerland AG 2020
J. Davis and K. Tabia (Eds.): SUM 2020, LNAI 12322, pp. 35–51, 2020.
https://doi.org/10.1007/978-3-030-58449-8_3

e.g., [42]. A perfect example of this situation is offered by Sugeno integrals which have both a (possibilistic) logic translation [25] and can be learnt in an extended version space setting [44].

An early example of the interface between reasoning and learning can be found with rule-based fuzzy logic controllers [43]. At the beginning, rules were obtained from experts and processed by a fuzzy logic inference machinery. Later on, the idea of learning the fuzzy rules became prominent, but, while the rules so obtained were good enough for control purposes, they were no longer interpretable by humans! Recently, Denœux [11] showed a striking parallel between the combination machinery of belief functions and neural nets. In [8], one may find a rich bibliography of recent works mixing logic and neural nets in a more or less empirical way.

In this paper, we use the setting of possibility theory for modeling uncertainty, which allows for the representation of incomplete or imprecise information. The discussion of the KRR and ML sides of a classification task is illustrated in the following by means of a simple example dealing with professions that can be recommended to people and associated salaries on the basis of their tastes and aspirations. We handle it in an expert system style in order to remain close to the if-then rule format of knowledge. As we shall see, it can be both interfaced with logic and then with statistics.

The paper first deals with the if-then rule reasoning side, before showing its relation with possibilistic logic. Then we review how possibility degrees can be extracted from statistics, thus providing a basis for a learning process. Lastly, we survey the existing works in possibilistic learning, and discuss them.

2 Possibilistic Handling of a Rule-Based System

We first introduce an illustrative example, then explain the possibilistic machinery necessary for handling it, and finally point out links with a functional view on the one hand and a purely logical approach on the other hand. We consider a simple example of a knowledge base, as it appears in the final version of [29] in French. The first four rules relate, in a very sketchy and very incomplete way, tastes and aspirations of people with professions that can be recommended to them. They form a set of parallel rules. The last three rules that relate professions with salary levels, are also parallel. They can be chained with the previous set. Intentionally, these rules present different peculiarities: combined condition, pair of rules expressing equivalence, presence of disjunctions.

Example 1. – R1: if a person likes meeting people, then recommended professions are professor or business man or lawyer or doctor
- R2: if a person is fond of creation/inventions, then recommended professions are engineer or public researcher or architect
- R'2: if a person is not fond of creation/invention, then he/she cannot be an engineer nor a researcher nor an architect
- R3: if a person looks for job security and is fond of intellectual speculation, then recommended professions are professor or public researcher

- R'4 if a person is a professor or a researcher, then her salary is rather low
- R''4 if a person is an engineer, a lawyer or an architect, her salary is average or high
- R'''4 if a person is a business man or a doctor, then her salary is high.   □

2.1   Possibilistic If-Then Rules

All these rules are pervaded with uncertainty. We choose here to represent it in the setting of possibility theory [16,22], using an inference machinery first described in [28,30], in the spirit of expert systems. All the if-then rules, as in the above example, are of the form "if p then q", or more generally, "if p_1 and ... and p_n then q", where p (resp. p_i) stands for $a(x) \in P$ (resp. $a_i(x) \in P_i$) and q for $b(x) \in Q$; a, a_i, b denote attributes applied to some item x, P, P_i, Q are subsets of the respective domains of these attributes. Thus, in our example, the domain of the attribute *profession* is $\mathcal{D}_{pro} = \{$business man, lawyer, doctor, professor, researcher, architect, engineer$\}$, and the domain of the attribute *salary* is $\mathcal{D}_{sal} = \{$low, average, high$\}$.

The rules are supposed to be applied to a base of facts (kept separated from rules in expert systems). The facts pertain to items x, which are here supposed to be described in terms of the four attributes *lik.-meet.-peo.*, *fond-creation*, *job-secur.*, *fond-intel.-spec.*. For simplicity, these attributes are supposed to have binary domains $\{$yes, no$\}$, i.e., $\{1,0\}$. These attribute values may be uncertain. The information about $a_i(x)$ is represented by a possibility distribution $\pi_{a_i(x)}$, a mapping from $\mathcal{D}_{a_i}$ to $[0,1]$. $\pi_{a_i(x)}$ is supposed to be normalized, i.e., $\exists u \in \mathcal{D}_{a_i}, \pi_{a_i(x)}(u) = 1$, which expresses the consistency of the information about a_i for x; $\pi_{a_i(x)}$ may take intermediary degrees in $[0,1]$ for some values in the attribute domain of a_i.

In possibility theory [16,22], uncertainty is assessed in terms of $[0,1]$-valued possibility (Π) and necessity (N) degrees of propositions, according to the duality $N(p) = 1 - \Pi(\neg p)$. On binary domains, i.e., for a true or false proposition p, it is equivalent to work with the possibility distribution defined by the pair $(\pi(p), \pi(\neg p))$. Normalization is assumed, i.e., $\max(\pi(p), \pi(\neg p)) = 1$. $(\pi(p), \pi(\neg p)) = (1,0), (0,1), (1,1)$ respectively mean that p is true, false, and unknown. The pair $(\pi(p), \pi(\neg p)) = (1, \lambda)$, with $0 < \lambda < 1$ means that it is certain at level $1 - \lambda$ that p is true.

The inference machinery includes five basic steps that are now described. We restate this forgotten approach, while highlighting what is the underlying functional machinery. The two first steps compute the compatibility between rule conditions and facts by possibilistic pattern matching. Then uncertainty is propagated and results of parallel rules are combined.

1. **Compatibility between an elementary condition and a fact.** The compatibility between condition p_i and the information is computed as the pair $(\pi(p_i), \pi(\neg p_i))$ ($\overline{P}$ denotes the complement of P):
   $$\pi(p_i) = \Pi(P_i) = \sup_{u \in P_i} \pi_{a_i(x)}(u); \qquad \pi(\neg p_i) = \Pi(\overline{P_i}) = \sup_{u \notin P_i} \pi_{a_i(x)}(u).$$
   The normalization of $\pi_{a_i(x)}$ ensures that $\max(\pi(p_i), \pi(\neg p_i)) = 1$.

2. **Compatibility between a compound condition and facts.** In case of a condition of the form 'p_1 and ... and p_n', the elementary compatibilities are combined conjunctively: $\pi(p) = \min_{i=1}^{n} \pi(p_i)$; $\pi(\neg p) = \max_{i=1}^{n} \pi(\neg p_i)$.
The first of the two formulas requires the logical independence of the attributes a_i, which means that the joint possibility distribution $\pi_{(a_1(x),...,a_n(x))}$ can be decomposed into $\min_{i=1}^{n} \pi_{a_i(x)}$ for any x; otherwise the formula provides an upper approximation of $\pi(p)$ [16]. It can be checked that these formulas preserve normalization. Observe also that as soon as $\exists i, (\pi(p_i), \pi(\neg p_i)) = (0, 1)$, expressing the falsity of p_i, then p is false (i.e., $(\pi(p), \pi(\neg p)) = (0, 1)$); besides, if $\exists j, (\pi(p_j), \pi(\neg p_j)) = (1, 1)$, expressing ignorance about p_j, while all the other conditions are at least somewhat certainly true (i.e., $\forall i \neq j, (\pi(p_i), \pi(\neg p_i)) = (1, \lambda_i)$), then $(\pi(p), \pi(\neg p)) = (1, 1)$, expressing ignorance about condition p.

3. **Uncertainty propagation.** The uncertainty attached to a rule "if p then q" is naturally assessed in terms of conditional distributions. Conditional possibility in the qualitative setting is defined by $\pi(q|p) = \begin{cases} 1 & \text{if } \pi(p) = \Pi(p \wedge q) \\ \Pi(p \wedge q) & \text{otherwise.} \end{cases}$. In the quantitative setting, $\pi(q|p) = \Pi(p \wedge q)/\pi(p)$ [16,22]. Thus, in the possibilistic setting the rule "if p then q" is represented by a pair $(\pi(q|p), \pi(\neg q|p)) = (1, r), r < 1$, and $(\pi(q|p), \pi(\neg q|p)) = (r', 1)$ corresponds to the rule "if p then $\neg q$", with certainty level $1 - r'$. The normalization condition $\max(\pi(q|p), \pi(\neg q|p)) = 1$ is assumed. Similarly, in context $\neg p$, we use the normalized pair $(\pi(q|\neg p), \pi(\neg q|\neg p))$.

This information can be put in a matrix form: $\begin{bmatrix} \pi(q|p) & \pi(q|\neg p) \\ \pi(\neg q|p) & \pi(\neg q|\neg p) \end{bmatrix}$. An uncertain rule like R1 of the form "if p then q" has a matrix of the form $\begin{bmatrix} 1 & 1 \\ r & 1 \end{bmatrix}$; a pair of rules like R2 and R'2 of the form "if p then q" and "if $\neg p$ then $\neg q$" is encoded by a matrix of the form $\begin{bmatrix} 1 & s \\ r & 1 \end{bmatrix}$; and the matrix $\begin{bmatrix} 1 & 0 \\ 0 & 1 \end{bmatrix}$ expresses a form of equivalence between p and q. The propagation of uncertainty is performed by the max-min composition:

$$\pi(q) = \max(\min(\pi(q|p), \pi(p)), \min(\pi(q|\neg p), \pi(\neg p))$$
$$\pi(\neg q) = \max(\min(\pi(\neg q|p), \pi(p)), \min(\pi(\neg q|\neg p), \pi(\neg p)),$$

i.e., by a matrix product $\otimes$ (min is replaced by product in the quantitative setting):[1]

$$\begin{bmatrix} \pi(q) \\ \pi(\neg q) \end{bmatrix} = \begin{bmatrix} \pi(q|p) & \pi(q|\neg p) \\ \pi(\neg q|p) & \pi(\neg q|\neg p) \end{bmatrix} \otimes \begin{bmatrix} \pi(p) \\ \pi(\neg p) \end{bmatrix}. \tag{1}$$

---

[1] For the max-min composition, $\begin{bmatrix} \pi(q) \\ \pi(\neg q) \end{bmatrix} = \begin{bmatrix} \Pi(p \wedge q) & \Pi(\neg p \wedge q) \\ \Pi(p \wedge \neg q) & \Pi(\neg p \wedge \neg q) \end{bmatrix} \otimes \begin{bmatrix} \pi(p) \\ \pi(\neg p) \end{bmatrix}$ (1'), taking advantage of the monotonicity and of the max-decomposability of possibility measures. In contrast with (1), there are inequality constraints between matrix terms and $(\pi(p), \pi(\neg p))$.

The result is normalized, i.e., $\max(\pi(q), \pi(\neg q)) = 1$, since the columns of the square matrix are normalized and $\max(\pi(p), \pi(\neg p)) = 1$.

In case of ignorance about p (i.e., $\pi(p) = \pi(\neg p) = 1$), one concludes that q is ignored as well ($\pi(q) = \pi(\neg q) = 1$), except if the knowledge says that q (or $\neg q$) is somewhat certain in both contexts p and $\neg p$. Namely, even if p is unknown, we have for instance, $\begin{bmatrix} 1 & 1 \\ r & t \end{bmatrix} \otimes \begin{bmatrix} 1 \\ 1 \end{bmatrix} = \begin{bmatrix} 1 \\ \max(r, t) \end{bmatrix}$, i.e., q is somewhat certain, unconditionally.

In the following, we assume that we are never in the above situation, and that we deal with rules, i.e., for a rule R_i, or a pair (R_i, R_i') such that $\begin{bmatrix} \pi(q_i|p_i) & \pi(q_i|\neg p_i) \\ \pi(\neg q_i|p_i) & \pi(\neg q_i|\neg p_i) \end{bmatrix} = \begin{bmatrix} 1 & s_i \\ r_i & 1 \end{bmatrix}$, with possibly s_i or r_i equal to 1. Letting $\begin{bmatrix} \pi(p_i) \\ \pi(\neg p_i) \end{bmatrix} = \begin{bmatrix} \lambda_i \\ \rho_i \end{bmatrix}$, we get $\begin{bmatrix} \pi(q_i) \\ \pi(\overline{q_i}) \end{bmatrix} = \begin{bmatrix} \alpha_i \\ \beta_i \end{bmatrix} = \begin{bmatrix} 1 & s_i \\ r_i & 1 \end{bmatrix} \otimes \begin{bmatrix} \lambda_i \\ \rho_i \end{bmatrix}$. Due to the normalisation of the inputs, we have $\alpha_i = \max(\lambda_i, s_i)$ and $\beta_i = \max(\rho_i, r_i)$., i.e., $\begin{bmatrix} \alpha_i \\ \beta_i \end{bmatrix} = \max(\begin{bmatrix} s_i \\ r_i \end{bmatrix}, \begin{bmatrix} \lambda_i \\ \rho_i \end{bmatrix})$.

4. **Possibility distribution associated to the conclusion of a rule.** If we have applied the rule "if p then q" where q stands for "$b(x) \in Q$", the resulting possibility distribution is $\pi_{b(x)}^*(u) = \begin{cases} \pi(q) & \text{if } u \in Q \\ \pi(\neg q) & \text{otherwise} \end{cases}$, which also writes ($\mu_Q$ is the characteristic function of $Q \subseteq \mathcal{D}_{b(x)}$), $\pi_{b(x)}^*(u) = \max(\min(\mu_Q(u), \pi(q)), \min(\mu_{\overline{Q}}(u), \pi(\neg q)))$.

5. **Combination of the conclusions obtained from several parallel rules.** When several rules pertain to the same attribute b for an item x, one applies a min-based conjunctive combination. In case of two rules with conclusions q^i ("$b(x) \in Q^i$"):

$$\pi_{b(x)}^*(u) = \min(\pi_{b(x)}^{*1}(u), \pi_{b(x)}^{*2}(u)) =$$
$$\max(\min(\mu_{Q^1 \cap Q^2}(u), \pi(q^1), \pi(q^2)), \min(\mu_{Q^1 \cap \overline{Q^2}}(u), \pi(q^1), \pi(\neg q^2)),$$
$$\min(\mu_{\overline{Q^1} \cap Q^2}(u), \pi(\neg q^1), \pi(q^2)), \min(\mu_{\overline{Q^1} \cap \overline{Q^2}}(u), \pi(\neg q^1), \pi(\neg q^2))).$$

Obviously, the normalization of $\pi_{b(x)}^{*1}$ and $\pi_{b(x)}^{*2}$ does not guarantee the normalization of $\pi_{b(x)}^*$. In this case the non normalization of the result reveals a lack of coherence of the rules which are then inconsistent with some inputs [18]. Since the subsets $Q^1 \cap Q^2$, $\overline{Q^1} \cap Q^2$, $Q^1 \cap \overline{Q^2}$, $\overline{Q^1} \cap \overline{Q^2}$ form a partition of $\mathcal{D}_{b(x)}$ (provided that they are not empty), $\pi_{b(x)}^*$ is the union of *disjoint* subsets with weights:

$$\begin{bmatrix} \Pi(Q_1 \cap Q_2) \\ \Pi(Q_1 \cap \overline{Q_2}) \\ \Pi(\overline{Q_1} \cap Q_2) \\ \Pi(\overline{Q_1} \cap \overline{Q_2}) \end{bmatrix} = \begin{bmatrix} \min(\pi(q^1), \pi(q^2)) \\ \min(\pi(q^1), \pi(\neg q^2)) \\ \min(\neg q^1), \pi(q^2)) \\ \min(\pi(\neg q^1), \pi(\neg q^2)) \end{bmatrix} = \begin{bmatrix} \min(\alpha_1, \alpha_2) \\ \min(\alpha_1, \beta_2) \\ \min(\beta_1, \alpha_2) \\ \min(\beta_1, \beta_2) \end{bmatrix}.$$

Using the notations of step 3 for the uncertainty propagation in rule R_i the computation of $\pi^*_{b(x)}$ can be computed as a min-max product, denoted $\boxdot$:

$$\begin{bmatrix} \Pi(Q_1 \cap Q_2) \\ \Pi(Q_1 \cap \overline{Q_2}) \\ \Pi(\overline{Q_1} \cap Q_2) \\ \Pi(\overline{Q_1} \cap \overline{Q_2}) \end{bmatrix} = \begin{bmatrix} s_1 & 1 & s_2 & 1 \\ s_1 & 1 & 1 & r_2 \\ 1 & r_1 & s_2 & 1 \\ 1 & r_1 & 1 & r_2 \end{bmatrix} \boxdot \begin{bmatrix} \lambda_1 \\ \rho_1 \\ \lambda_2 \\ \rho_2 \end{bmatrix},$$

and

$$\pi^*_{b(x)}(u) = \min(\pi^{*1}_{b(x)}(u), \pi^{*2}_{b(x)}(u)) = \begin{bmatrix} \min(\max(s_1,\lambda_1),\max(s_2,\lambda_2)) \\ \min(\max(s_1,\lambda_1),\max(r_2,\rho_2)) \\ \min(\max(r_1,\rho_1),\max(s_2,\lambda_2)) \\ \min(\max(r_1,\rho_1),\max(r_2,\rho_2)) \end{bmatrix} \begin{array}{l} \text{if } u \in Q_1 \cap Q_2 \\ \text{if } u \in Q_1 \cap \overline{Q_2} \\ \text{if } u \in \overline{Q_1} \cap Q_2 \\ \text{if } u \in \overline{Q_1} \cap \overline{Q_2} \end{array}.$$

This is for two rules, but it straightforwardly extends to n rules.

2.2 Example

In the example, the output of rules $R1$, $(R2, R'2)$ handled together and $R3$ correspond to the 3 subsets $Q_1 = \{\text{"professor"}, \text{"businessman"}, \text{"lawyer"}, \text{"doctor"}\}$; $Q_2 = \{\text{"engineer"}, \text{"researcher"}, \text{"architect"}\}$; $Q_3 = \{\text{"professor"}, \text{"researcher"}\}$. Note that $Q_1 \cap Q_2 = \emptyset$; $Q_3 \subseteq Q_1 \cup Q_2$.

In order to compute the output $\pi^*_{pro(x)}$, applying step 5 to 3 rules, we have to consider $8 = 2^3$ subsets $Q_1^* \cap Q_2^* \cap Q_3^*$ where Q_i^* stands for Q_i or $\overline{Q_i}$. We have $Q_1 \cap Q_2 \cap Q_3 = Q_1 \cap Q_2 \cap \overline{Q_3} = \emptyset$; $Q_1 \cap \overline{Q_2} \cap Q_3 = Q_1 \cap Q_3 = \{\text{"professor"}\}$; $Q_1 \cap \overline{Q_2} \cap \overline{Q_3} = Q_1 \cap \overline{Q_3} = \{\text{"businessman"}, \text{"lawyer"}, \text{"doctor"}\}$; $\overline{Q_1} \cap Q_2 \cap Q_3 = Q_2 \cap Q_3 = \{\text{"researcher"}\}$; $\overline{Q_1} \cap Q_2 \cap \overline{Q_3} = Q_2 \cap \overline{Q_3} = \{\text{"architect"}, \text{"engineer"}\}$; $\overline{Q_1} \cap \overline{Q_2} \cap Q_3 = \emptyset$; $\overline{Q_1} \cap \overline{Q_2} \cap \overline{Q_3} = \overline{Q_1} \cap \overline{Q_2} = \{\text{"others"}\}$ (where $\mathcal{D}_{pro} = Q_1 \cup Q_2 \cup Q_3 \cup \{others\}$).

Since the two first subsets and the second last one are empty, we have only 5 lines in the matrix product

$$\begin{bmatrix} \Pi(Q_1 \cap \overline{Q_3}) \\ \Pi(Q_1 \cap Q_3) \\ \Pi(Q_2 \cap Q_3) \\ \Pi(Q_2 \cap \overline{Q_3}) \\ \Pi(\overline{Q_1} \cap \overline{Q_2}) \end{bmatrix} = \begin{bmatrix} \min(\alpha_1,\beta_2,\beta_3) \\ \min(\alpha_1,\beta_2,\alpha_3) \\ \min(\beta_1,\alpha_2,\alpha_3) \\ \min(\beta_1,\alpha_2\,\beta_3) \\ \min(\beta_1,\beta_2,\beta_3) \end{bmatrix} = \begin{bmatrix} s_1 & 1 & 1 & r_2 & 1 & r_3 \\ s_1 & 1 & 1 & r_2 & s_3 & 1 \\ 1 & r_1 & s_2 & 1 & s_3 & 1 \\ 1 & r_1 & s_2 & 1 & 1 & r_3 \\ 1 & r_1 & 1 & r_2 & 1 & r_3 \end{bmatrix} \boxdot \begin{bmatrix} \lambda_1 \\ \rho_1 \\ \lambda_2 \\ \rho_2 \\ \lambda_3 \\ \rho_3 \end{bmatrix}.$$

which gives:

$$\pi^*_{pro(x)}(u) = \begin{bmatrix} \min(\max(s_1,\lambda_1),\max(r_2,\rho_2),\max(r_3,\rho_3)) \\ \min(\max(s_1,\lambda_1),\max(r_2,\rho_2),\max(s_3,\lambda_3)) \\ \min(\max(r_1,\rho_1),\max(s_2,\lambda_2),\max(s_3,\lambda_3)) \\ \min(\max(r_1,\rho_1),\max(s_2,\lambda_2),\max(r_3,\rho_3)) \\ \min(\max(r_1,\rho_1),\max(r_2,\rho_2),\max(r_3,\rho_3)) \end{bmatrix} \begin{array}{l} \text{if } u \in Q_1 \cap \overline{Q_3} \\ \text{if } u \in Q_1 \cap Q_3 \\ \text{if } u \in Q_2 \cap Q_3 \\ \text{if } u \in Q_2 \cap \overline{Q_3} \\ \text{if } u \in \overline{Q_1} \cap \overline{Q_2} \end{array}.$$

2.3   From a Functional View to a Logical View

The above approach strictly adheres to possibility theory, following the different steps of a rule-based inference system. Thus it amounts to the computation of a function. Each layer of parallel rules can be viewed as the result of a $\min - \max$ matrix product. Several layers (two in our example) of rules, used in forward chaining, correspond to a cascade of such matrix products, the inputs of which are computed as a pattern matching step. These products define an input-output (non-linear) function. It has some structural and computational resemblance with a $\min - \max$ multiple layer neural net [9]. As proposed in [29], this matrix calculus can be used as a basis for explanation purposes, by leaving uninstantiated some of the weights and observing their impact on the result. In that respect, the possibilistic handling of symbolic weights presents some similarity with the idea underlying provenance calculus [32] in databases.

The above approach, which can be understood as the description of an input-output function, can be also put into a logical format. Indeed, the rule "if p then q is somewhat certain" is represented by the pair $(\pi(q|p), \pi(\neg q|p)) = (1, r)$. Using possibility-necessity duality, we get $N(q|p) = 1 - \pi(\neg q|p) = 1 - r$ and $N(\neg q|p) = 1 - \pi(q|p) = 0$. The conditional necessity and the necessity of a material implication $N(p \to q)$ are distinct notions, although they are closely related: We have $N(q|p) = 0$ if $N(\neg p) = N(p \to q)$, and $N(q|p) = N(p \to q)$ otherwise.

In possibilistic logic [22,26], classical logic formulas are associated with lower bounds of their necessity degrees, and material implication is used instead of conditioning. When formulas are associated with degree 1, we retrieve classical logic.

The inference can be also be put under the form of a max-min matrix calculus [16], This is structurally similar to (1), but note the inequality:

$$\begin{bmatrix} N(q) \\ N(\neg q) \end{bmatrix} \geq \begin{bmatrix} N(p \to q) & N(\neg p \to q) \\ N(p \to \neg q) & N(\neg p \to \neg q) \end{bmatrix} \otimes \begin{bmatrix} N(p) \\ N(\neg p) \end{bmatrix}. \tag{2}$$

Even if (1) and (2) look different, they do the same job for propagating uncertainty.[2]

This can be checked by careful examination. Take (1) as $\begin{bmatrix} \pi(q) \\ \pi(\neg q) \end{bmatrix} = \begin{bmatrix} \alpha \\ \beta \end{bmatrix} = \begin{bmatrix} 1 & s \\ r & 1 \end{bmatrix} \otimes \begin{bmatrix} \lambda \\ \rho \end{bmatrix}$. Then due to input normalization, $\alpha = \max(s, \lambda)$; $\beta = \max(r, \rho)$

---

[2] By duality the inequality (2) writes $\begin{bmatrix} \pi(q) \\ \pi(\neg q) \end{bmatrix} \leq \begin{bmatrix} \Pi(p \wedge q) & \Pi(\neg p \wedge q) \\ \Pi(p \wedge \neg q) & \Pi(\neg p \wedge \neg q) \end{bmatrix} \square$ $\begin{bmatrix} \pi(\neg p) \\ \pi(p) \end{bmatrix}$ (2') where $\square$ is the $\min - \max$ product. Note that (2') provides an upper bound for $\begin{bmatrix} \pi(q) \\ \pi(\neg q) \end{bmatrix}$ obtained by (1'). It can be checked that this upper bound coincides with this latter vector if $\Pi(p \wedge q) = \Pi(\neg p \wedge \neg q) = 1$ or if $\Pi(\neg p \wedge q) = \Pi(p \wedge \neg q) = 1$.

(this remains true for product-based conditioning). Besides, we should have $\min(N(p \to q), N(p \to \neg q)) = 0 = \min(N(\neg p \to q), N(\neg p \to \neg q))$ in (2), otherwise it would constrain the input with $N(\neg p) > 0$ or $N(p) > 0$. Consider the case where (2) is of the form

$$\begin{bmatrix} 1 - \beta \\ 1 - \alpha \end{bmatrix} \geq \begin{bmatrix} N(p \to q) & 0 \\ 0 & N(\neg p \to \neg q) \end{bmatrix} \otimes \begin{bmatrix} 1 - \rho \\ 1 - \lambda \end{bmatrix}$$

in agreement with conditional necessity.

The max-min product reduces to $\begin{bmatrix} 1 - \beta \\ 1 - \alpha \end{bmatrix} \geq \min(\begin{bmatrix} N(p \to q) \\ N(\neg p \to \neg q) \end{bmatrix}, \begin{bmatrix} 1 - \rho \\ 1 - \lambda \end{bmatrix})$.
This is the dual of expression (1) that reduces to the max of two vectors (up to the inequality). Thus, $\alpha \leq \max(1 - N(\neg p \to \neg q), \lambda) = \max(\Pi(\neg p \wedge q), \lambda)$ is to be compared with $\alpha = \max(s, \lambda) = \max(\pi(q|\neg p), \lambda)$. In general, $\pi(q|\neg p) = \Pi(\neg p \wedge q)$ except if $\pi(\neg p) = \Pi(\neg p \wedge q)$ where $\pi(q|\neg p) = 1$; in this latter case the inequality writes $\alpha = \max(\pi(q|\neg p), \lambda) = 1 \leq \max(\pi(\neg p), \lambda) = \max(\pi(\neg p), \pi(p)) = 1$, which clearly holds. A similar computation can de made for β.

A set of (uncertain) if-then rules, represented by conditional tables, can be translated into a possibilistic logic base. In contrast with logic, the expression in terms of conditional possibility captures the directed nature of if-then rules (contraposition does not hold: $\pi(q|p) \neq \pi(\neg p|\neg q)$ while $p \to q \equiv \neg q \to \neg p$).

The rule-based approach has also some similarity with possibilistic Bayesian networks [1] where conditioning is also used, but the max-min matrix product only performs local propagation, while the chain rule in possibilistic nets compute a joint possibility distribution over variables [1]. Possibilistic networks, using either min-based or product-based conditioning, can be translated exactly under the form of a set of possibilistic logic formulas and vice-versa [1]. Thus, we have three representation formats (rule-based, logic, possibilistic net) that can be translated into one another. Min-based and product-based possibilistic conditionings are based on counterparts of Bayes formula; see [21] for a discussion of possibilistic Bayes rules in an abductive perspective and [20] in a fusion perspective.

3   Possibilistic Learning

In the previous section, we have seen that a set of uncertain rules can be read either as a function induced by a cascade of $\min - \max$ matrix compositions, agreeing with possibility theory and resembling a $\min - \max$ multiple layer neural net, or a possibilistic logic base, or yet a possibilistic Bayesian-like net. If the rules are not provided by some expert, the question of learning one of these representations arises.

In our example, we may imagine having a Boolean table where people are described in terms of their aspirations, the profession they practice, and the level of their salary. Under a functional view, one may learn a $\min - \max$ neural network [51]. With a probabilistic approach, if we view the salary level as classes, using, e.g., a naive Bayesian classifier one may compute the probability of classifying a new item in a particular class. One may also learn the concept of, e.g., "high" salary, and associate the rule(s) with a probability. But, how to get a possibility distribution over the salaries, or over the professions? How to get certainty levels associated to rules?

Before going further, it is important to keep in mind several distinctive features of possibility theory. First, possibility theory is dedicated to the representation of epistemic uncertainty, since the pair of dual measures of possibility and of necessity can distinguish between what is possible without being certain at all and what is certain to some extent. So, the possibilistic setting is specially appropriate in case of missing information, of imprecise information, or of poor information (when information is too sparse for statistical purposes). Second, possibility theory has a qualitative version (where typically a finite linearly ordered scale is used and conditioning is min-based) and a quantitative version (where the scale is the unit interval $[0, 1]$ and conditioning is defined by the product). In any case, possibility theory is not a substitute to probability theory, and if good statistics are available one should privilege probabilistic approaches for learning purposes. Third, the possibilistic framework is highly compatible with classical logic, through the use of standard possibilistic logic. In that respect, one should remember that a belief base in classical logic usually has a subset of models that represents an imprecise view of the world inasmuch there are propositions whose truth value remains unknown, while the specification of a join probability distribution does not leave room for incomplete information, nor for inconsistent information.

In the following, we first recall possible ways of extracting a possibility distribution and certainty levels from data, before surveying the variety of works existing in possibilistic learning, which are not always aware of each other.

3.1 Building Possibility Distributions and Eliciting Certainty

A survey of practical methods for obtaining possibility distributions, both in quantitative and in qualitative possibility theory, can be found in [23]. In the qualitative case, the task is made easier since only the relative values of the possibility degrees matter when expressing strict inequalities between them. Then one may exploit constraints induced by certainty qualification, or representing conditional statements. We now focus on the quantitative case. Roughly speaking, one may either read statistical information in a possibilistic way, or learn from imprecise or incomplete data. We first review the statistical approach.

From Statistical Data to Possibility Distributions. There are two main methods for changing probability into possibility: one named "antipignistic" and another obeying minimum specificity principle, that we recall successively. They are useful when having to combine pieces of uncertain information received in different formats.

Let us start with a motivating example [14]. Consider a game of coin tossing with an unfair coin. Assume the coin is biased so that heads are the most frequent outcome. The set of events is $X = \{x_1 = \text{head}, x_2 = \text{tail}\}$. We have $p(x_1) > 1/2 > p(x_2)$ with $p(x_1) + p(x_2) = 1$ in terms of probability. A natural measurement of this necessity (certainty) $n(x_1)$ of getting head is the *excess of probability* in favor of head, namely $n(x_1) = p(x_1) - p(x_2)$, while $n(x_2) = 0$. Indeed, if $p(x_2) = 0$ we are fully certain that heads will be the outcome. The larger $p(x_2) < 0.5$ the less certain we are about the head output.

More generally, in a finite setting, let $X = \{x_1, \cdots, x_n\}$ with $p(x_i) = p_i$ and $\Sigma_{i=1}^n p_i = 1$. Then one can define the set function N given by

$$N(A) = \Sigma_{x_i \in A} \max(p_i - \max_{x_k \notin A} p_k, 0) = \max(P(A) - \max_{x_k \notin A} p_k, 0).$$

Assuming $p_1 \geq \cdots \geq p_i \geq \cdots \geq p_n$ and letting $A_i = \{x_1, \cdots, x_i\}$, we have $N(A_i) = \Sigma_{j=1}^i p_j - p_{i+1}$, $\forall i = 1, \cdots, n$ with p_{n+1} by convention. Then it can be shown that $N(A) = \max_{A_k \subseteq A} N(A_k)$ and N is a necessity measure (it is min decomposable, and $\min(N(A), N(\overline{A})) = 0, \forall A$); moreover, $N(A) \geq Prob(A), \forall A \subseteq X$ [14]. Let $\pi(x_i) = \pi_i$. The possibility distribution π underlying N (such that $N(A) = \min_{x \notin A} 1 - \pi(x)$) is $\pi_i = \Sigma_{j=1}^n \min(p_j, p_i), i = 1, \ldots, n$.

It defines a bijective mapping which turns probability measures into possibility measures [14]. In particular we have $\pi_i > \pi_{i+1} \Leftrightarrow p_i > p_{i+1}$. This approach provides a statistical interpretation of certainty levels that experts attach to rules. The idea of excess of probability provides an intuitive ground to the perception of the idea of certainty.

This transformation has other intuitive grounds. The converse transformation is given by $p_i = \Sigma_{j=i}^n \frac{1}{j}(\pi_j - \pi_{j+1})$. Especially, this converse mapping turns a uniform possibility distributions into a uniform probability distribution when no information is available. It is also known as Smets' *pignistic* transformation and the Shapley value; see [12,50] for more justifications.

Another transformation of statistical probability into possibility is based on a totally different rationale. This is the *most specific* possibility distribution π^* whose associated possibility measure dominates the probability measure [10,13]. In a finite setting, if $p_1 > \cdots > p_i, \cdots > p_n$, then the most specific possibility distribution is defined by: $\pi_i^* = \Sigma_{j \geq i} p_j$.

It can be checked that $\forall i = 1, \cdots n, \pi_i^* \leq \pi_i$, i.e., π^* is more specific than the antipignistic transform. This transformation is motivated by the concern to keep as much information as possible.

Whatever the transformation used, an issue is then how to apply it for conditional possibility. Indeed consider the simplest network linking two variables, one has $p(x, y) = p(y|x) \cdot p(x)$ in probability, while in possibility we have $\pi(x, y) = \pi(y|x) \cdot \pi(x)$ (or $\pi(x, y) = \min(\pi(y|x), \pi(x))$ in a qualitative setting). It can be checked that whatever the transformation, whatever how conditioning is defined for possibility, there is no well-behaved correspondence when changing simultaneously the joint distribution, the conditional distribution and the marginal distribution from probability to possibility; see [33] for a discussion and experiments. If we apply a transformation, it should be done on the *joint distribution* (from which the other distributions can be derived).

Handling Imprecise Data. As already said, it may be questionable, when good statistical probabilities are available, to turn them into possibilities. Possibilities makes much more sense when learning from incomplete data or poor data.

In case of epistemic uncertainty, one may have imprecise information under the form of interval data, thus leading to a random set, which may in turn be viewed as the basic probability assignment underlying a belief function [49]. In general, this set of intervals is not nested, and thus does not directly give birth to a necessity measure in place of the belief function. Then one may look for a nested approximation of the set of intervals (e.g., by enlarging some of them) [15], and more generally by computing the best upper or lower consonant approximation of a belief function [17]. See [23] for a detailed overview. It has been also proved that the least informative belief function having a given pignistic probability transform is unique and consonant [27] and is thus a necessity measure (it is the antipignistic transformation).

However, a straightforward way of deriving a possibility distribution from statistical interval data is to consider the *contour function* [49] of the basic probability assignment m (i.e., the one-point coverage function of the random set) $\pi_*(u) = \Sigma_{u \in I} m(I)$, where $m(I)$ is the relative number of observations of interval I. Clearly, this is only a partial view of the data, as it is generally not possible to reconstruct m from π_*. This view of possibility distributions as induced by random sets was very early pointed out by Kampé de Fériet [38]. Even if π_* is not generally normalized, and once renormalized leads to a quite imprecise representation of the belief function [17], it is also useful for defining a possibilistic likelihood-based score function.

Indeed, as log-likelihood functions are used to measure the adequateness of a probability distribution with respect to a set of data, a new likelihood function based on contour functions has been proposed in the possibilistic framework [35] in order to learn possibilistic network structures from imprecise data. The idea is to define a random set likelihood (replacing the probability distribution by the basic probability assignment) and then to approximate it in terms of contour function. It can be shown that the possibilititic transform of the basic probability assignment maximizing the random set likelihood maximizes the possibilititic likelihood, which indicates that the two likelihood agree from a maximization point of view. The maximization takes place under the constraint that the sum of the possibility degrees is equal to some fixed value S called imprecision degree where $S > 1$ in order to control the imprecision of data sets [34].

Besides, an informational distance function for possibility distributions that agrees with the view of possibility distributions as families of probabilities that are upper and lower bounded by the associated possibility and necessity measures, has been proposed in [47]. The authors show that, given a set of data following a probability distribution, the optimal possibility distribution with respect to this informational distance is the distribution obtained as the result of the probability-possibility transformation that agrees with the maximal specificity principle. It is also shown that when the optimal distribution is not available due to representation bias, maximizing this possibilistic informational distance provides more faithful results than approximating the probability distribution and then applying the probability-possibility transformation.

3.2 Possibilistic Learning: A Survey

Existing works in possibilistic learning can be roughly clustered in four groups. We can learn: i) default rules, ii) possibilistic networks and naive possibilistic classifiers, iii) possibilistic logic theories, and iv) decision trees.

Learning Default Rules. It is desirable to guarantee that any new default rule deducible from a set of default rules extracted from a data set is also valid with respect to this data set. It has been shown [3] that default rules of the form "if p then generally q", denoted by $p \rightsquigarrow q$, where $\rightsquigarrow$ obeys the postulates of preferential inference [39], have both

- a possibilistic semantics expressed by the constraint $\Pi(p \wedge q) > \Pi(p \wedge \neg q)$, for some possibility measure Π.
- a probabilistic semantics expressed by the constraint $Prob(p \wedge q) > Prob(p \wedge \neg q)$ for some *big-stepped probability Prob*.

This is a very special kind of probability measure such that if $p_1 > p_2 > \ldots > p_{n-1} \geq p_n$, the following inequalities hold $\forall i = 1, \ldots, n-1$, $p_i > \Sigma_{j=i,n}\ p_j$. Note that here we retrieve the idea of an excess of probability which is present whatever the conditioning. Then, one can safely infer a new default $p \rightsquigarrow q$ from a set of K defaults $\Delta = \{p_k \rightsquigarrow q_k | k = 1, K\}$ if and only if the constraints modeling Δ entail the constraints modeling $p \rightsquigarrow q$. Thus, extracting defaults amounts to looking for big-stepped probabilities, by clustering lines describing items in Boolean tables, so as to find default rules, see [2] for details. Then the rules discovered are genuine default rules, and can be encoded in possibilistic logic (assuming rational monotony for the inference relation).

Learning Possibilistic Networks and Possibilistic Classifiers. We first focus on the use of possibilistic networks for classification. Based on the quantitative counterpart of Bayes rule, possibilistic naive possibilistic classifiers have been proposed. In [4], the authors handle uncertain inputs in binary possibilistic-based classification, using a possibilistic counterpart of Jeffrey's rule, and present an efficient algorithm in polynomial time. The interest of possibilistic classifiers for handling uncertainty is also advocated in [7] in the case of numerical data where possibility distributions are used for encoding families of Gaussian probabilistic distributions compatible with the data set.

The learning of possibilistic networks, as graphical models suitable for representing and reasoning with uncertain and incomplete information, has been pioneered by R. Kruse and its co-authors, e.g., [5,6,31] emphasizing their interest for dealing with imprecise data. In [36], two types of approaches are compared for learning possibilistic networks from data, one based on probability-possibility transformation, the other using contour functions and possibilistic likelihood. The experimental comparison gives a slight advantage to the second type of approach.

Learning Possibilistic Logic Theories. A new formalization of inductive logic programming (ILP) in first-order possibilistic logic that allows us to handle exceptions by means of prioritized rules has been proposed in [46]. The possibilistic formalization provides a sound encoding of non-monotonic reasoning [3] that copes with rules with exceptions and prevents an example from being classified in more than one class.

A different approach to the induction of possibilistic logic theories is proposed in [41]. It relies on the fact that any set of Markov logic [45] formulas can be exactly translated into possibilistic logic formulas [26,40]. So we can use the learning machinery of Markov logic to learn possibilistic logic theories. The approach allows for representing joint relational models in such a way that each weighted formula can be interpreted in isolation.

Learning Decision Trees. An extension of classification trees to deal with uncertain information where uncertainty is encoded in possibility theory is presented in [37] to cope with imprecise class labels in the training set. Class labels in data sets are no longer singletons but are given in the form of possibility distributions. The approach uses an information closeness index when evaluating the entropy of a given training partition in the attribute selection step. Besides, taking advantage of a cumulative entropy function derived from the informational distance function for possibility distributions [47], a procedure is presented in [48], which performs significant choices of split and provides a statistically relevant stopping criterion that allows the learning of trees whose size is well-suited w.r.t. the available data.

4   Concluding Remarks - Final Discussion

Possibilistic learning has been investigated in a rather scattered way until now and has not yet raised much attention. In order to learn a possibilistic model one may think of learning first a probabilistic model and to transform probabilities into possibility and necessity values. But this may create difficulties about the agreement of the possibilistic reasoning machinery with the probabilities that are learnt. A better strategy would be to directly learn a possibilistic model.

We might also think of taking advantage of the $\min - \max$ matrix composition view of a set of if-then rules. In the example used in the first part of the paper, the attributes have a clear meaning, it would be natural to look for a two layer model, although on might argue here that both aspirations and salary levels determine the profession one looks for. Once the number of the layers is fixed, and the variables of each layer determined, one may directly compute the relevant possibility degrees from the set of examples. But this is under the non very realistic assumption of knowing the variables in each layer. One may try to use a neural net approach, but this will raise the problem of its interpretation. Yet another type of method might be to learn rules using a version space approach, whose extension to a possibilistic setting, for coping with outliers, makes sense [24]. These are clearly topics for further research.

This paper has aimed at showing that rule-based systems and learning are not just the past and the future of artificial intelligence respectively, but they have connections and may be made more compatible. The paper has used the possibilistic setting which is especially appropriate as a bridge between logic and numerical approaches. This paper can be also related to the more general concern of unifying logic-based machine learning techniques, a topic that has been forgotten or neglected [24].

References

1. Benferhat, S., Dubois, D., Garcia, L., Prade, H.: On the transformation between possibilistic logic bases and possibilistic causal networks. Int. J. Approx. Reas. **29**(2), 135–173 (2002)
2. Benferhat, S., Dubois, D., Lagrue, S., Prade, H.: A big-stepped probability approach for discovering default rules. Int. J. Uncert. Fuzz. Knowl.-Based Syst. **11**(Suppl. 1), 1–14 (2003)
3. Benferhat, S., Dubois, D., Prade, H.: Possibilistic and standard probabilistic semantics of conditional knowledge bases. J. Log. Comput. **9**(6), 873–895 (1999)
4. Benferhat, S., Tabia, K.: Binary naive possibilistic classifiers: handling uncertain inputs. Int. J. Intell. Syst. **24**(12), 1203–1229 (2009)
5. Borgelt, C., Kruse, R.: Learning from imprecise data: possibilistic graphical models. Comput. Stat. Data Anal. **38**(4), 449–463 (2002)
6. Borgelt, C., Steinbrecher, M., Kruse, R.: Graphical Models: Representations for Learning, Reasoning and Data Mining, 2nd edn. Wiley (2009)
7. Bounhas, M., Ghasemi Hamed, M., Prade, H., Serrurier, M., Mellouli, K.: Naive possibilistic classifiers for imprecise or uncertain numerical data. Fuzzy Sets Syst. **239**, 137–156 (2014)
8. Bouraoui, Z., et al.: From shallow to deep interactions between knowledge representation, reasoning and machine learning (Kay R. Amel group). CoRR abs/1912.06612 (2019)
9. Buckley, J.J., Hayashi, Y.: Fuzzy neural networks: a survey. Fuzzy Sets Syst. **66**, 1–13 (1994)
10. Delgado, M., Moral, S.: On the concept of possibility-probability consistency. Fuzzy Sets Syst. **21**, 311–318 (1987)
11. Denœux, T.: Logistic regression, neural networks and Dempster-Shafer theory: a new perspective. Knowl.-Based Syst. **176**, 54–67 (2019)
12. Dubois, D., Foulloy, L., Mauris, G., Prade, H.: Probability-possibility transformations, triangular fuzzy sets, and probabilistic inequalities. Reliable Comput. **10**, 273–297 (2004)
13. Dubois, D., Prade, H.: On several representations of an uncertain body of evidence. In: Gupta, M., Sanchez, E. (eds.) Fuzzy Information and Decision Processes, pp. 167–181. North-Holland, Amsterdam (1982)

14. Dubois, D., Prade, H.: Unfair coins and necessity measures: a possibilistic interpretation of histograms. Fuzzy Sets Syst. **10**(1), 15–20 (1983)
15. Dubois, D., Prade, H.: Fuzzy sets and statistical data. Europ. J. Oper. Res. **25**, 345–356 (1986)
16. Dubois, D., Prade, H.: Possibility Theory - An Approach to Computerized Processing of Uncertainty. Plenum Press (1988)
17. Dubois, D., Prade, H.: Consonant approximations of belief functions. Int. J. Approximate Reasoning **4**, 419–449 (1990)
18. Dubois, D., Prade, H.: On the validation of fuzzy knowledge bases. In: Tzafestas, S.G., Venetsanopoulos, A.N. (eds.) Fuzzy Reasoning in Information, Decision and Control Systems, pp. 31–49. Kluwer Acad. Publ. (1994)
19. Dubois, D., Prade, H.: Fuzzy criteria and fuzzy rules in subjective evaluation - a general discussion. In: Proceedings of the 5th European Congress on Intelligent Techniques and Soft Computing (EUFIT 1997), Aachen, vol. 1, pp. 975–979 (1997)
20. Dubois, D., Prade, H., Yager, R.R.: Merging fuzzy information. In: Bezdek, J., Dubois, D., Prade, H. (eds.) Fuzzy Sets in Approximate Reasoning and Information Systems. The Handbooks of Fuzzy Sets Series, pp. 335–401. Kluwer, Boston (1999)
21. Dubois, D., Prade, H.: An overview of ordinal and numerical approaches to causal diagnostic problem solving. In: Gabbay, D.M., Kruse, R. (eds.) Abductive Reasoning and Uncertainty Management Systems. Handbook of Defeasible Reasoning and Uncertainty Management Systems, vol. 4, pp. 231–280. Kluwer Acad. Publ. (2000). Gabbay, D.M., Smets, Ph. (series eds.)
22. Dubois, D., Prade, H.: Possibility theory and its applications: where do we stand? In: Kacprzyk, J., Pedrycz, W. (eds.) Springer Handbook of Computational Intelligence, pp. 31–60. Springer, Heidelberg (2015). https://doi.org/10.1007/978-3-662-43505-2_3
23. Dubois, D., Prade, H.: Practical methods for constructing possibility distributions. Int. J. Intell. Syst. **31**(3), 215–239 (2016)
24. Dubois, D., Prade, H.: Towards a logic-based view of some approaches to classification tasks. In: Lesot, M.-J., et al. (eds.) IPMU 2020. CCIS, vol. 1239, pp. 697–711. Springer, Cham (2020). https://doi.org/10.1007/978-3-030-50153-2_51
25. Dubois, D., Prade, H., Rico, A.: The logical encoding of Sugeno integrals. Fuzzy Sets Syst. **241**, 61–75 (2014)
26. Dubois, D., Prade, H., Schockaert, S.: Generalized possibilistic logic: foundations and applications to qualitative reasoning about uncertainty. Artif. Intell. **252**, 139–174 (2017)
27. Dubois, D., Prade, H., Smets, P.: A definition of subjective possibility. Int. J. Approx. Reasoning **48**(2), 352–364 (2008)
28. Farreny, H., Prade, H.: Default and inexact reasoning with possibility degrees. IEEE Trans. Syst. Man Cybern. **16**(2), 270–276 (1986)
29. Farreny, H., Prade, H.: Positive and negative explanations of uncertain reasoning in the framework of possibility theory. In: Proceedings of the 5th Conference on Uncertainty in Artificial Intelligence (UAI 1989), Windsor, ON, August 18–20, pp. 95–101 (1989). Expanded version: explications de raisonnements dans l'incertain. Rev. d'Intell. Artif. **4**(2), 43–75 (1990)
30. Farreny, H., Prade, H., Wyss, E.: Approximate reasoning in a rule-based expert system using possibility theory: a case study. In: Kugler, H.J. (ed.) Proceedings of the 10th World IFIP Congress on Information Processing 1986, Dublin, September 1–5, pp. 407–413. North-Holland (1986)

31. Gebhardt, J., Kruse, R.: Learning possibilistic networks from data. In: Fisher, D., Lenz, H.-J. (eds.) Learning from Data - Proceedings of the 5th International Workshop on Artificial Intelligence and Statistics, Key West, January 1995, pp. 143–153. Springer, New York (1996)
32. Green, T.J., Karvounarakis, G., Tannen, V.: Provenance semirings. In: Libkin, L. (ed.) Proceedings of the 26th ACM SIGACT-SIGMOD-SIGART Symposium on Principles of Database Systems, Beijing, June 11–13, pp. 31–40 (2007)
33. Haddad, M., Leray, P., Ben Amor, N.: Apprentissage des réseaux possibilistes à partir de données: un survol. Rev. d'Intell. Artif. **29**(2), 229–252 (2015)
34. Haddad, M., Leray, Ph., Ben Amor, N.: Possibilistic networks: parameters learning from imprecise data and evaluation strategy. arXiv:1607.03705 [cs.AI] (2016)
35. Haddad, M., Leray, Ph., Ben Amor, N.: Possibilistic MDL: a new possibilistic likelihood based score function for imprecise data. In: ECSQARU 2017, pp. 435–445 (2017)
36. Haddad, M., Leray, Ph., Levray, A., Tabia, K.: Learning the parameters of possibilistic networks from data: empirical comparison. In: Proceedings of the FLAIRS Conference, pp. 736–741 (2017)
37. Jenhani, I., Ben Amor, N., Benferhat, S., Elouedi, Z.: SIM-PDT: a similarity based possibilistic decision tree approach. In: Hartmann, S., Kern-Isberner, G. (eds.) FoIKS 2008. LNCS, vol. 4932, pp. 348–364. Springer, Heidelberg (2008). https://doi.org/10.1007/978-3-540-77684-0_23
38. Kampé de Fériet, J.: Interpretation of membership functions of fuzzy sets in terms of plausibility and belief. In: Gupta, M., Sanchez, E. (eds.) Fuzzy Information and Decision Processes, pp. 93–98. North-Holland, Amsterdam (1982)
39. Kraus, S., Lehmann, D., Magidor, M.: Nonmonotonic reasoning, preferential models and cumulative logics. Artif. Intell. **44**, 167–207 (1990)
40. Kuzelka, O., Davis, J., Schockaert, S.: Encoding Markov logic networks in possibilistic logic. In: Meila, M., Heskes, T. (eds.) Proceedings of the 31st Conference on Uncertainty in Artificial Intelligence (UAI 2015), Amsterdam, July 12–16, pp. 454–463. AUAI Press (2015)
41. Kuzelka, O., Davis, J., Schockaert, S.: Induction of interpretable possibilistic logic theories from relational data. In: Proceedings of the IJCAI 2017, Melbourne, pp. 1153–1159 (2017)
42. Labreuche, C.: A general framework for explaining the results of a multi-attribute preference model. Artif. Intell. **175**(7–8), 1410–1448 (2011)
43. Mamdani, E.H., Assilian, S.: An experiment in linguistic synthesis with a fuzzy logic controller. Int. J. Man-Mach. Stud. **7**(1), 1–13 (1975)
44. Prade, H., Rico, A., Serrurier, M.: Elicitation of sugeno integrals: a version space learning perspective. In: Rauch, J., Raś, Z.W., Berka, P., Elomaa, T. (eds.) ISMIS 2009. LNCS (LNAI), vol. 5722, pp. 392–401. Springer, Heidelberg (2009). https://doi.org/10.1007/978-3-642-04125-9_42
45. Richardson, M., Domingos, P.M.: Markov logic networks. Mach. Learn. **62**(1–2), 107–136 (2006)
46. Serrurier, M., Prade, H.: Introducing possibilistic logic in ILP for dealing with exceptions. Artif. Intell. **171**(16–17), 939–950 (2007)
47. Serrurier, M., Prade, H.: An informational distance for estimating the faithfulness of a possibility distribution, viewed as a family of probability distributions, with respect to data. Int. J. Approx. Reason. **54**(7), 919–933 (2013)
48. Serrurier, M., Prade, H.: Entropy evaluation based on confidence intervals of frequency estimates: application to the learning of decision trees. In: Proceedings of the 32nd International Conference on Machine Learning, Lille, July 6–11, pp. 1576–1584 (2015)

49. Shafer, G.: A Mathematical Theory of Evidence. Princeton University Press (1976)
50. Sudkamp, T.: On probability-possibility transformations. Fuzzy Sets Syst. **51**, 73–81 (1992)
51. Teow, L.-N., Loe, K.-F.: An effective learning method for max-min neural networks. In: Proceedings of the IJCAI 1997, Nagoya, pp. 1134–1139 (1997)

Logic, Probability and Action: A Situation Calculus Perspective

Vaishak Belle[1,2(⊠)]

[1] University of Edinburgh, Edinburgh, UK
[2] Alan Turing Institute, London, UK
vaishak@ed.ac.uk

Abstract. The unification of logic and probability is a long-standing concern in AI, and more generally, in the philosophy of science. In essence, logic provides an easy way to specify properties that must hold in every possible world, and probability allows us to further quantify the weight and ratio of the worlds that must satisfy a property. To that end, numerous developments have been undertaken, culminating in proposals such as probabilistic relational models. While this progress has been notable, a general-purpose first-order knowledge representation language to reason about probabilities and dynamics, including in continuous settings, is still to emerge. In this paper, we survey recent results pertaining to the integration of logic, probability and actions in the situation calculus, which is arguably one of the oldest and most well-known formalisms. We then explore reduction theorems and programming interfaces for the language. These results are motivated in the context of cognitive robotics (as envisioned by Reiter and his colleagues) for the sake of concreteness. Overall, the advantage of proving results for such a general language is that it becomes possible to adapt them to any special-purpose fragment, including but not limited to popular probabilistic relational models.

1 Introduction

The unification of logic and probability is a long-standing concern in AI [72], and more generally, in the philosophy of science [31]. The motivation stems from the observation that (human and agent) knowledge is almost always incomplete. It is then not enough to say that some formula ϕ is unknown. One must also know which of ϕ or $\neg\phi$ is the more likely, and by how much. On the more pragmatic side, when reasoning about uncertain propositions and statements, it is beneficial to be able to leverage the underlying relational structure. Basically, logic provides an easy way to specify properties that must hold in every possible world, and probability allows us to further quantify the weight and ratio of the worlds that must satisfy a property. For example, the sibling relation is symmetric in every possible world, whereas the influence of smoking among siblings can be considered a statistical property, perhaps only true in 80% of the worlds.

The author was supported by a Royal Society University Research Fellowship.

Another argument increasingly made in favor of unifying logic and probability is that perhaps it would help us enable an apparatus analogous to Kahneman's so-called *System 1* versus *System 2* processing in human cognition [43]. That is, we want to interface experiential and reactive processing (assumed to be handled by some data-driven probabilistic learning methodology) with cogitative processing (assumed to be handled by a deliberative reasoning methodology).

To that end, numerous developments have been undertaken in AI. Closely following Bayesian networks [52,70], and particle filters [30,35], the areas of statistical relational learning and probabilistic relational modeling [28,36] emerged, and have been very successful. Since the world is rarely static, the application of such proposals to dynamic worlds has also seen many successes, e.g., [68,80]. However, these closely follow propositional representations, such as Bayesian networks, using logic purely for templating purposes (i.e., syntactic sugar in programming language parlance). So, although the progress has been notable, a general-purpose first-order knowledge representation language to reason about probabilities and dynamics, including in continuous settings, is still to emerge.

In the early days of the field, approaches such as [33] provided a logical language that allowed one to reason about the probabilities of atoms, which could be further combined over logical connectives. That work has inspired numerous extensions for reasoning about dynamics. But this has been primarily in the propositional setting [40,45,82], or with discrete probabilistic models [78]. (See [15] for extended discussions.) In this paper, we survey recent results pertaining to the integration of logic, probability and actions in the situation calculus [65, 71]. The situation calculus is one of the oldest and most well-known knowledge representation formalisms. In that regard, the results illustrate that we obtain perhaps the most expressive formalism for reasoning about degrees of belief in the presence of noisy sensing and acting. For that language, we then explore reduction theorems and programming interfaces. Of course, the advantage of proving results for such a general language is that it becomes possible to adapt them to any special-purpose fragment, including but not limited to popular probabilistic relational models.

To make the discussion below concrete, we motivate one possible application of such a language: *cognitive robotics*, as envisioned by Reiter [71] and further discussed in [48]. This is clearly not the only application of a language such as the situation calculus, which has found applications in areas such as service composition, databases, automated planning, decision-theoretic reasoning and multi-agent systems [71,83].

2 Motivation: Cognitive Robotics

The design and control of autonomous agents, such as robots, has been a major concern in artificial intelligence since the early days [65]. Robots can be viewed as systems that need to act purposefully in open-ended environments, and so are required to exhibit everyday commonsensical behavior. For the most part, however, traditional robotics has taken a "bottom-up" approach [79] focusing on low-level sensor-effector feedback. Perhaps the most dominant reason for this is that controllers for physical robots need to address the noise in effectors and sensors, often characterized by continuous probability distributions, which significantly complicates the reasoning and planning problems faced by a robot. While the simplicity of Bayesian statistics, defined over a fixed number of (propositional) random variables, has enabled the successful handling of probabilistic information in robotics modules, the flip side is that the applicability of contemporary methods is at the mercy of the roboticist's ingenuity. It is also unclear how precisely commonsensical knowledge can be specified using conditional independences between random variables while also accounting for how these dependencies further change as the result of actions.

Cognitive robotics [48], as envisioned by Reiter and his colleagues [71], follows closely in the footsteps of McCarthy's seminal ideas [64]: it takes the view that understanding the relationships between the beliefs of the agent and the actions at its disposal is key to a commonsensical robot that can operate purposefully in uncertain, dynamic worlds. In particular, it considers the study of knowledge representation and reasoning problems faced by the agent when attempting to answer questions such as [53]:

- *to execute a program, what information does a robot need to have at the outset vs. the information that it can acquire en route by perceptual means?*
- *what does the robot need to know about its environment vs. what need only be known by the designer?*
- *when should a robot use perception to find out if something is true as opposed to reasoning about what it knows was true in the past?*

The goal, in other words, is to develop a theory of high-level control that maps the knowledge, ignorance, intention and desires of the agent to appropriate actions. In this sense, cognitive robotics not only aims to connect to traditional robotics, which already leverages probabilistic reasoning, vision and learning for stochastic control, but also to relate to many other areas of AI, including automated planning, agent-oriented programming, belief-desire-intention architectures, and formal epistemology.

In lieu of this agenda, many sophisticated control methodologies and formal accounts have emerged, summarized in the following section. Unfortunately, despite the richness of these proposals, one criticism leveled at much of the work in cognitive robotics is that the theory is far removed from the kind of continuous uncertainty and noise seen in typical robotic applications. That is, the formal machinery of GOLOG to date does not address the complications due to noise and uncertainty in realistic robotic applications, at least in a way that relates these complications to what the robot believes, and how that changes over actions. The assumptions under which real-time behavior can be expected is also left open. For example, can standard probabilistic projection methodologies, such as Kalman and particle filters, be subsumed as part of a general logical framework?

The results discussed in this article can be viewed as a research agenda that attempts to bridge the gap between knowledge representation advances and robotic systems. By generalizing logic-based knowledge representation languages to reason about discrete and continuous probability distributions in the specification of both the initial beliefs of the agent and the noise in the sensors and effectors, the idea is to contribute to commonsensical and provably correct high-level controllers for agents in noisy worlds.

3   Tools of the Trade

To represent the beliefs and the actions, efforts in cognitive robotics would need to rely on a formal language of suitable expressiveness. Reiter's variant of the situation calculus has perhaps enjoyed the most success among first-order formalisms, although related proposals offer attractive properties of their own.[1] Reiter's variant was also the language considered in a recent survey on cognitive robotics [48], and so the reported results can easily be put into context.[2]

In this section, we will briefly recap some of the main foundational results discussed in [48]. In a few cases, we report on recent developments expanding on those results.

3.1   Language

Intuitively, the language $\mathcal{L}$ of the situation calculus [65] is a many-sorted dialect of predicate calculus, with sorts for *actions*, *situations* and *objects* (for everything else, and includes the set of reals $\mathbb{R}$ as a subsort). A situation represents a world history as a sequence of actions. A set of initial situations correspond to the ways the world might be initially. Successor situations are the result of doing actions, where the term $do(a, s)$ denotes the unique situation obtained on doing a in s. The term $do(\bar{a}, s)$, where $\bar{a}$ is the sequence $[a_1, \ldots, a_n]$ abbreviates $do(a_n, do(\ldots, do(a_1, s) \ldots))$. Initial situations are defined as those without a predecessor, and we let the constant S_0 denote the actual initial situation. See [71] for a comprehensive treatment.

The picture that emerges from the above is a set of trees, each rooted at an initial situation and whose edges are actions. In general, we want the values of predicates and functions to vary from situation to situation. For this purpose, $\mathcal{L}$ includes *fluents* whose last argument is always a situation.

Following [71], dynamic domains in $\mathcal{L}$ are modeled by means of a *basic action theory* $\mathcal{D}$, which consists domain-independent foundational axioms, and

---

[1] For example, the fluent calculus [77] offers an intuitive and simple state update mechanism in a first-order setting, and extensions of propositional dynamic logic [41] offer decidable formalisms.

[2] There has been considerable debate on why a quantified relational language is crucial for knowledge representation and commonsense reasoning; see references in [26,57], for example. Moreover, owing to the generality of the underlying language, decidable variants can be developed (e.g., [20,38]).

a domain-dependent first-order initial theory $\mathcal{D}_0$ (standing for what is true initially), and domain-dependent precondition and effect axioms, the latter taking the form of so-called successor state axioms that incorporates a monotonic solution to the frame problem [71].

To represent knowledge, and how that changes, one appeals to the possible-worlds approach [34]. The idea is that there many different ways the world can be, where each world stands for a complete state of affairs. Some of these are considered possible by a putative agent, and they determine what the agent knows and does not know. Essentially, situations can be viewed as possible worlds [74]: a special binary fluent K, taking two situation arguments determines the accessibility relation between worlds. So, $K(s', s)$ says that when the agent is at s, he considers s' possible. *Knowledge*, then, is simply truth at accessible worlds: $Knows(\phi, s) \doteq \forall s'. K(s', s) \supset \phi[s']$.

Sensing axioms additionally capture the discovery of the truth values of fluents. For example, to check whether f is true at s, we would use: $SF(sensetrue_f, s) \equiv f(s) = 1$. A successor state axiom formalizes the incorporation of these sensed values in the agent's mental state: $K(s', do(a, s)) \equiv \exists s''[K(s'', s) \wedge s' = do(a, s'') \wedge Poss(a, s'') \wedge (SF(a, s'') \equiv SF(a, s))]$. This says that if s'' is the predecessor of s', such that s'' was considered possible at s, then s' would be considered possible from $do(a, s)$ contingent on sensing outcomes.

3.2   Reasoning Problems

A fundamental problem underlying almost all applications involving basic action theories is *projection*. Given a sequence of actions a_1 through a_n, denoted $\bar{a} = [a_1, \ldots, a_n]$, we are often interested in asking whether ϕ holds after these via entailment: $\mathcal{D} \models \phi[do(\bar{a}, S_0)]$? One of the main results by Reiter is the existence of a reduction operator called *regression* that eliminates the actions: $\mathcal{D} \models \phi[do(\bar{a}, S_0)]$ iff $\mathcal{D}_{una} \cup \mathcal{D}_0 \models \mathcal{R}[\phi[do(\bar{a}, S_0)]]$. Here, $\mathcal{D}_{una}$ is an axiom that declares that all named actions are unique, and $\mathcal{R}[\phi[do(\bar{a}, S_0)]]$ mentions only a single situation term, S_0.

In the worst case, regressed formulas are exponentially long in the length of the action sequence [71], and so it has been argued that for long-lived agents like robots, continually updating the current view of the state of the world, is perhaps better suited. Lin and Reiter [60] proposed a theory of progression that satisfies: $\mathcal{D} \models \phi[do(\bar{a}, S_0)]$ iff $\mathcal{D}_{una} \cup \mathcal{P}(\mathcal{D}_0, \bar{a}) \models \phi[S_0]$. Here $\mathcal{P}(\mathcal{D}_0, \bar{a})$ is the updated initial theory that denotes the state of the world on doing $\bar{a}$. In general, progression requires second-order logic, but many special cases that are definable in first-order logic have since been identified (e.g., [61]).

3.3   Closed vs Open Worlds

$\mathcal{D}_0$ is assumed to be any set of first-order formulas, but then computing its entailments, regardless of whether we appeal to regression or progression, would be undecidable. Thus, restricting the theory to be equivalent to a relational database is one possible tractable fragment, but this makes the closed world assumption which is not really desirable for robotics. A second possibility is to assume that at the time of query evaluation, the agent has complete knowledge about the predicates mentioned in the query. This leads to a notion of local completeness [27]. A third possibility is to provide some control over the computational power of the evaluation scheme, leading to a form of *limited reasoning*. First-order fragments such as *proper* and *proper*$^+$ [47,56], which correspond to an infinite set of ground literals and clauses respectively, have been shown to work well with projection schemes for restricted classes of action theories [61,62].

An altogether different and more general strategy for reasoning about incomplete knowledge is to utilize the epistemic situation calculus. A regression theorem was already proved in early work [74], and a progression theorem has been considered in [63]. However, since propositional reasoning in epistemic logic is already intractable [34], results such as the *representation theorem* [57] that shows how epistemic operators can be eliminated under *epistemic closure* (i.e., knowing what one knows as well as what one does not know) needs to be leveraged at least. Alternatively, one could perhaps appeal to limited reasoning in the epistemic setting [46].

3.4   High-Level Control

To program agents whose actions are interpreted over a basic action theory, *high-level programming languages* such as GOLOG emerged [54]. These languages contained the usual programming features like sequence, conditional, iteration, recursive procedures, and concurrency but the key difference was that the primitive instruction was an action from a basic action theory. The execution of the program was then understood as $\mathcal{D} \models Do(\delta, S_0, do(\bar{a}, S_0))$ where δ is a GOLOG program, and on starting from S_0, the program successfully terminates in $do(\bar{a}, S_0)$. So, from S_0, executing the program leads to performing actions $\bar{a}$.

As argued in [48], GOLOG programs can range from a fully deterministic instruction $a_1; \ldots; a_n$ to a general search **while** $\neg\phi$ **do** πa. a: the former instructs the agent to perform action a_1, then a_2, and so on until a_n in sequence, and the latter instructs to try every possible action (sequence) until the goal is satisfied. It is between these two extremes where GOLOG is most powerful: it enables a partial specification of programs that can perform guided search for sub-goals in the presence of other loopy or conditional plans.

To guide search in the presence of nondeterminism, rewards can be stipulated on situations leading to a decision-theoretic machinery [17]. Alternatively, if the nondeterminism is a result of not knowing the true state of the world, sensing actions can be incorporated during program execution, leading to an *online* semantics for GOLOG execution [73].

4   Tools Revisited

In this section, we revisit the results from the previous section and discuss how these have been generalized to account for realistic, continuous, models of noise.

Perhaps the most general formalism for dealing with *degrees of belief* in formulas, and in particular, with how degrees of belief should evolve in the presence of noisy sensing and acting is the account proposed by Bacchus et al. [1], henceforth BHL. Among its many properties, the BHL model shows precisely how beliefs can be made less certain by acting with noisy effectors, but made more certain by sensing (even when the sensors themselves are noisy). Not only is it embedded in the rich theory of the situation calculus, including the use of Reiter's successor state axioms, it is also a stochastic extension to the categorical epistemic situation calculus. The main advantage of a logical account like BHL is that it allows a specification of belief that can be partial or incomplete, in keeping with whatever information is available about the application domain. It does not require specifying a prior distribution over some random variables from which posterior distributions are then calculated, as in Kalman filters, for example [79]. Nor does it require specifying the conditional independences among random variables and how these dependencies change as the result of actions, as in the temporal extensions to Bayesian networks [70]. In the BHL model, some logical constraints are imposed on the initial state of belief. These constraints may be compatible with one or very many initial distributions and sets of independence assumptions. (See [15] for extensive discussions.) All the properties of belief will then follow at a corresponding level of specificity.

4.1 Language

The BHL model makes use of two distinguished binary fluents p and l [9]. The p fluent determines a probability distribution on situations, by associating situations with *weights*. More precisely, the term $p(s', s)$ denotes the relative *weight* accorded to situation s' when the agent happens to be in situation s. Of course, p can be seen as a companion to K. As one would for K, the properties of p in initial states, which vary from domain to domain, are specified with axioms as part of $\mathcal{D}_0$. The term $l(a, s)$ is intended to denote the likelihood of action a in situation s to capture noisy sensors and effectors. For example, think of a sonar aimed at the wall, which gives a reading for the true value of a fluent f that corresponds to the distance between the robot and the wall. Supposing the sonar's readings are subject to additive Gaussian noise. If now a reading of z were observed on the sonar, intuitively, those situations where $f = z$ should be considered more probable than those where $f \neq z$. Then we would have: $l(sense_f(z), s) = u \equiv u = \mathcal{N}(z - f(s); 0, 1))$. Here, a standard normal is assumed, where the mean is 0, and the variance is 1.[3] Analogously, noisy effectors can be modeled using actions with double the arguments: $l(move(x, y), s) = u \equiv u = \mathcal{N}(y - x; 0, 1)$. This says the difference between actual distance moved and the intended amount is normally distributed, corresponding to additive Gaussian noise. Such noise models can also be made context

---

[3] If the specification of the p-axiom or the l-axiom includes disjunctions and existential quantifiers, we will then be dealing with uncertainty about distributions. See [14], for example.

dependent (e.g., specifying the sensor's error profile to be worse for lower temperatures, where the temperature value is situation-dependent). In the case of noisy effectors, the successor state axioms have to be defined to use the second argument, as this is what actually happens at a situation [15].

Analogous to the notion of knowledge, the *degree of belief* in ϕ in situation s is defined as the weight of accessible worlds where ϕ is true:

$$Bel(\phi, s) \doteq \frac{1}{\gamma} \sum_{\{s' : \phi[s']\}} p(s', s).$$

Here, γ is the normalization factor and corresponds to the numerator but with ϕ replaced by *true*. The change in p values over actions is specified using a successor state axiom, analogous to the one for K: $p(s', do(a, s)) = u \equiv \exists s'' \; [s' = do(a, s'') \wedge Poss(a, s'') \wedge \; u = p(s'', s) \times l(a, s'')] \vee \neg \exists s'' \; [s' = do(a, s'') \wedge Poss(a, s'') \wedge u = 0]$. This axioms determines how l affects the p-value of successor situations.

As the BHL model is defined as a sum over possible worlds, it cannot actually handle Gaussians and other continuous distributions involving π, e, exponentiation, and so on. Therefore, BHL always consider discrete probability distributions that *approximate* the continuous ones. However, this limitation was lifted in [15], which shows how *Bel* is defined in continuous domains.

4.2   Reasoning Problems

The projection problem in this setting is geared for reasoning about formulas that now mention *Bel*. In particular, we might be interested in knowing whether $\mathcal{D} \models Bel(\phi, do(\bar{a}, S_0)) \geq r$ for a real number r.

One reductive approach would be to translate both $\mathcal{D}$ and ϕ, which would mention *Bel*, into a predicate logic formula. This approach, however, presents a serious computational problem because belief formulas expand into a large number of sentences, resulting in an enormous search space with initial and successor situations. The other issue with this approach is that sums (and integrals in the continuous case) reduce to complicated second-order formulas.

In [10], it is shown how Reiter's regression operator can be generalized to operate directly on *Bel*-terms. This involves appealing to the likelihood axioms. For example, imagine a robot that is uncertain about its distance d to the wall, and the prior is a uniform distribution on the interval $[2, 12]$. Assume the robot (noise-free) moves away by 2 units and is now interested in the belief about $d \leq 5$. Regression would tell the robot that this is equivalent to its initial beliefs about $d \leq 3$ which here would lead to a value of .1. Imagine then the robot is also equipped with a sonar unit with additive Gaussian noise. After moving away by 2 units, if the sonar were now to provide a reading of 8, then regression would derive that belief about $d \leq 5$ is equivalent to $1/\gamma \times \int_2^3 .1 \times \mathcal{N}(6 - x; 0, 1) \; dx$. Essentially, the posterior belief about $d \leq 5$ is reformulated as the product of the prior belief about $d \leq 3$ and the likelihood of $d \leq 3$ given an observation of

6. That is, observing 8 after moving away by 2 units is equated here to observing 6 initially. (Here, γ is the normalization factor.)

Progression too could potentially be addressed by expanding formulas involving Bel-terms, but it is far from clear what precisely this would look like. In particular, given initial beliefs about fluents (such as the one about d earlier), we intuit that a progression account would inform us how this distribution changed. For example, on moving away from the wall by 2 units, we would now expect d to be uniformly distributed on the interval $[4, 14]$. However, this leads to a complication: because if the robot had instead moved towards the wall by 4 units, then those points where $d \in [2, 4]$ initially are mapped to a single point $d = 0$ that should then obtain a probability mass of .2, while the other points retain their initial density of .1. In [11], it is shown that for a certain class of basic action theories called *invertible theories*, such complications are avoidable, and moreover, the progressed database can be specified by means of simple syntactic manipulations.

4.3   Closed vs Open Worlds

The closed vs open world discussion does not seem immediately interesting here, because, after all, the language is clearly open in the sense of not knowing the values of fluents, and according a distribution to these values. However, consider that the closed-world assumption was also motivated previously by computational concerns. In that regard, the above regression and progression results already studied special cases involving conjugate distributions [18], such as Gaussians which admit attractive analytical simplifications. For example, efficient Kalman filters [79] often make the assumption that the initial prior and the noise models are Gaussians, in which case the posterior would also be a Gaussian. In [12], it is further shown that when the initial belief is a Bayesian network, by way of regression, projection can be handled effectively by sampling. (That is, once the formula is regressed, the network is sampled and the samples are evaluated against the regressed formula.)

In the context of probabilistic specifications, the notion of "open"-ness can perhaps be interpreted differently. We can take this to mean that we do not know the distribution of the random variables, or even that the set of random variables is not known in advance. As argued earlier, this is precisely the motivation for the BHL scheme, and a recent modal reformulation of BHL illustrates the properties of such a language in detail [7]. A detailed demonstration of how such specifications would work in the context of *robot localization* was given in [14].

The question of how to effectively compute beliefs in such rich settings is not clear, however. We remark that various static frameworks have emerged for handling imprecision or uncertainty in probabilistic specifications [24,58,66]. For example, when we have finitely many random variables but there is uncertainty about the underlying distribution, credal representations are of interest [24], and under certain conditions, they can be learned and reasoned with in an efficient manner [58]. On the other hand, when we have infinitely many random variables (but with a single underlying distribution), proposal such as [72] and [2] are of interest, the latter being a weighted representation inspired by *proper*[+] knowledge bases. Extending these to allow uncertainty about the underlying distribution may also be possible. Despite being static, by means of regression or progression, perhaps such open knowledge bases can be exploited for cognitive robotics applications, but that remains to be seen.

4.4 High-Level Control

A high-level programming language that deals with noise has to reason about two kinds of complications. First, when a noisy physical or sensing action in the program is performed, we must condition the next instruction on how the belief has changed as a result of that action. Second, because sensing actions in the language are of the form *sense(z)* that expects an input z, an *offline* execution would simulate possible values for z whereas an *online* execution would expect an external source to provide z (e.g., reading off the value of a sonar). We also would not want the designer to be needlessly encumbered by the error profiles of the various effectors and sensors, so she has to be encouraged to program around *sense-act* loops; that is, every action sequence should be accompanied with a suitable number of sensing readings so that the agent is "confident" (i.e., the distribution of the fluent in question is *narrow*) before performing more actions. In [13], such a desiderata was realized to yield a stochastic version of knowledge-based programming [71]. Primitive instructions are *dummy* versions of noisy actions and sensors; e.g., *move(x,y)* is simply *move(x)* and *sonar(z)* is simply *sonar*. The idea then is that the modeler simply uses these dummy versions as she would with noise-free actions, but the execution semantics incorporates the change in belief. It is further shown that program execution can be realized by means of a *particle filtering* [79] strategy: weighted samples are drawn from the initial beliefs, which correspond to initial situations, and on performing actions, fluent values at these situations are updated by means of the successor state axioms. The degree of belief in ϕ corresponds to summing up the weights of samples where ϕ is true.

Such an approach can be contrasted with notable probabilistic relational modelling proposals such as [68]: the difference mainly pertains to three sources of generality. First, a language like the situation calculus allows knowledge bases to be arbitrary quantificational theories, and BHL further allows uncertainty about the distributions defined for these theories. Second, the situation calculus, and by extension, GOLOG and the paradigm in [13] allows us to reason about non-terminating and unbounded behavior [23]. Third, since an explicit belief state is allowed, it becomes possible to provide a systematic and generic treatment for multiple agents [6,44].

On the issue of tractable reasoning, an interesting observation is that because these programs require reasoning with an explicit belief state [34], one might wonder whether the programs can be "compiled" to a *reactive plan*, possibly with loops, where the next action to be performed depends only on the sensing information received in the current state. This relates knowledge-based programming to *generalized planning* [55,76], and of course, the advantage is also that numerous strategies have been identified to synthesize such loopy, reactive plans. Such plans are also shown to be sufficient for goal achievability [59]; however, knowledge-based programs are known to be exponentially more succinct than loopy, reactive plans [49]. In [42], a generic algorithmic framework was proposed to synthesize such plans in noise-free environments. How the correctness of such plans should be generalized to noisy environments was considered in [3,8]. The algorithmic synthesis problem was then considered in [81].

5 Related Work and Discussions

There are many threads of research in AI, automated planning and robotics that are close in spirit to what is reported here. For example, belief update via the incorporation of sensor information has been considered in probabilistic formalisms such as Bayesian networks [52,70], Kalman and particle filters [79]. But these have difficulties handling strict uncertainty. Moreover, since rich models of actions are rarely incorporated, shifting conditional dependencies and distributions are hard to address in a general way. While there are graphical formalisms with an account of actions, such as [25,39], they too have difficulties handling strict uncertainty and quantification. To the best of our knowledge, no existing probabilistic formalism handles changes in state variables like those possible in the BHL scheme. Related to these are relational probabilistic models [21,32,66,67]. Although limited accounts for dynamic domains are common here [50,69], explicit actions are seldom addressed in a general way. We refer interested readers to discussions in [15], where differences are also drawn to prior developments in reasoning about actions, including stochastic but non-epistemic GOLOG dialects [37].

Arguably, many of the linguistic restrictions of such frameworks is often motivated by computational considerations. So what is to be gained by a general approach? This question is especially significant when we take into account that numerous "hybrid" approaches have emerged over the years that provide a bridge between a high-level language and a low-level operation [19,51]. Our sense is that while these and other approaches are noteworthy, and are extended in a modular manner to keep things tractable and workable on an actual physical robot, it still leaves a lot at the mercy of the roboticist's ingenuity. For example, extending an image recognition algorithm to reason about a structured world is indeed possible, but it is more likely than not that this ontology is also useful for a number of other components, such as the robot's grasping arm; moreover, changes to one must mean changes to all. Abstracting a complex behavior module of a robot is a painstaking effort: often the robot's modules are written in different programming languages with varying levels of abstraction, and to reduce these interactions to *atoms* in the high-level language would require considerable know-how of the system. Moreover, although a roboticist can abstract probabilistic sensors in terms of high-level categorical ones, there is loss in detail, as it is not clear at the outset which aspect of the sensor data is being approximated and by how much. Thus, all of these "bottom-up" approaches ultimately challenge the claim that the underlying theory is a genuine characterization of the agent.

In service of that, the contributions reported in this work attempt to express all the (inner and outer) workings of a robot in a *single mathematical language*: a mathematical language that can capture rich structure as well as natively reason about the probabilistic uncertainty plaguing a robot; a mathematical language that can reason with all available information, some of which may be probabilistic, and some categorical; a mathematical language that can reason about the physical world at different levels of abstraction, in terms of objects, atoms, and whatever else physicists determine best describes our view of the world. Undoubtedly, given this glaring expressiveness, the agenda will raise significant challenges for the applicability of the proposal in contemporary robots, but our view is that, it will also engender novel extensions to existing algorithms to cope with the expressiveness. Identifying tractable fragments, for example, will engender novel theoretical work. As already discussed, many proposals from the statistical relational learning community are very promising in this regard, and are making steady progress towards the overall ambition. (But as discussed, they fall short in terms of being able to reason about non-terminating behavior, arbitrary first-order quantification, among other things, and so identifying richer fragments is a worthwhile direction.) It is also worth remarking that the tractability of reasoning (and planning) has been the primary focus of much of the research in knowledge representation. The broader question of how to learn models has received lesser attention, and this is precisely where statistical relational learning and related paradigms will prove useful [4]. (It would be especially interesting to consider relational learning with neural modules [29].) Indeed, in addition to approaches such as [22,75], there have been a number of advances recently on learning dynamic representations (e.g., [68]), which might provide fertile ground to lift such ideas for cognitive robotics. Computability results for qualitative learning in dynamic epistemic logic has been studied in [16]. Recently, proper$^+$ knowledge bases were shown to be polynomial-time learnable for querying tasks [5]. Ultimately, learning may provide a means to coherently arrive at action descriptions at different levels of granularity from data [26]. In the long term, the science of building a robot, which currently is more of an art, can perhaps be approached systematically. More significantly, through the agenda of cognitive robotics, we might gain deep insights on how commonsense knowledge and actions interact for general-purpose, open-ended robots. In that regard, the integration of logic, probability and actions will play a key role.

References

1. Bacchus, F., Halpern, J.Y., Levesque, H.J.: Reasoning about noisy sensors and effectors in the situation calculus. Artif. Intell. **111**(1–2), 171–208 (1999)
2. Belle, V.: Weighted model counting with function symbols. In: UAI (2017)
3. Belle, V.: On plans with loops and noise. In: AAMAS (2018)
4. Belle, V.: Symbolic logic meets machine learning: a brief survey in infinite domains (2020)

5. Belle, V., Juba, B.: Implicitly learning to reason in first-order logic. In: NeurIPS (2019)
6. Belle, V., Lakemeyer, G.: Multiagent only knowing in dynamic systems. J. Artif. Intell. Res. **49** (2014)
7. Belle, V., Lakemeyer, G.: Reasoning about probabilities in unbounded first-order dynamical domains. In: IJCAI (2017)
8. Belle, V., Levesque, H.: Foundations for generalized planning in unbounded stochastic domains. In: KR (2016)
9. Belle, V., Levesque, H.J.: Reasoning about continuous uncertainty in the situation calculus. In: Proceedings of the IJCAI (2013)
10. Belle, V., Levesque, H.J.: Reasoning about probabilities in dynamic systems using goal regression. In: Proceedings of the UAI (2013)
11. Belle, V., Levesque, H.J.: How to progress beliefs in continuous domains. In: KR (2014)
12. Belle, V., Levesque, H.J.: PREGO: an action language for belief-based cognitive robotics in continuous domains. In: Proceedings of the AAAI (2014)
13. Belle, V., Levesque, H.J.: Allegro: belief-based programming in stochastic dynamical domains. In: IJCAI (2015)
14. Belle, V., Levesque, H.J.: A logical theory of localization. Studia Logica **104**, 741–772 (2015)
15. Belle, V., Levesque, H.J.: Reasoning about discrete and continuous noisy sensors and effectors in dynamical systems. Artif. Intell. **262**, 189–221 (2018)
16. Bolander, T., Gierasimczuk, N.: Learning actions models: qualitative approach. In: van der Hoek, W., Holliday, W.H., Wang, W. (eds.) LORI 2015. LNCS, vol. 9394, pp. 40–52. Springer, Heidelberg (2015). https://doi.org/10.1007/978-3-662-48561-3_4
17. Boutilier, C., Reiter, R., Soutchanski, M., Thrun, S.: Decision-theoretic, high-level agent programming in the situation calculus. In: Proceedings of the AAAI, pp. 355–362 (2000)
18. Box, G.E.P., Tiao, G.C.: Bayesian Inference in Statistical Analysis. Addison-Wesley (1973)
19. Burgard, W., Cremers, A.B., Fox, D., Hähnel, D., Lakemeyer, G., Schulz, D., Steiner, W., Thrun, S.: Experiences with an interactive museum tour-guide robot. Artif. Intell. **114**(1–2), 3–55 (1999)
20. Calvanese, D., De Giacomo, G., Montali, M., Patrizi, F.: First-order μ-calculus over generic transition systems and applications to the situation calculus. Inf. Comput. **259**(3), 328–347 (2018)
21. Choi, J., Amir, E., Hill, D.J.: Lifted inference for relational continuous models. In: Proceedings of the UAI, pp. 126–134 (2010)
22. Choi, J., Guzman-Rivera, A., Amir, E.: Lifted relational Kalman filtering. In: Proceedings of the IJCAI, pp. 2092–2099 (2011)
23. Claßen, J., Lakemeyer, G.: A logic for non-terminating Golog programs. In: KR, pp. 589–599 (2008)
24. Cozman, F.G.: Credal networks. Artif. Intell. **120**(2), 199–233 (2000)
25. Darwiche, A., Goldszmidt, M.: Action networks: a framework for reasoning about actions and change under uncertainty. In: Proceedings of the UAI, pp. 136–144 (1994)
26. Davis, E., Marcus, G.: Commonsense reasoning and commonsense knowledge in artificial intelligence. Commun. ACM **58**(9), 92–103 (2015)
27. De Giacomo, G., Levesque, H.J.: Projection using regression and sensors. In: IJCAI (1999)

28. De Raedt, L., Kersting, K.: Statistical relational learning. In: Encyclopedia of Machine Learning, pp. 916–924. Springer (2011)
29. De Raedt, L., Manhaeve, R., Dumancic, S., Demeester, T., Kimmig, A.: Neuro-symbolic = neural + logical + probabilistic. In: NeSy 2019 @ IJCAI (2019)
30. Dean, T., Wellman, M.: Planning and Control. Morgan Kaufmann Publishers Inc. (1991)
31. Demey, L., Kooi, B., Sack, J.: Logic and Probability (2013)
32. Domingos, P., Kok, S., Poon, H., Richardson, M., Singla, P.: Unifying logical and statistical AI. In: Proceedings of the AAAI, pp. 2–7 (2006)
33. Fagin, R., Halpern, J.Y.: Reasoning about knowledge and probability. J. ACM 41(2), 340–367 (1994)
34. Fagin, R., Halpern, J.Y., Moses, Y., Vardi, M.Y.: Reasoning About Knowledge. MIT Press, Cambridge (1995)
35. Fox, D., Hightower, J., Liao, L., Schulz, D., Borriello, G.: Bayesian filtering for location estimation. IEEE Pervasive Comput. 2(3), 24–33 (2003)
36. Getoor, L., Taskar, B.: Introduction to statistical relational learning (adaptive computation and machine learning) (2007)
37. Grosskreutz, H., Lakemeyer, G.: ccGolog - a logical language dealing with continuous change. Logic J. IGPL 11(2), 179–221 (2003)
38. Gu, Y., Soutchanski, M.: A description logic based situation calculus. Ann. Math. Artif. Intell. 58(1–2), 3–83 (2010)
39. Hajishirzi, H., Amir, E.: Reasoning about deterministic actions with probabilistic prior and application to stochastic filtering. In: Proceedings of the KR (2010)
40. Halpern, J.Y., Tuttle, M.R.: Knowledge, probability, and adversaries. J. ACM 40, 917–960 (1993)
41. Herzig, A., Lang, J., Longin, D., Polacsek, T.: A logic for planning under partial observability. In: Proceedings of the AAAI/IAAI, pp. 768–773 (2000)
42. Hu, Y., De Giacomo, G.: A generic technique for synthesizing bounded finite-state controllers. In: ICAPS (2013)
43. Kahneman, D.: Thinking, Fast and Slow. Macmillan (2011)
44. Kelly, R.F., Pearce, A.R.: Complex epistemic modalities in the situation calculus. In: KR (2008)
45. Kushmerick, N., Hanks, S., Weld, D.: An algorithm for probabilistic planning. Artif. Intell. 76(1), 239–286 (1995)
46. Lakemeyer, G., Lespérance, Y.: Efficient reasoning in multiagent epistemic logics. In: Proceedings of the ECAI, pp. 498–503 (2012)
47. Lakemeyer, G., Levesque, H.J.: Evaluation-based reasoning with disjunctive information in first-order knowledge bases. In: Proceedings of the KR, pp. 73–81 (2002)
48. Lakemeyer, G., Levesque, H.J.: Cognitive robotics. In: Handbook of Knowledge Representation, pp. 869–886. Elsevier (2007)
49. Lang, J., Zanuttini, B.: Probabilistic knowledge-based programs. In: Twenty-Fourth International Joint Conference on Artificial Intelligence (2015)
50. Lang, T., Toussaint, M., Kersting, K.: Exploration in relational domains for model-based reinforcement learning. J. Mach. Learn. Res. (JMLR) 13, 3691–3734 (2012)
51. Lemaignan, S., Ros, R., Mösenlechner, L., Alami, R., Beetz, M.: Oro, a knowledge management platform for cognitive architectures in robotics. In: IROS (2010)
52. Lerner, U., Moses, B., Scott, M., McIlraith, S., Koller, D.: Monitoring a complex physical system using a hybrid dynamic Bayes net. In: Proceedings of the UAI, pp. 301–310 (2002)
53. Levesque, H., Reiter, R.: High-level robotic control: beyond planning. Position paper at AAAI Spring Symposium on Integrating Robotics Research (1998)

54. Levesque, H., Reiter, R., Lespérance, Y., Lin, F., Scherl, R.: Golog: a logic programming language for dynamic domains. J. Logic Program. **31**, 59–84 (1997)
55. Levesque, H.J.: What is planning in the presence of sensing? In: Proceedings of the AAAI/IAAI, pp. 1139–1146 (1996)
56. Levesque, H.J.: A completeness result for reasoning with incomplete first-order knowledge bases. In: Proceedings of the KR, pp. 14–23 (1998)
57. Levesque, H.J., Lakemeyer, G.: The Logic of Knowledge Bases. The MIT Press (2001)
58. Levray, A., Belle, V.: Learning tractable credal networks. In: AKBC (2020)
59. Lin, F., Levesque, H.J.: What robots can do: robot programs and effective achievability. Artif. Intell. **101**(1–2), 201–226 (1998)
60. Lin, F., Reiter, R.: How to progress a database. Artif. Intell. **92**(1–2), 131–167 (1997)
61. Liu, Y., Lakemeyer, G.: On first-order definability and computability of progression for local-effect actions and beyond. In: Proceedings of the IJCAI, pp. 860–866 (2009)
62. Liu, Y., Levesque, H.: Tractable reasoning with incomplete first-order knowledge in dynamic systems with context-dependent actions. In: Proceedings of the IJCAI, pp. 522–527 (2005)
63. Liu, Y., Wen, X.: On the progression of knowledge in the situation calculus. In: IJCAI (2011)
64. McCarthy, J.: Programs with common sense. In: Semantic Information Processing, pp. 403–418. MIT Press (1968)
65. McCarthy, J., Hayes, P.J.: Some philosophical problems from the standpoint of artificial intelligence. In: Machine Intelligence, pp. 463–502 (1969)
66. Milch, B., Marthi, B., Russell, S.J., Sontag, D., Ong, D.L., Kolobov, A.: BLOG: probabilistic models with unknown objects. In: Proceedings of the IJCAI, pp. 1352–1359 (2005)
67. Ng, R., Subrahmanian, V.: Probabilistic logic programming. Inf. Comput. **101**(2), 150–201 (1992)
68. Nitti, D.: Hybrid Probabilistic Logic Programming. Ph.D. Thesis. KU Leuven (2016)
69. Nitti, D., Belle, V., Raedt, L.D.: Planning in discrete and continuous Markov decision processes by probabilistic programming. In: ECML (2015)
70. Pearl, J.: Probabilistic Reasoning in Intelligent Systems: Networks of Plausible Inference. Morgan Kaufmann (1988)
71. Reiter, R.: Knowledge in Action: Logical Foundations for Specifying and Implementing Dynamical Systems. MIT Press (2001)
72. Russell, S.J.: Unifying logic and probability. Commun. ACM **58**(7), 88–97 (2015)
73. Sardina, S., De Giacomo, G., Lespérance, Y., Levesque, H.J.: On the semantics of deliberation in indigolog–from theory to implementation. Ann. Math. Artif. Intell. **41**(2–4), 259–299 (2004)
74. Scherl, R.B., Levesque, H.J.: Knowledge, action, and the frame problem. Artif. Intell. **144**(1–2), 1–39 (2003)
75. Shirazi, A., Amir, E.: First-order logical filtering. In: Proceedings of the IJCAI, pp. 589–595 (2005)
76. Srivastava, S.: Foundations and Applications of Generalized Planning. Ph.D. thesis, Department of Computer Science, University of Massachusetts Amherst (2010)
77. Thielscher, M.: From situation calculus to fluent calculus: state update axioms as a solution to the inferential frame problem. Artif. Intell. **111**(1–2), 277–299 (1999)

78. Thielscher, M.: Planning with noisy actions (preliminary report). In: Proceedings of the Australian Joint Conference on Artificial Intelligence, pp. 27–45 (2001)
79. Thrun, S., Burgard, W., Fox, D.: Probabilistic Robotics. MIT Press (2005)
80. Tran, S.D., Davis, L.S.: Event modeling and recognition using Markov logic networks. In: Forsyth, D., Torr, P., Zisserman, A. (eds.) ECCV 2008. LNCS, vol. 5303, pp. 610–623. Springer, Heidelberg (2008). https://doi.org/10.1007/978-3-540-88688-4_45
81. Treszkai, L., Belle, V.: A correctness result for synthesizing plans with loops in stochastic domains. Int. J. Approximate Reasoning (2020)
82. Van Benthem, J., Gerbrandy, J., Kooi, B.: Dynamic update with probabilities. Stud. Logica **93**(1), 67–96 (2009)
83. Van Harmelen, F., Lifschitz, V., Porter, B.: Handbook of Knowledge Representation. Elsevier (2008)

When Nominal Analogical Proportions Do Not Fail

Miguel Couceiro[1], Erkko Lehtonen[2], Laurent Miclet[3], Henri Prade[4(✉)], and Gilles Richard[4]

[1] Université de Lorraine, CNRS, Inria N.G.E., Nancy, France
`miguel.couceiro@loria.fr`
[2] Centro de Matemática e Aplicações, Faculdade de Ciências e Tecnologia, Universidade Nova de Lisboa, Lisbon, Portugal
`e.lehtonen@fct.unl.pt`
[3] IRISA, University of Rennes, Rennes, France
`laurent.miclet@gmail.com`
[4] IRIT, CNRS & Université Paul Sabatier, Toulouse, France
`{prade,richard}@irit.fr`

Abstract. Analogical proportions are statements of the form "**a** is to **b** as **c** is to **d**", where **a**, **b**, **c**, **d** are tuples of attribute values describing items. The mechanism of analogical inference, empirically proved to be efficient in classification and reasoning tasks, started to be better understood when the characterization of the class of classification functions with which the analogical inference always agrees was established for Boolean attributes. The purpose of this paper is to study the case of finite attribute domains that are not necessarily two-valued, i.e., when attributes are nominal. In particular, we describe the more stringent class of "hard" analogy preserving (HAP) functions $f: X_1 \times \ldots \times X_m \to X$ over finite domains $X_1, \ldots, X_m, X$ for binary classification purposes. This description is obtained in two steps. First we observe that such AP functions are *almost affine*, that is, their restriction to any $S_1 \times \cdots \times S_m$, where $S_i \subseteq X_i$ and $|S_i| \leq 2$ $(1 \leq i \leq m)$, can be turned into an affine function by renaming variable and function values. We then use this result together with some universal algebraic tools to show that they are essentially unary or *quasi-linear*, which provides a general representation of HAP functions. As a by-product, in the case when $X_1 = \cdots = X_m = X$, it follows that this class of HAP functions constitutes a clone on X, thus generalizing several results by some of the authors in the Boolean case.

1 Introduction

An analogy establishes a parallel between two situations, which are similar in many respects and dissimilar in others. If such a parallel holds to some extent,

The authors acknowledge a partial support of ANR-11-LABX-0040-CIMI (Cent. Int. de Math. et d'Informat.) within program ANR-11-IDEX-0002-02, project ISIPA, and a partial support of Fundação para a Ciência e a Tecnologia (Portuguese Foundation for Science and Technology) through project UID/MAT/00297/2019 (Centro de Matemática e Aplicações) and project PTDC/MAT-PUR/31174/2017.

J. Davis and K. Tabia (Eds.): SUM 2020, LNAI 12322, pp. 68–83, 2020.
https://doi.org/10.1007/978-3-030-58449-8_5

there are pairs (a, b) and (c, d) such as "a is to b in situation 1 as c is to d in situation 2" [7]. Analogical proportions are of the form "a is to b as c is to d". It is only recently that researchers have provided representations of this quaternary relation in different settings [10, 12], or algorithms for finding an item d, given a, b, and c, for building an analogical proportion from these three items when it is possible [9]. The items considered in this paper are supposed to be represented by vectors of attribute values.

Analogical inference relies on the idea that if four items a, b, c, d are in analogical proportion for each of the n attributes describing them, it may still be the case for another attribute. For instance, if class labels are known for a, b, c and unknown for d, then one may infer the label for d as a solution of an analogical proportion equation [16]. Obviously, analogical inference rule is not a sound rule, and the effectiveness of analogical classifiers [2,11] looks quite mysterious. From a theoretical viewpoint it is quite challenging to find and characterize situations where such an inference rule can be applied in a sound way. In case of Boolean attributes, a first step for explaining this state of facts was to characterize the set of functions for which analogical inference is sound, i.e., no error occurs, no matter which triplets of examples are used. In [4], it was proved that these so-called "analogy-preserving" (AP) functions coincide exactly with the set of affine Boolean functions. Moreover, when the function is close to being affine, it was also shown that the prediction accuracy remains high [5]. When attributes are valued on finite domains, which we refer to as the "*nominal case*" (it includes the Boolean case), the problem of identifying the AP functions had remained a challenging open problem until now. This paper aims to solving this problem in the context of binary classification problems and to providing a complete description of the more stringent class of "hard" AP functions.

The paper is organized as follows. Section 2 provides the necessary background on analogical proportions and analogical inference in the Boolean and in the nominal cases. Then we introduce the notion of analogy-preserving functions on which analogical inference never fails, and discusses an illustrative example in the nominal case, which emphasizes the linkage of analogical proportions with trees cataloguing items according to the values of the attributes used for describing them. Later, a local description of hard analogy preserving functions is given in terms of almost affine functions, which is then extended to a global description given in terms of the notion of quasi-linear functions.

2   Background

Postulates. An analogical proportion is a 4-ary relation, denoted $a : b :: c : d$, between items a, b, c, d, supposed to obey the following 3 postulates (e.g., [10]):

- $\forall a, b,\ a : b :: a : b$ (*reflexivity*)
- $\forall a, b, c, d,\ a : b :: c : d \rightarrow c : d :: a : b$ (*symmetry*)
- $\forall a, b, c, d,\ a : b :: c : d \rightarrow a : c :: b : d$ (*central permutation*)

The repeated and alternate application of the two last postulates entail that an analogical proportion $a : b :: c : d$ has 8 equivalent forms: $a : b :: c : d = c : d :: a : b = c : a :: d : b = d : b :: c : a = d : c :: b : a = b : a :: d : c = b : d :: a : c = a : c :: b : d$. Some immediate consequences can be observed:

i) $\forall a, b, \ a : a :: b : b$ (*identity*)
ii) $\forall a, b, c, d, \ a : b :: c : d \to d : b :: c : a$ (*extreme permutation*)
iii) $\forall a, b, c, d, \ a : b :: c : d \to b : a :: d : c$ (*inside pair reversing*)
iv) $\forall a, b, c, d, \ a : b :: c : d \to d : c :: b : a$ (*complete reversal*)

Boolean Case. Let us now assume for a while that a, b, c, d denote Boolean variables, i.e., their values belong to the set $\{0, 1\}$. This may be thought of as encoding the fact that a given property is true or false for the considered item. Since items are usually described in terms of several properties, this modeling of analogical proportions is then extended to tuples in a component-wise manner as recalled later. As shown in [13], the minimal Boolean model obeying the analogical proportion postulates makes $a : b :: c : d$ true only for the six patterns

$$(a, b, c, d) \in \{(0,0,0,0), (1,1,1,1), (0,0,1,1), (1,1,0,0), (0,1,0,1), (1,0,1,0)\},$$

while $a : b :: c : d$ is false for the other ten patterns of values for the four variables a, b, c, d. This is the truth table of a quaternary connective that can be logically expressed as $a : b :: c : d = ((a \wedge \neg b) \equiv (c \wedge \neg d)) \wedge ((\neg a \wedge b) \equiv (\neg c \wedge d))$ [12].

It can be seen on this expression that the analogical proportion "a is to b as c is to d" formally states that "a differs from b as c differs from d and b differs from a as d differs from c". It means that $a = b \Leftrightarrow c = d$, and that $a \neq b \Leftrightarrow c \neq d$ (with the further requirement that both truth value changes are in the same direction (either from 1 to 0, or from 0 to 1, when going from a to b, and from c to d). So, the analogy is as much a matter of dissimilarity as a matter of similarity.

Nominal Case. In the nominal case, attributes are supposed to take their values on finite domains (which are not necessarily ordered). For instance, the attribute domain of `color` may be the set $\{blue, red, yellow\}$. Let s and t be two values in such a finite domain X. It follows from reflexivity and central permutation postulates that $s : t :: s : t$ and $s : s :: t : t$ should hold. By the symmetry postulate, s and t play the same role. Note also that s and t are not necessarily distinct. This leads to a minimal model of analogical proportion for nominal values, which can be stated as follows: $a : b :: c : d$ holds if and only if

$$(a, b, c, d) \in \{(s, t, s, t), (s, s, t, t) \mid s, t \in X\}.$$

This clearly covers the Boolean case as a particular case, leading to the 6 lines seen in the Boolean truth table. If $|X| = n$, we obviously have n^4 tuples (a, b, c, d). Among them, we have i) n valid analogies of type $s : s :: s : s$, ii) $n(n-1)$ of type $s : t :: s : t$ with $s \neq t$, and iii) $n(n-1)$ of type $s : s :: t : t$ with $s \neq t$.

Hence, a total of $n(2n - 1)$, which shows that the number of valid analogies increases with the square of the cardinality of the underlying set. For instance, for an attribute such as `color` whose values belong to, e.g., $X = \{blue, red, yellow\}$, we have only 15 valid analogies among 81 combinations. For instance, *red* : *yellow* :: *red* : *yellow* holds, but it is not the case for *red* : *yellow* :: *red* : *blue*. Following the definition, an analogical proportion that holds with nominal attribute values involves *at most two distinct values*. This remark will have important consequences in the theoretical part of this paper. So, in the nominal case, $a : b :: c : d$ is false if and only if $|\{a, b, c, d\}| \geq 3$ or $(a, b, c, d) \in Neg$, where

$$Neg = \{(s, t, t, s), (s, s, s, t), (s, s, t, s), (s, t, s, s), (t, s, s, s) \mid s, t \in X, s \neq t\}.$$

Representing objects with a single Boolean or nominal attribute is usually not expressive enough. In general, items are represented by *tuples* of values, each component being the value of an attribute, either Boolean or nominal. Extending the definition of analogy to nominal tuples of the form $\mathbf{a} = (a_1, \ldots, a_m)$ belonging to a Cartesian product $X = X_1 \times \ldots \times X_m$ can be done component-wise as follows:

$$\mathbf{a} : \mathbf{b} :: \mathbf{c} : \mathbf{d} \text{ holds} \Leftrightarrow \forall i \in [1, m], a_i : b_i :: c_i : d_i \text{ holds.}$$

Throughout the paper, the attribute domains $X_1, \ldots, X_m$ are assumed to be finite sets with at least two elements each.

Analogical Inference. In the Boolean case, the problem of finding an $x \in \{0, 1\}$ such that $a : b :: c : x$ holds, does not always have a solution. Indeed, neither $0 : 1 :: 1 : x$ nor $1 : 0 :: 0 : x$ has a solution (since 0111, 0110, 1000, 1001 are not valid patterns for an analogical proportion). In fact, a solution exists if and only if $(a \equiv b) \vee (a \equiv c)$ holds. When a solution exists, it is unique and is given by $x = c \equiv (a \equiv b)$. This corresponds to the original view advocated by S. Klein [8], who however applied the latter formula even to the cases $0 : 1 :: 1 : x$ and $1 : 0 :: 0 : x$, where it yields $x = 0$ and $x = 1$ respectively.

In the nominal case, the situation is similar. The analogical proportion $a : b :: c : x$ may have no solution ($s : t :: t : x$ has no solution as soon as $s \neq t$), and otherwise (if $a = b$ or $a = c$) the solution is unique, and is given by $x = b$ if $a = c$ and $x = c$ if $a = b$. Namely, the solutions of $s : t :: s : x$, $s : s :: t : x$, and $s : s :: s : x$ are $x = t$, $x = t$, and $x = s$, respectively.

This motivates the following inference pattern first proposed in [16]

$$\frac{\forall i \in \{1, \ldots, m\}, \quad a_i : b_i :: c_i : d_i \text{ holds}}{a_{m+1} : b_{m+1} :: c_{m+1} : d_{m+1} \text{ holds}}$$

which enables us to compute d_{m+1}, provided that $a_{m+1} : b_{m+1} :: c_{m+1} : x$ has a solution. This pattern expresses a rather bold inference which amounts to saying that if the representations of four items are in analogical proportion on m attributes, they should remain in analogical proportion with respect to their labels. Note that, we can restrict ourselves to binary labels, since a multiple class prediction can be obtained by solving a series of binary class problems.

In this paper, we adopt a completely different viewpoint: instead of adding constraints for ensuring the soundness of analogical inference, we want to characterize contexts where this inference is valid, without adding any further constraints. In the next section, we proceed with a discussion on analogical inference, and we present the notion of analogy-preserving functions.

3   Analogy-Preserving Functions

In the analogical inference pattern that was introduced in the previous section, we implicitly assume that there is a dependency that links labels to the values of the m attributes. More precisely, there is some unknown function f such that $e_{m+1} = f(e_1, \ldots, e_m)$, for any item $\mathbf{e} = (e_1, \ldots, e_m)$. Such a function f can be thought of as a classifier that associates to each item, a (unique) class based on the values of the m attribute values describing it.

Since the solutions of analogical equations (when they exist) are unique, the previous pattern can be also written as follows:

$$
\begin{array}{ccccc}
a_1 & \cdots & a_i & \cdots & a_m & f(\mathbf{a}) \\
b_1 & \cdots & b_i & \cdots & b_m & f(\mathbf{b}) \\
c_1 & \cdots & c_i & \cdots & c_m & f(\mathbf{c}) \\
\hline
d_1 & \cdots & d_i & \cdots & d_m & f(\mathbf{d})
\end{array}
$$

where $\mathbf{a} = (a_1, \ldots, a_m)$, $\mathbf{b} = (b_1, \ldots, b_m)$, $\mathbf{c} = (c_1, \ldots, c_m)$ and $\mathbf{d} = (d_1, \ldots, d_m)$.

Remark 3.1. Note that in the nominal case, each column i has at most two distinct elements belonging to the attribute domain X_i of attribute i.

As previously highlighted, the conclusions obtained by analogical inference are brittle. This means here that for a given $\mathbf{d} = (d_1, \ldots, d_i, \ldots, d_m)$ for which we want to evaluate $f(\mathbf{d})$, there may exist several triplets $(\mathbf{a}, \mathbf{b}, \mathbf{c})$ such that $f(\mathbf{a}) : f(\mathbf{b}) :: f(\mathbf{c}) : x$ is solvable, maybe leading to different solutions. In that case, at least from a theoretical viewpoint, it is clear that applying the analogical inference principle for a given $\mathbf{d}$ will not give a unique value to predict $f(\mathbf{d})$. To cope with real-life situations, one generally uses a majority vote for computing a plausible $f(\mathbf{d})$. But an interesting particular case is when all the analogical predictions are the same whatever the triplets. This will be the case as soon as the function f is analogy-preserving, a notion we now formally define.

3.1   Basic Notions and Motivation

In the following, $X_1, \ldots, X_m, X$ denote finite sets with cardinality at least 2.

Definition 3.2. *Let* $\mathbf{X} = X_1 \times \cdots \times X_m$. *A function* $f \colon \mathbf{X} \to X$ *is analogy-preserving (AP for short) if for every* $\mathbf{a}, \mathbf{b}, \mathbf{c}, \mathbf{d} \in \mathbf{X}$,

$$\mathbf{a} : \mathbf{b} :: \mathbf{c} : \mathbf{d} \text{ and } solvable(f(\mathbf{a}), f(\mathbf{b}), f(\mathbf{c})) \Rightarrow sol(f(\mathbf{a}), f(\mathbf{b}), f(\mathbf{c})) = f(\mathbf{d}),$$

where $solvable(f(\mathbf{a}), f(\mathbf{b}), f(\mathbf{c}))$ *means that there exists an* x *such that* $f(\mathbf{a}) : f(\mathbf{b}) :: f(\mathbf{c}) : x$ *holds, and* $sol(f(\mathbf{a}), f(\mathbf{b}), f(\mathbf{c}))$ *is the unique solution for* x.

Note that if f is AP, there cannot exist $\mathbf{a}, \mathbf{b}, \mathbf{c}, \mathbf{d}, \mathbf{a}', \mathbf{b}', \mathbf{c}'$ such that

1. $\mathbf{a} : \mathbf{b} :: \mathbf{c} : \mathbf{d}$ and $\mathbf{a}' : \mathbf{b}' :: \mathbf{c}' : \mathbf{d}$,
2. $solvable(f(\mathbf{a}), f(\mathbf{b}), f(\mathbf{c}))$ and $solvable(f(\mathbf{a}'), f(\mathbf{b}'), f(\mathbf{c}'))$, and
3. $sol(f(\mathbf{a}), f(\mathbf{b}), f(\mathbf{c})) \neq sol(f(\mathbf{a}'), f(\mathbf{b}'), f(\mathbf{c}'))$.

In other words, AP functions are exactly those for which analogical inference never fails. Let us denote by **AP** the set of all AP functions. The following examples suggest that the class of AP functions is quite large and diverse.

Example 3.3. Consider the class of essentially unary[1] functions $f \colon \mathbf{X} \to X$, i.e., of the form $f(\mathbf{x}) = \varphi(x_i)$, for some map $\varphi \colon X_i \to X$. It is not difficult to see that such functions are AP. This corresponds to the simplest example of classifiers since the predicted classes are then determined by the value of a single attribute.

Example 3.4. Consider now the class of *injective* functions $f \colon \mathbf{X} \to X$, i.e., of functions that satisfy the condition: if $\mathbf{x} \neq \mathbf{y}$, then $f(\mathbf{x}) \neq f(\mathbf{y})$. Again, such functions are AP. The key observation is that if $\mathbf{a}, \mathbf{b}, \mathbf{c}$ are pairwise distinct, then so are $f(\mathbf{a}), f(\mathbf{b}), f(\mathbf{c})$ and the condition $solvable(f(\mathbf{a}), f(\mathbf{b}), f(\mathbf{c}))$ in the definition of AP functions does not hold. Thus injective functions are AP. However, injective functions are of little interest in classification since the number of class labels is expected to be smaller than the number of items.

Example 3.5. Using the same argument, we can relax the previous example to obtain other classes of AP functions. For instance, let $\mathbf{1} = (1, \ldots, 1) \in \{0, 1\}^m$ and consider the class of *pseudo-Boolean functions* $f \colon \{0, 1\}^m \to X$ satisfying the following condition: the kernel[2] of f is the form $\ker f = \{(\mathbf{a}, \mathbf{a} \oplus \mathbf{1}) \mid \mathbf{a} \in \{0, 1\}^m\}$, where $\oplus$ denotes addition modulo 2. Again, it is not difficult to verify that such functions are AP. There are examples of "reflexive" functions [6], i.e., functions satisfying the condition that for every $\mathbf{x} \in \{0, 1\}^m$, $f(\mathbf{x} \oplus \mathbf{1}) = f(\mathbf{x})$.

Example 3.6. Examples 3.4 and 3.5 can be generalized as follows. Recall that the set $\mathbb{B} := \{0, 1\}$ constitutes a 2-element field with the operations $\oplus$ (addition modulo 2) and $\otimes$ (multiplication modulo 2). For any natural number m, the set $\mathbb{B}^m$, equipped with scalar multiplication and addition of vectors, is a vector space over $\mathbb{B}$. Let V be a fixed subspace of the vector space $\mathbb{B}^m$. Any function $f \colon \mathbb{B}^m \to X$ such that $\ker f$ is the set of affine spaces that are translations of V is AP. Examples 3.4 and 3.5 correspond to the cases where V is the trivial subspace and the 1-dimensional subspace $\{\mathbf{0}, \mathbf{1}\}$, respectively.

In view of Remark 3.1 we will focus on the following subclass of AP functions.

Definition 3.7. *An AP function $f \colon \mathbf{X} \to X$ is called a* hard AP *(HAP) function if for all $S_i \subseteq X_i$ with $|S_i| \leq 2$ $(1 \leq i \leq m)$ it holds that $|\mathrm{Im} f|_S| \leq 2$, where $S := S_1 \times \cdots \times S_m$. We denote the class of HAP functions by* **HAP**.

Remark 3.8. Observe that **HAP** contains all essentially unary functions. If X is a 2-element set, then **HAP** = **AP**.

---

[1] An argument x_i is said to be *inessential* in $f \colon \mathbf{X} \to X$ if for all $(a_1, \ldots, a_m) \in \mathbf{X}$, $a_i' \in X_i$, we have $f(a_1, \ldots, a_m) = f(a_1, \ldots, a_{i-1}, a_i', a_{i+1}, \ldots, a_m)$. Otherwise, x_i is said to be *essential* in f. The number of essential arguments of f is called the *essential arity* of f.

[2] Recall that the *kernel* of f is $\ker f := \{(\mathbf{a}, \mathbf{b}) \in \{0, 1\}^m \times \{0, 1\}^m \mid f(\mathbf{a}) = f(\mathbf{b})\}$.

3.2 ANF Representations and Affine Functions

In this section we recall some well-known facts about the simplest interesting case of functions, namely, the Boolean functions.

There are several formalisms to represent Boolean functions, such as the classical DNF and CNF representations. However, in the analogical framework the algebraic representation of Boolean functions turns out be more relevant than the former classical representations based on the standard logical operators $\vee$ (disjunction) and $\wedge$ (conjunction). Following [15,17], each Boolean function $f\colon \{0,1\}^m \to \{0,1\}$ (of *arity* m) is uniquely represented by a multilinear polynomial called the algebraic normal form of f that we recall below.

Let $\mathbb{B}$ be the 2-element field over $\{0,1\}$ with its 2 usual operators $\oplus$ (addition modulo 2) and $\otimes$ (multiplication modulo 2). Note that they correspond respectively to the *exclusive or* and to the *conjunction* in logical terms. Equipped with scalar multiplication (which coincides here with the multiplication modulo 2) over $\mathbb{B}$ and addition, the set $\mathbb{B}[x_1,\ldots,x_m]$ of polynomials on the m indeterminates $x_1,\ldots,x_m$ is a vector space over $\mathbb{B}$.

A (*multilinear*) *monomial* is a term of the form $\mathbf{x}_I := \prod_{i\in I} x_i$, for some (possibly empty) finite set of positive integers I with the convention that 1 is the empty monomial $\mathbf{x}_\emptyset$. The size $|I|$ is called the *degree* of $\mathbf{x}_I$, denoted $d(\mathbf{x}_I)$. A (*multilinear*) *polynomial* is a sum of monomials

$$\sum_{I \subseteq \{1,\ldots,m\}} \omega_I \cdot \mathbf{x}_I$$

where each ω_I belongs to $\mathbb{B}$ (addition is understood as addition modulo 2). Note that the monomials 0 and 1 are just $0 \cdot \mathbf{x}_\emptyset$ and $1 \cdot \mathbf{x}_\emptyset$, respectively. The *degree* of a polynomial is then the maximum degree among the degrees of its monomials.

An *algebraic normal form* (ANF) of a Boolean function f of arity m is simply a multilinear polynomial in $\mathbb{B}[x_1,\ldots,x_m]$ that represents it:

$$f(x_1,\ldots,x_m) = \sum_{I \subseteq \{1,\ldots,m\}} \omega_I \cdot \mathbf{x}_I.$$

It is well known that the ANF representation of a Boolean function is unique (see, e.g., [6]), and thus we can define the *degree* $d(f)$ of a Boolean function f as the degree of the polynomial that represents it. Note that the constant 0 and 1 functions are the only Boolean functions of degree 0, whereas *projections* (that correspond to the selection of a single attribute and that are represented by variables x_i) and their negations (that are represented by polynomials of the form $x_i \oplus 1$) are the only functions of degree 1.

A Boolean function $f\colon \mathbb{B}^m \to \mathbb{B}$ is said to be *affine* if $d(f) \leq 1$, i.e., there exist $\omega_0, \omega_1, \ldots, \omega_m \in \mathbb{B}$ such that

$$f(x_1,\ldots,x_m) = \sum_{i=1}^{m} \omega_i \cdot x_i + \omega_0.$$

The set of affine functions of arity m is denoted by $\mathcal{L}_m$, so that $\mathcal{L} = \bigcup_{m \geq 0} \mathcal{L}_m$ is the set of all affine functions. If $\omega_0 = 0$, then such an affine function is said to be *linear*. Thus, affine functions are either linear functions or their negations.

Our interest in this class of affine functions is motivated by the characterization of AP Boolean functions (i.e., in case when items are described by Boolean attributes) [4]: AP Boolean functions are exactly those Boolean functions that are affine, i.e., $\mathbf{AP} = \mathcal{L}$.

In the following sections, we will generalize this result to the case of nominal attributes. This is not a straightforward extension of the Boolean case as we shall see. Before doing that, we provide an illustrative example that puts nominal analogical proportions in another perspective, and that reveals the close relationship of analogical proportions and taxonomic trees, as recently suggested in the Boolean case [1].

3.3   An Illustrative Example

In the illustrative example below, items are assumed to be described by means of three attributes (i.e., $m = 3$), numbered from 1 to 3, namely: $1 =$ shape, $2 =$ color, and $3 =$ weight, where respectively $X_1 = \{circle\ (c),\ square\ (s)\}$, $X_2 = \{blue\ (b),\ red\ (r),\ yellow\ (y)\}$, and $X_3 = \{light\ (l),\ heavy\ (h)\}$. Due to space constraints, we chose a small example, with a non-Boolean nominal attribute, namely, X_2 with $|X_2| = 3$. There are two class labels referred to by 0 and 1.

Table 1 enumerates the 12 items, $\mathbf{a}, \mathbf{a'}, \mathbf{b}, \ldots, \mathbf{f'}$, that can be distinguished on the basis of the three attributes above. Moreover, we consider 4 ways of classifying them into the two classes, each of which corresponding to the 4 functions g_1, g_2, g_3, g_4.

Table 1. Items, attributes, and classifications.

Items	shape	color	weight	g_1	g_2	g_3	g_4
a	c	b	l	1	1	1	1
a'	c	b	h	1	1	0	0
b	c	r	l	1	0	0	1
b'	c	r	h	0	0	0	0
c	c	y	l	0	1	0	1
c'	c	y	h	0	1	0	0
d	s	b	l	1	0	1	1
d'	s	b	h	1	0	0	0
e	s	r	l	1	1	0	1
e'	s	r	h	0	1	0	0
f	s	y	l	0	0	0	1
f'	s	y	h	0	0	0	0

Clearly, g_1 is not an AP function since: i) $\mathbf{a} : \mathbf{a}' :: \mathbf{b} : \mathbf{b}'$ holds, ii) $1 : 1 :: 1 : x$ is solvable (just take $x = 1$), iii) but $g_1(\mathbf{b}') = 0$ is not the solution of $1 : 1 :: 1 : x$.

The function g_2 looks more promising, since the 4-tuple $\mathbf{a}, \mathbf{a}', \mathbf{b}, \mathbf{b}'$ is associated with $1 : 1 :: 0 : 0$, which holds as an analogical proportion. However, looking at $\mathbf{a}, \mathbf{b}', \mathbf{d}, \mathbf{e}'$ we again have an analogical proportion on the three attributes. However this is associated with $1 : 0 :: 0 : 1$ which is not an analogical proportion. Nonetheless, $1 : 0 :: 0 : x$ is not a solvable proportion: as such, it cannot be considered as a counter-example for proving that g_2 is not AP. In fact, by an exhaustive search we can see that there is no counter-example in the table showing that g_2 is not AP. Thus, g_2 is AP: class 1 can be described by the equivalence $c \equiv \neg r$ (since $\neg r = b \vee y$ and $\neg c = s$ here), and class 0 can be described by $c \oplus \neg r$.

For g_3, we can consider the following four tuples: $\mathbf{c} = (c, y, l)$ with $g_3(\mathbf{c}) = 0$, $\mathbf{c}' = (c, y, h)$ with $g_3(\mathbf{c}') = 0$, $\mathbf{d} = (s, b, l)$ with $g_3(\mathbf{d}) = 1$, and $\mathbf{d}' = (s, b, h)$ with $g_3(\mathbf{d}') = 0$. In this case, we have $\mathbf{c} : \mathbf{c}' :: \mathbf{d} : \mathbf{d}'$ and $solvable(g_3(\mathbf{c}), g_3(\mathbf{c}'), g_3(\mathbf{d}))$, but $g_3(\mathbf{c}) : g_3(\mathbf{c}') :: g_3(\mathbf{d}) : g_3(\mathbf{d}')$ does not hold. This shows that g_3 is not AP. For g_3, class 1 corresponds to the blue light objects, which clearly corresponds to a monomial of degree 2.

The situation is simpler for g_4, where class 1 corresponds to the light objects. It is not difficult to see that it is essentially unary, and thus an AP function.

What happens with these different classification functions is better understood by looking at classification trees, which is the topic of the next subsection.

3.4   Taxonomic Trees

A table describing all the possible items that can be distinguished in terms of a set of nominal attributes can be straightforwardly associated with a taxonomic tree, taking the attributes in some order. The tree corresponding to Table 1, with two binary attributes and one ternary one, is given in Fig. 1. At the third level, we retrieve the $2 \cdot 2 \cdot 3 = 12$ items from $\mathbf{a}$ to $\mathbf{f}'$. They can be encoded by following the path from the root, using a standard convention: at each level the edges are numbered from the left from 0 to 1, or to 2. Thus, for example, $\mathbf{b}'$ is associated with 011, corresponding to attribute values c, r, h; see Fig. 1.

A large number of analogical proportions are hidden between the leaves of such a taxonomic tree. Namely, in our example with 12 items, there are exactly 30 *distinct* analogical proportions on the three attributes (where all the elements in the 4-tuples are distinct). For instance, we have $\mathbf{a} : \mathbf{a}' :: \mathbf{b} : \mathbf{b}'$, or $\mathbf{c} : \mathbf{c}' :: \mathbf{d} : \mathbf{d}'$. This can be checked by observing that here these analogical proportions involving 3 attributes are either

- of the form $uxt : uyt :: vxt : vyt$ (with one constant attribute value), or
- of the form $uxt : vxz :: uyt : vyz$ (with no constant attribute value),

where t, u, v, x, y, z are attribute values, one by attribute in each tuple (such as tux), since an analogical proportion can involve at most 2 distinct values for each attribute. The ordering of attributes has no special meaning, but is the same in each tuple. The 2 above patterns remain the same under symmetry. Note that

$$uxt : uxt :: vxt : vxt \quad \text{and} \quad uxt : vyt :: uxt : vyt$$

are not considered, since they hold trivially by identity or reflexivity. Note also that the form $uxt : vxz :: uyt : vyz$ is the same as $uxt : uyt :: vxz : vyz$ by central permutation (even if the number of constant attributes in the first and second pairs of tuples vary from two to one).

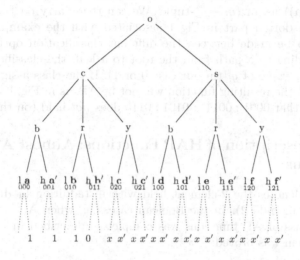

Fig. 1. Example of classification tree.

Reading this taxonomic tree horizontally, there is quite a number of analogical proportions that hold between 4-tuples of items. Assume for a short while that we only have a tree induced by 3 *binary* attributes. Then each of the two forms $uxt : uyt :: vxt : vyt$ and $uxt : vxz :: uyt : vyz$ yields 6 analogical proportions, i.e., in this case we have 12 distinct non-trivial analogical proportions. Indeed, considering the first form $uxt : uyt :: vxt : vyt$, there are $3 \cdot 2 = 6$ possible choices of value for t in case of 3 binary attributes, the possible instantiations of ux, uy, vx, vy being all equivalent due to analogical proportion postulates. For the second form $uxt : vxz :: uyt : vyz$, we can observe that uxt and uyt (as well as vxz and vyz) differ only on one attribute value. There are 6 possible instantiations for this attribute in case of 3 binary attributes, the possible instantiations of the two remaining binary attributes being all equivalent.

In case of two binary attributes and one ternary attribute as in the example, a similar counting can be made. For the first form, we now have $1 \cdot 3 + 2 \cdot 2 = 7$ possible instantiations for t. Moreover when t is not a value of the ternary attribute, we have 3 possible ways of instantiating ux, uy, vx, vy. Altogether the first form then yields $1 \cdot 3 + 2 \cdot 2 \cdot 3 = 15$ analogical proportions. For the second form there are 3 ways of instantiating the "changing attribute" if it is the ternary one; and $2 \cdot 2$ if it is a binary attribute, in this later case there are 3 possible ways of instantiating the remaining attributes. Again, we get 15 analogical proportions, and a total of 30 distinct analogical proportions.

As suggested by the above example, the number of analogical proportions increases rapidly with the cardinalities of the attribute domains and with the number of levels in the tree. This suggests how important the presence of analogical proportions in a classification process is.

Given one of our functions g_i, nothing forbids to consider its value as another attribute (an n-ary one if there are n classes). So instead of considering $\mathbf{a}$, we consider $(\mathbf{a}, g_i(\mathbf{a}))$ as an $(m + 1)$-tuple. We can then carry on the building of the tree as the dotted part in Fig. 1 associated with the example. But now, a choice has to be made between the different classification options for each item corresponding to a path from the root to a leaf: the classification option is related to the value of g_i (in our case 0 or 1). If the class assignment is not "well-balanced" the resulting function will not be AP, as in Fig. 1 where we can clearly observe that $0001 : 0011 :: 0101 : 0110$ does *not* hold (on the 4th digit).

4   Local Description of HAP Functions: Almost Affine Functions

Recall that $\mathbb{B}$ denotes the 2-element field with $\oplus$ (addition modulo 2) and $\otimes$ (multiplication modulo 2). In the Boolean case, i.e., when $X_1 = \cdots = X_m = X = \mathbb{B}$, the class of AP functions was completely described in [4], where the following theorem was proved:

Theorem 4.1. *A function $f : \mathbb{B}^m \to \mathbb{B}$ is AP if and only if it is affine. In particular, the class of AP functions constitutes a clone of Boolean functions, i.e., it contains all projections and it is closed under compositions.*

To extend this result to HAP functions in the arbitrary nominal case (see Subsect. 3.1), we shall make use of the following useful observations. From the definition of HAP functions it follows that the restriction $f|_S$ of a HAP function f to any subset $S := S_1 \times \cdots \times S_m \subseteq \mathbf{X} = X_1 \times \cdots \times X_m$ with $|S_i| \leq 2$ $(1 \leq i \leq m)$ must have at most 2 distinct values. Therefore, every such function can be thought of as a Boolean function by a suitable renaming of variable and function values. Thus, from Theorem 4.1, we have the following corollary.

Corollary 4.2. *Let X be a set, let $S_1, \ldots, S_m$ be 2-element sets, and set $S := S_1 \times \cdots \times S_m$. A function $f : S \to X$ is HAP if and only if there exist $\varphi^f : \{0, 1\} \to X$ and $\varphi_i^f : S_i \to \{0, 1\}$ such that*

$$f|_S = \varphi^f(\varphi_1^{f,S}(x_1) \oplus \cdots \oplus \varphi_m^{f,S}(x_m) \oplus c). \tag{1}$$

Remark 4.3. Note that the term $\oplus c$ can be encoded into φ^f so that (1) can be simplified into

$$f|_S = \varphi^f(\varphi_1^{f,S}(x_1) \oplus \cdots \oplus \varphi_m^{f,S}(x_m)). \tag{2}$$

We will generalize these results by introducing the notion of "almost affine" functions, and show that HAP functions are exactly the almost affine functions.

Definition 4.4. *A function* $f\colon \mathbf{X} \to X$ *is* almost affine *if for any* $S_i \subseteq X_i$ *with* $|S_i| \leq 2$, *and* $S := S_1 \times \cdots \times S_m$, *there exist* $\varphi^{f,S}\colon \{0,1\} \to \mathrm{Im}(f|_S)$ *and* $\varphi_i^{f,S}\colon S_i \to \{0,1\}$ *such that* $f|_S = \varphi^{f,S}(\varphi_1^{f,S}(x_1) \oplus \cdots \oplus \varphi_m^{f,S}(x_m))$.

Theorem 4.5. *A function* $f\colon \mathbf{X} \to X$ *is HAP if and only if it is almost affine.*

Proof. By Corollary 4.2, we know that every HAP function is almost affine. Moreover, by definition, every restriction $f|_S$ of an almost affine function f is AP on $S \subseteq \mathbf{X}$. Therefore, f is HAP. □

As we will see, this description is quite useful. However, it has some limitations since it requires a local inspection on each subset

$$S = S_1 \times \cdots \times S_m \subseteq \mathbf{X} = X_1 \times \cdots \times X_m.$$

In the next section we provide a global description of almost affine functions, and thus a description of HAP functions.

5 Global Description of HAP Functions: Quasi-Linear Functions

In the previous section, we showed that the class of HAP functions $f\colon \mathbf{X} \to X$ coincides exactly with the class of almost affine functions. In this section we will show that the HAP functions are either essentially unary or quasilinear.

Definition 5.1. *A function* $f\colon \mathbf{X} \to X$ *is* quasilinear *if there exist* $\varphi\colon \{0,1\} \to X$ *and* $\varphi_i\colon X_i \to \{0,1\}$ $(1 \leq i \leq m)$ *such that* $f = \varphi(\varphi_1(x_1) \oplus \cdots \oplus \varphi_m(x_m))$.

We are going to make use of Jablonski's Fundamental Lemma (see, e.g. [14]).

Lemma 5.2 (Jablonski's Fundamental Lemma).

1. *Let* $f\colon X^m \to X$ *be a function that has at least two essential arguments and* $|\mathrm{Im}(f)| = \ell > 2$. *Then there exist sets* $S_i \subseteq X$ $(1 \leq i \leq m)$ *with* $|S_i| \leq 2$ *such that for* $S := S_1 \times \cdots \times S_m$, $|\mathrm{Im}(f|_S)| \geq 3$.
2. *More generally, let* $f\colon \mathbf{X} \to X$ *be a function that has at least two essential arguments and* $|\mathrm{Im}(f)| = \ell > 2$. *Then for any* k *with* $2 < k \leq \ell$, *there exist sets* $S_i \subseteq X_i$ $(1 \leq i \leq m)$ *with* $|S_i| \leq k-1$ *such that for* $S := S_1 \times \cdots \times S_m$, $|\mathrm{Im}(f|_S)| \geq k$.

Lemma 5.3. *Let* $f\colon \mathbf{X} \to X$. *Assume that for all* $S_i \subseteq X_i$ *with* $|S_i| \leq 2$ $(1 \leq i \leq m)$, *and* $S := S_1 \times \cdots \times S_m$, *we have that* $|\mathrm{Im}(f|_S)| \leq 2$. *Then* f *is essentially unary or* $|\mathrm{Im}(f)| \leq 2$.

Proof. Suppose, to the contrary, that f has at least two essential arguments and $|\mathrm{Im}(f)| = \ell \geq 3$. By Lemma 5.2, item 2., there exist sets $S_i \subseteq X_i$ $(1 \leq i \leq m)$ with $|S_i| \leq 2$ such that for $S := S_1 \times \cdots \times S_m$, $|\mathrm{Im}(f|_S)| \geq 3$. We have reached a contradiction. $\square$

In other words, Lemma 5.3 asserts that an almost affine function f is either essentially unary or has a range of at most two elements.

Proposition 5.4. *A function $f \colon \mathbf{X} \to X$ is almost affine if and only if it is essentially unary or quasilinear.*

Proof. Assume that $f \colon \mathbf{X} \to X$ is almost affine and has at least two essential arguments. Then $|\mathrm{Im}(f)| \leq 2$ by Lemma 5.3.

We are going to show that for all $S_i \subseteq X_i$ $(1 \leq i \leq m)$ and $S := S_1 \times \cdots \times S_m$, there exist maps $\varphi^{f,S} \colon \{0,1\} \to X$ and $\varphi_i^{f,S} \colon S_i \to \{0,1\}$ such that

$$f|_S = \varphi^{f,S}\big(\varphi_1^{f,S}(x_1) \oplus \cdots \oplus \varphi_m^{f,S}(x_m)\big).$$

The claim holds whenever $|S_i| \leq 2$ $(1 \leq i \leq m)$ by definition.

We proceed with an inductive argument. Assume that the claim holds for all sets $S_i \subseteq X_i$ $(1 \leq i \leq m)$ with $|S_i| \leq k_i$ for some $k_1, \ldots, k_m$ with $2 \leq k_i \leq |X_i|$. We will show that if $j \in \{1, \ldots, m\}$ and $k_j < |X_j|$, then the claim holds also for all sets $S_i \subseteq X_i$ $(1 \leq i \leq n)$ with $|S_i| \leq k_i$ $(i \neq j)$ and $|S_j| = k_j + 1$.

So, let $S_i \subseteq X_i$ $(1 \leq i \leq m)$ with $|S_i| \leq k_i$ $(i \neq j)$ and $|S_j| = k_j + 1$, and write $S := S_1 \times \cdots \times S_m$. Assume that $a, b \in S_j$, $a \neq b$. Let $S_j' := S_j \setminus \{b\}$ and $S_j^* := \{a, b\}$, and let

$$S' := S_1 \times \cdots \times S_{j-1} \times S_j' \times S_{j+1} \times \cdots \times S_m,$$
$$S^* := S_1 \times \cdots \times S_{j-1} \times S_j^* \times S_{j+1} \times \cdots \times S_m,$$
$$T := S_1 \times \cdots \times S_{j-1} \times \{a\} \times S_{j+1} \times \cdots \times S_m = S' \cap S^*,$$
$$T' := S_1 \times \cdots \times S_{j-1} \times \{b\} \times S_{j+1} \times \cdots \times S_m = S \setminus S'.$$

By the inductive hypothesis, there exist maps $\varphi^{f,S'}$, $\varphi_i^{f,S'}$ $(1 \leq i \leq m)$ and φ^{f,S^*}, φ_i^{f,S^*} $(1 \leq i \leq m)$ such that $f|_{S'} = \varphi^{f,S'}\big(\varphi_1^{f,S'}(x_1) \oplus \cdots \oplus \varphi_m^{f,S'}(x_m)\big)$, and $f|_{S^*} = \varphi^{f,S^*}\big(\varphi_1^{f,S^*}(x_1) \oplus \cdots \oplus \varphi_m^{f,S^*}(x_m)\big)$.

Let now $\varphi^{f,S} := \varphi^{f,S'}$, $\varphi_i^{f,S} := \varphi_i^{f,S'}$ for $i \neq j$, and define $\varphi_j^{f,S} \colon S_j \to \{0,1\}$ as the extension of $\varphi_j^{f,S'} \colon S_j' \to \{0,1\}$ that satisfies the condition $\varphi_j^{f,S}(a) = \varphi_j^{f,S}(b)$ if and only if $\varphi_j^{f,S^*}(a) = \varphi_j^{f,S^*}(b)$.

It remains to show that $f|_S = \varphi^{f,S}\big(\varphi_1^{f,S}(x_1) \oplus \cdots \oplus \varphi_m^{f,S}(x_m)\big)$. Let $\mathbf{x} = (x_1, \ldots, x_m) \in S$. If $\mathbf{x} \in S'$ then

$$\varphi^{f,S}\big(\varphi_1^{f,S}(x_1) \oplus \cdots \oplus \varphi_m^{f,S}(x_m)\big) = \varphi^{f,S'}\big(\varphi_1^{f,S'}(x_1) \oplus \cdots \oplus \varphi_m^{f,S'}(x_m)\big) = f(\mathbf{x}).$$

Assume now that $\mathbf{x} \in S \setminus S' = T'$. Then $x_j = b$, so $\mathbf{x} \in S^*$. If $\varphi_j^{f,S^*}(a) = \varphi_j^{f,S^*}(b)$, then also $\varphi_j^{f,S}(a) = \varphi_j^{f,S}(b)$, and we have

$$\varphi^{f,S}\left(\varphi_1^{f,S}(x_1) \oplus \cdots \oplus \varphi_j^{f,S}(b) \oplus \cdots \oplus \varphi_m^{f,S}(x_m)\right)$$
$$= \varphi^{f,S}\left(\varphi_1^{f,S}(x_1) \oplus \cdots \oplus \varphi_j^{f,S}(a) \oplus \cdots \oplus \varphi_m^{f,S}(x_m)\right)$$
$$= \varphi^{f,S^*}\left(\varphi_1^{f,S^*}(x_1) \oplus \cdots \oplus \varphi_j^{f,S^*}(a) \oplus \cdots \oplus \varphi_m^{f,S^*}(x_m)\right)$$
$$= \varphi^{f,S^*}\left(\varphi_1^{f,S^*}(x_1) \oplus \cdots \oplus \varphi_j^{f,S^*}(b) \oplus \cdots \oplus \varphi_m^{f,S^*}(x_m)\right) f(\mathbf{x}).$$

If $\varphi_j^{f,S^*}(a) \neq \varphi_j^{f,S^*}(b)$, i.e., $\varphi_j^{f,S^*}(a) \oplus 1 = \varphi_j^{f,S^*}(b)$, then also $\varphi_j^{f,S}(a) \oplus 1 = \varphi_j^{f,S}(b)$, and we have

$$\varphi^{f,S}\left(\varphi_1^{f,S}(x_1) \oplus \cdots \oplus \varphi_j^{f,S}(b) \oplus \cdots \oplus \varphi_m^{f,S}(x_m)\right)$$
$$= \varphi^{f,S}\left(\varphi_1^{f,S}(x_1) \oplus \cdots \oplus \varphi_j^{f,S}(a) \oplus 1 \oplus \cdots \oplus \varphi_m^{f,S}(x_m)\right)$$
$$= \varphi^{f,S^*}\left(\varphi_1^{f,S^*}(x_1) \oplus \cdots \oplus \varphi_j^{f,S^*}(a) \oplus 1 \oplus \cdots \oplus \varphi_m^{f,S^*}(x_m)\right)$$
$$= \varphi^{f,S^*}\left(\varphi_1^{f,S^*}(x_1) \oplus \cdots \oplus \varphi_j^{f,S^*}(b) \oplus \cdots \oplus \varphi_m^{f,S^*}(x_m)\right) f(\mathbf{x}).$$

Therefore $\varphi^{f,S}\left(\varphi_1^{f,S}(x_1) \oplus \cdots \oplus \varphi_m^{f,S}(x_m)\right) = f(\mathbf{x})$ for all $\mathbf{x} \in S$ ⊔

Example 5.5. Note that both g_2 and g_4 of Subsect. 3.3 are in fact HAP functions, since both are quasilinear. Indeed, g_2 and g_4 can be represented as

$$g_2(x_1, x_2, x_3) = \varphi(\varphi_1(x_1) \oplus \varphi_2(x_2) \oplus \varphi_3(x_3))$$

with

$$
\begin{aligned}
&\varphi_1 \colon \{c, s\} \to \{0, 1\}, &&c \mapsto 0, &&s \mapsto 1, \\
&\varphi_2 \colon \{b, r, y\} \to \{0, 1\}, &&b \mapsto 1, &&r \mapsto 0, &&y \mapsto 1, \\
&\varphi_3 \colon \{l, h\} \to \{0, 1\}, &&l \mapsto 0, &&h \mapsto 0, \\
&\varphi \colon \{0, 1\} \to \{0, 1\}, &&0 \mapsto 0, &&1 \mapsto 1
\end{aligned}
$$

and

$$g_4(x_1, x_2, x_3) = \psi(\psi_1(x_1) \oplus \psi_2(x_2) \oplus \psi_3(x_3))$$

with

$$
\begin{aligned}
&\psi_1 \colon \{c, s\} \to \{0, 1\}, &&c \mapsto 0, &&s \mapsto 0, \\
&\psi_2 \colon \{b, r, y\} \to \{0, 1\}, &&b \mapsto 0, &&r \mapsto 0, &&y \mapsto 0, \\
&\psi_3 \colon \{l, h\} \to \{0, 1\}, &&l \mapsto 1, &&h \mapsto 0, \\
&\psi \colon \{0, 1\} \to \{0, 1\}, &&0 \mapsto 0, &&1 \mapsto 1.
\end{aligned}
$$

Moreover, g_4 is essentially unary because

$$g_4(x_1, x_2, x_3) = \gamma(x_3)$$

with $\gamma \colon \{l, h\} \to \{0, 1\}, l \mapsto 1, h \mapsto 0$.

We have seen that the class of HAP functions on X coincides with the class of almost affine functions on X. In the case when $X_1 = \cdots = X_m = X$, this class is exactly Burle's clone of quasilinear functions [3], thus generalizing the last assertion of Theorem 4.1.

Proposition 5.6. *For every finite X, the class of HAP functions on X constitutes a clone, i.e., it contains every projection on X and it is closed under forming compositions: if $f\colon X^n \to X$ and each $g_i\colon X^m \to X$ is HAP on X, then so is $f' = f(g_1, \ldots, g_n)\colon X^m \to X$.*

6   Conclusion

The above results describe the class of hard analogy-preserving functions over finite domains, including the Boolean case as a particular case. The case of non-finite domains remains open. Still it is an important step towards a better understanding of analogical inference. The analogy-preserving functions are those for which analogical inference never fails for predicting their values. Still the set of situations where analogical inference gives good predictions is much larger, as shown by experiments, since a good prediction does not require that *all* predictions given by triplets are the same, but that a majority of triplets give the good prediction. However, these theoretical results contribute to a better understanding of analogical inference and show that applying analogical proportion-based inference amounts to enforcing linearity as much as possible, at least in a local manner. Analogical proportions are pervasive, as shown by their abundance in taxonomic trees, and are thus an important notion for reasoning from data.

References

1. Barbot, N., Miclet, L., Prade, H., Richard, G.: A new perspective on analogical proportions. In: Kern-Isberner, G., Ognjanović, Z. (eds.) ECSQARU 2019. LNCS (LNAI), vol. 11726, pp. 163–174. Springer, Cham (2019). https://doi.org/10.1007/978-3-030-29765-7_14
2. Bounhas, M., Prade, H., Richard, G.: Analogy-based classifiers for nominal or numerical data. Int. J. Approx. Reason. **91**, 36–55 (2017)
3. Burle, G.A.: Classes of k-valued logic which contain all functions of a single variable. Diskret. Analiz **10**, 3–7 (1967). (Russian)
4. Couceiro, M., Hug, N., Prade, H., Richard, G.: Analogy-preserving functions: a way to extend Boolean samples. In: Proceedings of the 26th International Joint Conference on Artificial Intelligence, IJCAI 2017, Melbourne, August 19–25, pp. 1575–1581 (2017)
5. Couceiro, M., Hug, N., Prade, H., Richard, G.: Behavior of analogical inference w.r.t. Boolean functions. In: Lang, J. (ed.) Proceedings of the 27th International Joint Conference on Artificial Intelligence, IJCAI 2018, Stockholm, July 13–19, pp. 2057–2063. ijcai.org (2018)
6. Crama, Y., Hammer, P.: Boolean Functions - Theory, Algorithms, and Applications. Encyclopedia of Mathematics and its Applications, vol. 142. Cambridge University Press (2011)

7. Hesse, M.: On defining analogy. Proc. Aristotelian Soc. **60**, 79–100 (1959)
8. Klein, S.: Analogy and mysticism and the structure of culture (and Comments and Reply). Curr. Anthropol. **24**(2), 151–180 (1983)
9. Langlais, P., Patry, A.: Translating unknown words by analogical learning. In: Joint Conference on Empirical Methods in Natural Language Processing (EMNLP) and Conference on Computational Natural Language Learning (CONLL), Prague, pp. 877–886 (2007)
10. Lepage, Y.: Analogy and formal languages. In: Proceedings of the FG/MOL, pp. 373–378 (2001)
11. Miclet, L., Bayoudh, S., Delhay, A.: Analogical dissimilarity: definition, algorithms and two experiments in machine learning. JAIR **32**, 793–824 (2008)
12. Miclet, L., Prade, H.: Handling analogical proportions in classical logic and fuzzy logics settings. In: Sossai, C., Chemello, G. (eds.) ECSQARU 2009. LNCS (LNAI), vol. 5590, pp. 638–650. Springer, Heidelberg (2009). https://doi.org/10.1007/978-3-642-02906-6_55
13. Prade, H., Richard, G.: Analogical proportions: from equality to inequality. Int. J. Approx. Reason. **101**, 234–254 (2018)
14. Salomaa, A.: On essential variables of functions, especially in the algebra of logic. Ann. Acad. Sci. Fenn. Ser. A I(339), 1–11 (1963)
15. Stone, M.H.: The theory of representation for Boolean algebras. Trans. Am. Math. Soc. **40**(1), 37–111 (1936)
16. Stroppa, N., Yvon, F.: Du quatrième de proportion comme principe inductif : une proposition et son application à l'apprentissage de la morphologie. Trait. Autom. Lang. **47**(2), 1–27 (2006)
17. Zhegalkin, I.I.: On the technique of calculating propositions in symbolic logic. Mat. Sb. **43**, 9–28 (1927)

Measuring Disagreement
with Interpolants

Jandson S. Ribeiro$^{(\boxtimes)}$, Viorica Sofronie-Stokkermans, and Matthias Thimm

University of Koblenz-Landau, Koblenz, Germany
{jandson,sofronie,thimm}@uni-koblenz.de

Abstract. We consider the problem of quantitatively assessing the conflict between knowledge bases in knowledge merging scenarios. Using the notion of Craig interpolation we define a series of disagreement measures and analyse their compliance with properties proposed in previous work by Potyka. We study basic complexity theoretic questions in that scenario and discuss the suitability of our approaches.

Keywords: Disagreement measure · Craig interpolation

1 Introduction

Inconsistencies arise easily whenever knowledge has to be aggregated from different sources [2,3,14]. Approaches to *belief merging* and *information fusion* address these issues by providing computational approaches for automatically resolving these issues in some sensible way. These fields bear a close relationship with the fields of *judgement* and *preference aggregation* [8,20] and also feature their own version of Arrow's impossibility result [1], insofar that there cannot be any "rational" belief merging approach [7]. This calls for semi-automatic methods that take human background knowledge into account when knowledge has to be merged in order not to remove important pieces of information. In order to support the task of semi-automatic merging, we investigate approaches to *analyse* belief merging settings, i.e., approaches that can explain reasons for inconsistency and assess their severity. More specifically, we investigate *disagreement measures* [17], i.e., functions that take a knowledge base profile $P = \langle \mathcal{K}_1, \ldots, \mathcal{K}_n \rangle$ as input and return a non-negative value that quantifies the severity of the disagreement between the different sources of information modelled by $\mathcal{K}_1, \ldots, \mathcal{K}_n$. Disagreement measures are closely related to *inconsistency measures* [9,18,19], which themselves are functions that assess the severity of inconsistency in a single knowledge base. Disagreement and inconsistency measures can be used to help in debugging inconsistencies in semi-automated settings [4,5,10].

In this paper, we develop novel disagreement measures based on the concept of Craig interpolation [6]. Given two knowledge bases $\mathcal{K}_1, \mathcal{K}_2$ with $\mathcal{K}_1 \cup \mathcal{K}_2$ being inconsistent, an *interpolant* is a formula that concisely characterises an aspect of this inconsistency (we will provide formal definitions later). Thus, interpolants

© Springer Nature Switzerland AG 2020
J. Davis and K. Tabia (Eds.): SUM 2020, LNAI 12322, pp. 84–97, 2020.
https://doi.org/10.1007/978-3-030-58449-8_6

play a similar role in analysing the disagreement between two knowledge bases as minimal inconsistent subsets do in analysing the inconsistency within a single knowledge base. As a matter of fact, minimal inconsistent subsets play a dominant role in many approaches to measuring inconsistency [12,15,21] and, therefore, it seems plausible to explore the use of interpolants in measuring disagreement. In order to do that, we consider the set of all interpolants (up to semantical equivalence) and define measures based on the size of that set and on the information content of the weakest/strongest interpolants (which are well-defined concepts due to the fact that the set of interpolants form a complete lattice). We show that our approaches provide meaningful results and comply with many of the rationality postulates introduced in [17]. We also undertake a small study of the computational complexity of several tasks relevant for our work, showing that (unsurprisingly) all of those are intractable. In summary, the contributions of this paper are as follows:

1. We make some general observations on interpolants in order to establish a framework suitable for measuring disagreement (Sect. 3).
2. We present novel disagreement measures based on interpolants (Sect. 4).
3. We investigate the compliance of these disagreement measures with rationality postulates from the literature (Sect. 5).
4. We investigate the computational complexity of several tasks pertaining to our disagreement measures (Sect. 6).

We introduce necessary preliminaries in Sect. 2 and conclude in Sect. 7.

We omit several proofs due to space restrictions. These can be found in an online appendix[1].

2   Preliminaries

Let At be some fixed propositional signature, i.e., a (possibly infinite) set of propositions, and let $\mathcal{L}(\mathsf{At})$ be the corresponding propositional language constructed using the usual connectives $\wedge$ (*conjunction*), $\vee$ (*disjunction*), $\rightarrow$ (*implication*), and $\neg$ (*negation*). A literal is a proposition p or a negated proposition $\neg p$.

Definition 1. *A knowledge base $\mathcal{K}$ is a finite set of formulas $\mathcal{K} \subseteq \mathcal{L}(\mathsf{At})$. Let $\mathbb{K}$ be the set of all knowledge bases.*

A clause is a disjunction of literals. A formula is in conjunctive normal form (CNF) if the formula is a conjunction of clauses. If Φ is a formula or a set of formulas we write $\mathsf{At}(\Phi)$ to denote the set of propositions appearing in Φ. For a set $\Phi = \{\phi_1, \dots, \phi_n\}$ let $\bigwedge \Phi = \phi_1 \wedge \dots \wedge \phi_n$ and $\neg\Phi = \{\neg\phi \mid \phi \in \Phi\}$.

Semantics for a propositional language is given by *interpretations* where an *interpretation* ω on At is a function $\omega : \mathsf{At} \rightarrow \{\mathsf{true}, \mathsf{false}\}$. Let $\Omega(\mathsf{At})$ denote the set of all interpretations for At. An interpretation ω *satisfies* (or is a *model* of)

---

[1] http://mthimm.de/misc/rst_dismes_proofs.pdf.

an atom $a \in \mathsf{At}$, denoted by $\omega \models a$, if and only if $\omega(a) = \mathsf{true}$. The satisfaction relation $\models$ is extended to formulas in the usual way. For $\Phi \subseteq \mathcal{L}(\mathsf{At})$ we also define $\omega \models \Phi$ if and only if $\omega \models \phi$ for every $\phi \in \Phi$.

In the following, let Φ, Φ_1, Φ_2 be formulas or sets of formulas. Define the set of models $\mathsf{Mod}(\Phi) = \{\omega \in \Omega(\mathsf{At}) \mid \omega \models \Phi\}$. We write $\Phi_1 \models \Phi_2$ if $\mathsf{Mod}(\Phi_1) \subseteq \mathsf{Mod}(\Phi_2)$. Φ_1, Φ_2 are *equivalent*, denoted by $\Phi_1 \equiv \Phi_2$, if and only if $\mathsf{Mod}(\Phi_1) = \mathsf{Mod}(\Phi_2)$. Define the closure $\mathsf{Cn}(\Phi)$ of a formula or set of formulas Φ via $\mathsf{Cn}(\Phi) = \{\phi \mid \Phi \models \phi\}$. If $\mathsf{Mod}(\Phi) = \emptyset$ we also write $\Phi \models \bot$ and say that Φ is *inconsistent* (or *unsatisfiable*).

3   Craig Interpolants

An important result in first-order logic is Craig's Interpolation Theorem [6].

Theorem 1 [6]. *Let ϕ, ψ be closed formulæ such that $\phi \models \psi$. Then there exists a closed formula I containing only predicate symbols, function symbols and constants occurring in both ϕ and ψ such that $\phi \models I$ and $I \models \psi$.*

Every formula I satisfying the property in Theorem 1 will be called an interpolant of ϕ and ψ. In the context of propositional logic, and of finite sets of propositional formulas, the concept of *interpolant* specializes as follows:

Definition 2. *Let Φ and Φ' be finite sets of propositional logic formulas. A formula ϕ is called an* interpolant of Φ wrt. Φ' *if*

1. $\Phi \models \phi$,
2. $\Phi' \cup \{\phi\} \models \bot$, and
3. $\mathsf{At}(\phi) \subseteq \mathsf{At}(\Phi) \cap \mathsf{At}(\Phi')$

Consider, for instance, two sets $\Phi_1 = \{r \vee \neg p, \neg r \vee \neg q\}$ and $\Phi_2 = \{p, q\}$. The formula $p \to \neg q$ is an interpolant of Φ_1 wrt Φ_2, as $\Phi_1 \models p \to \neg q$, $\Phi_2 \cup \{p \to \neg q\} \models \bot$ and $\mathsf{At}(\{p \to \neg q\}) = \{p, q\} \subseteq \mathsf{At}(\Phi_1) \cap \mathsf{At}(\Phi_2)$.

Clearly, two finite sets Φ and Φ' of formulas in propositional logic have an interpolant if and only if $\Phi \cup \Phi'$ is unsatisfiable. Let $\mathbb{I}(\Phi, \Phi')$ denote the set of interpolants of Φ and Φ'.

Let φ be a formula and x a propositional symbol, we write $\varphi[x \mapsto \top]$ to denote that all occurrences of x in φ are replaced by $\top$. Analogously, $\varphi[x \mapsto \bot]$ means that all ocurrences of x are replaced by $\bot$. For instance, for the formula $\varphi = p \vee \neg q$, we get $\varphi[p \mapsto \top] \equiv \top \vee \neg q$.

Definition 3. *Let ϕ be a propositional formula and $x \in \mathsf{At}(\phi)$. We use the following notation:*

- $\exists x\, \phi := \phi[x \mapsto \bot] \vee \phi[x \mapsto \top]$;
- $\forall x\, \phi := \phi[x \mapsto \bot] \wedge \phi[x \mapsto \top]$.

Let Φ be a finite set of propositional formulæ, and let ϕ be the conjunction of all the formulæ in Φ. For every $x \in \mathsf{At}(\phi)$, we use the following notation $\exists x \Phi := \exists x \phi$, $\forall x \Phi := \forall x \phi$.

Propositional logic allows uniform interpolation: For every formula ϕ and every set $\{y_1, \ldots, y_m\} \subseteq \mathsf{At}(\phi)$ there exists a propositional formula I_Φ such that $\mathsf{At}(I_\Phi) \subseteq \{y_1, \ldots, y_m\}$ with the property that for every formula ϕ' such that $\phi \wedge \phi' \models \perp$ and $\{y_1, \ldots, y_m\} = \mathsf{At}(\phi) \cap \mathsf{At}(\phi')$, I_Φ is an interpolant of ϕ and ϕ'. This follows from the following result (here formulated for finite sets of propositional formulæ).

Proposition 1. *Let Φ be a finite set of propositional formulas. Assume that $\mathsf{At}(\Phi) = \{x_1, \ldots, x_n, y_1, \ldots, y_m\}$. Then the following hold:*

1. $\Phi \models \exists x_1 \ldots \exists x_n \Phi$.
2. *Let ψ be a propositional formula with $\mathsf{At}(\psi) \subseteq \{y_1, \ldots, y_m\}$ such that $\Phi \models \psi$.
   Then $\exists x_1 \ldots \exists x_n \Phi \models \psi$.*
3. $\forall x_1 \ldots \forall x_n \Phi \models \Phi$ *(i.e. $\forall x_1 \ldots \forall x_n \Phi \models \phi_i$ for every formula $\phi_i \in \Phi$).*
4. *Let ψ be a formula with $\mathsf{At}(\psi) \subseteq \{y_1, \ldots, y_m\}$ such that $\psi \models \Phi$.
   Then $\psi \models \forall x_1 \ldots \forall x_n \Phi$.*

In the following, we define some auxiliary notions for interpolants and make some first observations regarding the structure of $\mathbb{I}(\Phi, \Phi')$.

Proposition 2. *Let Φ and Φ' be finite sets of formulas.*

1. *If $\Phi \cup \Phi' \not\models \perp$ then $\mathbb{I}(\Phi, \Phi') = \emptyset$.*
2. *If $\Phi \models \perp$ and $\Phi' \models \perp$ then $\mathbb{I}(\Phi, \Phi') = \mathcal{L}(\mathsf{At}(\Phi) \cap \mathsf{At}(\Phi'))$.*
3. *If $\Phi \models \perp$ and $\Phi' \not\models \perp$ then $\mathbb{I}(\Phi, \Phi') = \mathcal{L}(\mathsf{At}(\Phi) \cap \mathsf{At}(\Phi')) \cap (\neg \mathsf{Cn}(\Phi') \cup \{\phi \mid \phi \models \perp\})$.*
4. *If $\Phi \not\models \perp$ and $\Phi' \models \perp$ then $\mathbb{I}(\Phi, \Phi') = \mathcal{L}(\mathsf{At}(\Phi) \cap \mathsf{At}(\Phi')) \cap \mathsf{Cn}(\Phi)$.*

Proof. (1) Assume that $\mathbb{I}(\Phi, \Phi') \neq \emptyset$. Then there exists a formula ϕ such that $\Phi \models \phi$ and $\Phi' \cup \{\phi\} \models \perp$. Hence, $\Phi \cup \Phi' \models \perp$.

(2) If $\Phi \models \perp$ and $\Phi' \models \perp$ then (i) $\Phi \models \phi$ for all $\phi \in \mathcal{L}(\mathsf{At}(\Phi) \cap \mathsf{At}(\Phi'))$ and (ii) $\Phi' \cup \{\phi\} \models \perp$ for all $\phi \in \mathcal{L}(\mathsf{At}(\Phi) \cap \mathsf{At}(\Phi'))$.

(3) If $\Phi \models \perp$ then $\Phi \models \phi$ for all $\phi \in \mathcal{L}(\mathsf{At}(\Phi) \cap \mathsf{At}(\Phi'))$. If $\Phi' \not\models \perp$ then $\Phi' \cup \{\phi\} \models \perp$ for all ϕ such that (i) $\phi = \neg \phi'$, where $\phi' \in \mathsf{Cn}(\Phi')$ or (ii) $\phi \models \perp$.

(4) If $\Phi \not\models \perp$ and $\Phi' \models \perp$ then a formula ϕ is an interpolant iff $\Phi \models \phi$ and $\phi \in \mathcal{L}(\mathsf{At}(\Phi) \cap \mathsf{At}(\Phi'))$.

As the notion of an interpolant is a syntactical one, the set $\mathbb{I}(\Phi, \Phi')$ also contains infinite syntactical variants for each interpolant (except for the case where $\Phi \cup \Phi'$ is consistent); in particular, if $\phi \in \mathbb{I}(\Phi, \Phi')$ then for every formula $\phi' \in \mathcal{L}(\mathsf{At}(\Phi) \cap \mathsf{At}(\Phi'))$, if $\phi \equiv \phi'$ then $\phi' \in \mathbb{I}(\Phi, \Phi')$. However, we will consider the following finite representation of $\mathbb{I}(\Phi, \Phi')$. For that, let $[\cdot] : \mathcal{L}(\mathsf{At}) \to \Omega(\mathsf{At})$ be a function that maps each formula to its *equivalence class* w.r.t. $\equiv$.

Definition 4. *Let Φ and Φ' be finite set of formulas. We denote by $\mathbb{SI}(\Phi, \Phi')$ the set of equivalence classes of interpolants wrt. semantical equivalence, i. c.,*

$$\mathbb{SI}(\Phi, \Phi') = \{[\phi] \mid \phi \in \mathbb{I}(\Phi, \Phi')\}$$

In the following, by abuse of notation we refer to $[\phi]$ also as an interpolant if ϕ is an interpolant.

Remark 1. If the sets Φ, Φ' of propositional formulæ are finite then the set $\mathsf{At}(\Phi) \cap \mathsf{At}(\Phi')$ is finite, so the set of all equivalence classes of formulæ over this set of atoms is finite and has at most $2^{2^{|\mathsf{At}(\Phi) \cap \mathsf{At}(\Phi')|}}$ elements, thus $\mathbb{SI}(\Phi, \Phi')$ is finite.

It can be easily seen that the elements of $\mathbb{SI}(\Phi, \Phi')$ form a lattice wrt. $\models$ and the operations $\wedge$ and $\vee$. Formally, we can make the following observations.

Proposition 3. *Let Φ and Φ' be finite set of formulas.*

1. *If $[\phi], [\phi'] \in \mathbb{SI}(\Phi, \Phi')$ then $[\phi \wedge \phi'] \in \mathbb{SI}(\Phi, \Phi')$.*
2. *If $[\phi], [\phi'] \in \mathbb{SI}(\Phi, \Phi')$ then $[\phi \vee \phi'] \in \mathbb{SI}(\Phi, \Phi')$.*
3. *There is a uniquely defined $[\phi_w] \in \mathbb{SI}(\Phi, \Phi')$ with $\phi' \models \phi_w$ for all $[\phi'] \in \mathbb{SI}(\Phi, \Phi')$.*
4. *There is a uniquely defined $[\phi_s] \in \mathbb{SI}(\Phi, \Phi')$ with $\phi_s \models \phi'$ for all $[\phi'] \in \mathbb{SI}(\Phi, \Phi')$.*

Conditions 1 and 2 of the proposition above are trivial. We illustrate the intuition of both conditions 3 and 4 with an example. Consider the knowledge bases $\mathcal{K}_1 = \{a, b\}$ and $\mathcal{K}_2 = \{a, \neg b\}$. We have that

$$\mathbb{SI}(\mathcal{K}_1, \mathcal{K}_2) = \{[a \wedge b], [b], [a \rightarrow b], [a \leftrightarrow b]\}.$$

Note that every formula in $\mathbb{SI}(\mathcal{K}_1, \mathcal{K}_2)$ implies $a \rightarrow b$ (the ϕ_w of condition 3). Similarly, the formula $a \wedge b$ implies all formulae in $\mathbb{SI}(\mathcal{K}_1, \mathcal{K}_2)$, which makes $a \wedge b$ the formula $[\phi_s]$ of condition 4.

Remark 2. If the sets Φ, Φ' of propositional formulæ are finite then as $\mathbb{SI}(\Phi, \Phi')$ is finite, $(\mathbb{SI}(\Phi, \Phi'), \wedge, \vee)$ is a complete lattice.

We call $[\phi_w]$ the *weakest* interpolant and $[\phi_s]$ the *strongest* interpolant. It can be easily seen that

$$[\phi_w] = \left[\bigvee_{[\phi] \in \mathbb{SI}(\Phi, \Phi')} \phi \right] \qquad [\phi_s] = \left[\bigwedge_{[\phi] \in \mathbb{SI}(\Phi, \Phi')} \phi \right]$$

We abbreviate the weakest interpolant of $\mathbb{SI}(\Phi, \Phi')$ by $\mathsf{Weakest}(\Phi, \Phi')$ and the strongest interpolant of $\mathbb{SI}(\Phi, \Phi')$ by $\mathsf{Strongest}(\Phi, \Phi')$. If $\mathbb{SI}(\Phi, \Phi') = \emptyset$ both notions are undefined. We conclude this section with some further observations that will be useful in the remainder of the paper.

Proposition 4. *Let Φ, Φ', and Φ'' be finite set of formulas.*

1. $\mathbb{I}(\Phi, \Phi') \subseteq \mathbb{I}(\Phi, \Phi' \cup \Phi'')$
2. $\mathbb{I}(\Phi, \Phi') \subseteq \mathbb{I}(\Phi \cup \Phi'', \Phi')$
3. $\mathbb{SI}(\Phi, \Phi') \subseteq \mathbb{SI}(\Phi, \Phi' \cup \Phi'')$

4. $\mathbb{SI}(\varPhi, \varPhi') \subseteq \mathbb{SI}(\varPhi \cup \varPhi'', \varPhi')$
5. Weakest$(\varPhi, \varPhi') \models$ Weakest$(\varPhi, \varPhi' \cup \varPhi'')$
6. Weakest$(\varPhi, \varPhi') \models$ Weakest$(\varPhi \cup \varPhi'', \varPhi')$
7. Strongest$(\varPhi, \varPhi' \cup \varPhi'') \models$ Strongest$(\varPhi, \varPhi')$
8. Strongest$(\varPhi \cup \varPhi'', \varPhi') \models$ Strongest$(\varPhi, \varPhi')$

Proof. Considering item 1, for $\phi \in \mathbb{I}(\varPhi, \varPhi')$, from $\varPhi' \cup \{\phi\} \models \perp$ it directly follows $\varPhi' \cup \varPhi'' \cup \{\phi\} \models \perp$. The other two conditions of Definition 2 remain valid as well. Item 2 is proven analogously and the remaining items follow directly from 2. $\square$

4   Disagreement Measures Based on Interpolation

We consider the scenario of measuring *disagreement* between multiple knowledge bases [17] using interpolation. For that, we denote a *knowledge base profile* by $P = \langle \mathcal{K}_1, \ldots, \mathcal{K}_n \rangle$ with $\mathcal{K}_1, \ldots, \mathcal{K}_n$ being knowledge bases. Furthermore, for P of this form and another knowledge base $\mathcal{K}$, we denote by $P \circ \mathcal{K}$ the concatenation of P with $\mathcal{K}$, i.e., $P \circ \mathcal{K} = \langle \mathcal{K}_1, \ldots, \mathcal{K}_n, \mathcal{K} \rangle$. Furthermore, for $k \in \mathbb{N}$ define $P \circ^k \mathcal{K} = (P \circ^{k-1} \mathcal{K}) \circ \mathcal{K}$ and $P \circ^1 \mathcal{K} = P \circ \mathcal{K}$. Let $\mathfrak{K}$ denote the set of all knowledge base profiles. Then Potyka [17] defines a disagreement measure as follows. Let $\mathbb{R}^\infty_{\geq 0}$ be the set of non-negative real values including ∞.

Definition 5. *A disagreement measure $\mathcal{D}$ is a function $\mathcal{D} : \mathfrak{K} \to \mathbb{R}^\infty_{\geq 0}$ that satisfies*

Consistency $\mathcal{D}(P) = 0$ *iff* $\bigcup P$ *is consistent.*
Symmetry $\mathcal{D}(\langle \mathcal{K}_1, \ldots, \mathcal{K}_n \rangle) = \mathcal{D}(\langle \mathcal{K}_{\sigma(1)}, \ldots, \mathcal{K}_{\sigma(n)} \rangle)$ *for each permutation σ of* $\{1, \ldots, n\}$.

We also write $\mathcal{D}(\mathcal{K}_1, \ldots, \mathcal{K}_n)$ for $\mathcal{D}(\langle \mathcal{K}_1, \ldots, \mathcal{K}_n \rangle)$ to ease notation.

In the following, we define a series of disagreement measures that work on knowledge base profiles with exactly *two* elements. In order to generalise these measures to arbitrary knowledge base profiles, we consider the following constructions.

Definition 6. *Let $\mathcal{D}$ be a function $\mathcal{D} : \mathbb{K} \times \mathbb{K} \to \mathbb{R}^\infty_{\geq 0}$. Then the induced sum-measure $\mathcal{D}^\Sigma : \mathfrak{K} \to \mathbb{R}^\infty_{\geq 0}$ and max-measure $\mathcal{D}^{\max} : \mathfrak{K} \to \mathbb{R}^\infty_{\geq 0}$ are defined via*

$$\mathcal{D}^\Sigma(P) = \sum_{\mathcal{K}, \mathcal{K}' \in P} \mathcal{D}(\mathcal{K}, \mathcal{K}')$$

$$\mathcal{D}^{\max}(P) = \max\{\mathcal{D}(\mathcal{K}, \mathcal{K}') \mid \mathcal{K}, \mathcal{K}' \in P\}$$

In order to obtain valid disagreement measures using these two constructions, we only need to require the *Consistency* property from the used two-place functions:

Consistency[2] $\mathcal{D}(\mathcal{K}_1, \mathcal{K}_2) = 0$ iff $\mathcal{K}_1 \cup \mathcal{K}_2$ is consistent.

Proposition 5. *Let* $\mathcal{D}$ *be a function* $\mathcal{D} : \mathbb{K} \times \mathbb{K} \rightarrow \mathbb{R}^{\infty}_{\geq 0}$ *that satisfies Consistency*^{2} *then both* $\mathcal{D}^{\Sigma}$ *and* $\mathcal{D}^{\max}$ *are disagreement measures.*

Proof. We first consider $\mathcal{D}^{\Sigma}$. If $\bigcup P$ is consistent then $\mathcal{K} \cup \mathcal{K}'$ is consistent as well for all $\mathcal{K}, \mathcal{K}' \in P$, implying $\mathcal{D}(\mathcal{K}, \mathcal{K}') = 0$. It follows $\mathcal{D}^{\Sigma}(P) = \mathcal{D}^{\max}(P) = 0$ showing Consistency. The property Symmetry is satisfied by construction.

Let us now define some concrete measure using interpolants.

Definition 7. *Let* $\mathcal{K}_1, \mathcal{K}_2$ *be finite and consistent set of formulas. Define the measure* $\mathcal{D}_{\mathrm{SI}} : \mathbb{K} \times \mathbb{K} \rightarrow \mathbb{R}^{\infty}_{\geq 0}$ *via*

$$\mathcal{D}_{\mathrm{SI}}(\mathcal{K}_1, \mathcal{K}_2) = |\mathbb{SI}(\mathcal{K}_1, \mathcal{K}_2)|$$

Observe that $\mathcal{D}_{\mathrm{SI}}$ satisfies *Consistency*^{2}, as $\mathbb{SI}(\mathcal{K}, \mathcal{K}') = \emptyset$ if $\mathcal{K} \cup \mathcal{K}'$ is consistent. So $\mathcal{D}^{\Sigma}_{\mathrm{SI}}$ and $\mathcal{D}^{\max}_{\mathrm{SI}}$ are disagreement measures according to Proposition 5.

Example 1. Consider the three following knowledge bases: $\mathcal{K}_1 = \{a, b\}, \mathcal{K}_2 = \{b, c\}$ and $\mathcal{K}_3 = \{a, \neg b\}$. Going for their interpolants we get:

- $\mathbb{SI}(\mathcal{K}_1, \mathcal{K}_2) = \mathbb{SI}(\mathcal{K}_2, \mathcal{K}_1) = \emptyset$, as $\mathcal{K}_1$ is consistent with $\mathcal{K}_2$;
- $\mathbb{SI}(\mathcal{K}_1, \mathcal{K}_3) = \{[b], [a \wedge b], [a \rightarrow b]\}$ and $\mathbb{SI}(\mathcal{K}_3, \mathcal{K}_1) = \{[\neg b], [a \rightarrow \neg b], [a \wedge \neg b]\}$;
- $\mathbb{SI}(\mathcal{K}_2, \mathcal{K}_3) = \{[b]\}$ and $\mathbb{SI}(\mathcal{K}_3, \mathcal{K}_2) = \{[\neg b]\}$, as $\mathsf{At}(\{\mathcal{K}_2, \mathcal{K}_3\}) = \{b\}$.

Therefore, $\mathcal{D}^{\Sigma}_{\mathrm{SI}}(\mathcal{K}_1, \mathcal{K}_2, \mathcal{K}_3) = 8$ and $\mathcal{D}^{\max}_{\mathrm{SI}}(\mathcal{K}_1, \mathcal{K}_2, \mathcal{K}_3) = 3$.

We consider another measure based on the information content of strongest (resp. weakest) interpolant. We use the following definition of an information measure, similar in spirit to the definition given in [11].

Definition 8. *An information measure* J *is a function* $J : \mathcal{L}(\mathsf{At}) \rightarrow \mathbb{R}^{\infty}_{\geq 0}$ *that satisfies the following four properties:*

1. $J(\top) = 0$.
2. $J(\bot) = \infty$.
3. *If* $\phi \models \phi'$ *then* $J(\phi) \geq J(\phi')$.
4. *If* $\phi \not\models \bot$ *then* $J(\phi) < \infty$.

Here is a simple example of an information measures:

$$J_M(\phi) = \begin{cases} 0 & \text{if } \top \models \phi \\ \infty & \text{if } \phi \models \bot \\ 1/|\mathsf{Mod}(\phi)| & \text{otherwise} \end{cases} \tag{1}$$

It is easy to verify that J_M is indeed an information measure according to Definition 8.

Then consider the following measure.

Definition 9. *Let $\mathcal{K}_1, \mathcal{K}_2$ be finite and consistent set of formulas and J some information measure. Define the following measure* $\mathcal{D}_J : \mathbb{K} \times \mathbb{K} \to \mathbb{R}^\infty_{\geq 0}$:

$$\mathcal{D}_J(\mathcal{K}_1, \mathcal{K}_2) = \begin{cases} 0 & \text{if } \mathcal{K}_1 \cup \mathcal{K}_2 \text{ is consistent} \\ \infty & \text{if } J(\text{Weakest}(\mathcal{K}_1, \mathcal{K}_2)) = 0 \\ \frac{J(\text{Strongest}(\mathcal{K}_1, \mathcal{K}_2))}{J(\text{Weakest}(\mathcal{K}_1, \mathcal{K}_2))} & \text{otherwise} \end{cases}$$

Again, observe that $\mathcal{D}_J$ satisfies *Consistency*^{2} (independently of J). So $\mathcal{D}_J^\Sigma$ and $\mathcal{D}_J^{\max}$ are disagreement measures according to Proposition 5.

Example 2. Consider the three knowledge bases from Example 1: $\mathcal{K}_1 = \{a, b\}, \mathcal{K}_2 = \{b, c\}$ and $\mathcal{K}_3 = \{a, \neg b\}$. In the table below, we show their strongest and weakest interpolants as well as their disagreement measure $\mathcal{D}_{J_M}$, where J_M is the information measure defined in (1) above.

	$(\mathcal{K}_1, \mathcal{K}_3)$	$(\mathcal{K}_3, \mathcal{K}_1)$	$(\mathcal{K}_2, \mathcal{K}_3),$	$(\mathcal{K}_3, \mathcal{K}_2)$	$(\mathcal{K}_1, \mathcal{K}_2)$	$(\mathcal{K}_1, \mathcal{K}_2)$
Weakest	$[\neg a \vee b]$	$[\neg a \vee \neg b]$	$[b]$	$[\neg b]$	-	-
Strongest	$[a \wedge b]$	$[a \wedge \neg b]$	$[b]$	$[\neg b]$	-	-
$\mathcal{D}_{J_M}$	3	3	1	1	0	0

As $\mathcal{K}_1$ is consistent with $\mathcal{K}_2$ they do not have any interpolant, and therefore their disagreement measure $\mathcal{D}_{J_M}$ is zero. Note that each of the weakest interpolants between $\mathcal{K}_1$ and $\mathcal{K}_3$ have each 6 models, while their strongest interpolants have each 2 models.

5 Analysis

Potyka [17] proposes some desirable properties for disagreement measures, inspired by similar properties from inconsistency measurement [18]. Let $\mathcal{D}$ be some disagreement measure. For $P = \langle \mathcal{K}_1, \ldots, \mathcal{K}_n \rangle$, we say that $\mathcal{K}_i, i \in \{1, \ldots, n\}$, is *involved in a conflict*, if there is $C \subseteq \{1, \ldots, n\}$ such that $\bigcup_{j \in C} \mathcal{K}_j$ is consistent but $\mathcal{K}_i \cup \bigcup_{j \in C} \mathcal{K}_j$ is inconsistent.

Monotony (MO) $\mathcal{D}(\mathcal{K}_1, \ldots, \mathcal{K}_n) \leq \mathcal{D}(\mathcal{K}_1 \cup \mathcal{K}', \ldots, \mathcal{K}_n)$.

Dominance (DO) For formulas ϕ, ψ with $\phi \models \psi$ and $\phi \not\models \bot$, $\mathcal{D}(\mathcal{K}_1 \cup \{\phi\}, \ldots, \mathcal{K}_n) \geq \mathcal{D}(\mathcal{K}_1 \cup \{\psi\}, \ldots, \mathcal{K}_n)$.

Safe Formula Independence (SFI) For a formula ϕ with $\phi \not\models \bot$ and $\text{At}(\phi) \cap \text{At}(\bigcup \mathcal{K}_i) = \emptyset$, $\mathcal{D}(\mathcal{K}_1 \cup \{\phi\}, \ldots, \mathcal{K}_n) = \mathcal{D}(\mathcal{K}_1, \ldots, \mathcal{K}_n)$.

Adjunction Invariance (AI) For formulas ϕ, ψ, $\mathcal{D}(\mathcal{K}_1 \cup \{\phi, \psi\}, \ldots, \mathcal{K}_n) = \mathcal{D}(\mathcal{K}_1 \cup \{\phi \wedge \psi\}, \ldots, \mathcal{K}_n)$.

Tautology (TA) If $\mathcal{K}$ is a knowledge base with $\top \models \mathcal{K}$ then $\mathcal{D}(P) \geq \mathcal{D}(P \circ \mathcal{K})$.

Contradiction (CO) If $\mathcal{K}$ is inconsistent then $\mathcal{D}(P) \leq \mathcal{D}(P \circ \mathcal{K})$.

Majority (MAJ) If $\mathcal{K} \in P$ is consistent and involved in a conflict, then there is $k \in \mathbb{N}$ with $\mathcal{D}(P \circ^k \mathcal{K}) < \mathcal{D}(P)$.

Majority Agreement in the Limit (MAJL) If M is a maximal consistent subset of $\bigcup P$ then $\lim_{k \to \infty} \mathcal{D}(P \circ^k M) = 0$.

Table 1. Compliance of investigated disagreement measures with postulates.

	MO	DO	SFI	AI	TA	CO	MAJ	MAJL
$\mathcal{D}_{\mathrm{SI}}^{\Sigma}$	✓	✗	✓	✓	✗	✓	✗	✗
$\mathcal{D}_{\mathrm{SI}}^{\max}$	✓	✗	✓	✓	✓	✓	✗	✗
$\mathcal{D}_{J}^{\Sigma}$	✓	✗	✓	✓	✓	✓	✗	✗
$\mathcal{D}_{J}^{\max}$	✓	✗	✓	✓	✓	✓	✗	✗

We check compliance of the disagreement measures $\mathcal{D}_{\mathrm{SI}}^{\Sigma}, \mathcal{D}_{\mathrm{SI}}^{\max}, \mathcal{D}_{J}^{\Sigma}$ and $\mathcal{D}_{J}^{\max}$ against the disagreement measures postulates mentioned above. For this, we introduce the concept of agreement between interpretations. We say that an interpretation w agrees with an interpretation w' modulo a set of propositional symbols X iff $w(p) = w'(p)$, for all $p \in X$. Note that the agreement relation is symmetric, that is, if w agrees with w' modulo X then w' agrees with w modulo X. We recall the Coincidence Lemma on propositional logic:

Lemma 1 (Coincidence Lemma). *If two interpretations w and w' agree with At(α), then $w \models \alpha$ iff $w' \models \alpha$.*

Proposition 6. *Let $\mathcal{K}$ be knowledge base, and φ a consistent formula. If At$(\mathcal{K}) \cap$ At$(\varphi) = \emptyset$, At$(\alpha) \subseteq$ At$(\mathcal{K})$ and $\mathcal{K} \cup \{\varphi\} \models \alpha$ then $\mathcal{K} \models \alpha$.*

Proof. The case that $\mathcal{K}$ is inconsistent is trivial. So we focus on the case that $\mathcal{K}$ is consistent. Let then $w \in \mathsf{Mod}(\mathcal{K})$ be an interpretation of $\mathcal{K}$, we will show that $w \in \mathsf{Mod}(\alpha)$. As φ is consistent, it has at least one interpretation. Let $w' \in \mathsf{Mod}(\varphi)$, and let w_0 be the following interpretation: $w_0(p) = w(p)$, if $p \in \mathsf{At}(\mathcal{K})$; and $w_0(p) = w'(p)$, otherwise. Note that w_0 agrees with w modulo $\mathsf{At}(\mathcal{K})$, which implies that $w_0 \models \mathcal{K}$. As $\mathsf{At}(\mathcal{K}) \cap \mathsf{At}(\varphi) = \emptyset$, we get that w_0 agrees with w' modulo $\mathsf{At}(\varphi)$. This implies that $w_0 \models \varphi$. Therefore, as $w_0 \models \mathcal{K}$ and $w_0 \models \varphi$, we have that $w_0 \models \mathcal{K} \cup \{\varphi\}$. Thus, as $\mathcal{K} \cup \{\varphi\} \models \alpha$, we get that $w_0 \models \alpha$. As, by hypothesis, $\mathsf{At}(\alpha) \subseteq \mathsf{At}(\mathcal{K})$, we get that w agrees with w_0 modulo $\mathsf{At}(\alpha)$. This implies, from Lemma 1, that $w \models \alpha$.

Theorem 2. *The compliance of the measures $\mathcal{D}_{\mathrm{SI}}^{\Sigma}, \mathcal{D}_{\mathrm{SI}}^{\max}, \mathcal{D}_{J}^{\Sigma}$, and $\mathcal{D}_{J}^{\max}$ is as shown in Table 1.*

Note again that the proof of the above theorem can be found online[2].

Potyka [17] has defined disagreement measures from inconsistency measures. The idea is that the degree of disagreement in a knowledge profile would correspond to measuring the degree of inconsistencies between the knowledge bases in a knowledge profile. Potyka then discusses about some principles that disagreement measures should satisfy, and show that many of the disagreement measures induced from inconsistency measures fail to satisfy some of these principles. He

---

[2] http://mthimm.de/misc/rst_dismes_proofs.pdf.

then proposes the disagreement measure D_η, based on the η-inconsistency measure [13], to capture some of these principles. Following this line, we analyse here some desirable properties that disagreement measures on interpolants satisfy. We show that Potyka's D_η measure has some issues that are better handled by disagreement measures based on interpolants.

We start by addressing the Adjunction Invariance axiom. Breach of this axiom can lead to unintuitive behaviours. For instance, both $\{\alpha \wedge \neg\alpha\}$ and $\{\alpha, \neg\alpha\}$ should have the same disagreement value, since they present the same conflicts. Most of the disagreement measures induced from inconsistency measures analysed by Potyka, including his D_η measure, is not adjunctive invariant. On the other hand, as shown above, disagreement measures based on interpolants are adjunctive invariant. This is because interpolants are not syntax sensitive, instead they consider formulæ that share the same signature.

Potyka's disagreement measure D_η plateaus when each knowledge base in a knowledge profile conflicts with each other. As he himself criticizes, this scenario makes the measurement purely dependable on the size of the knowledge profile. Precisely, in that case $D_\eta(P) = (|P| - 1)/|P|$. In this scenario, a disagreement measure should still be able to distinguish an increase of conflicts, even if two knowledge profiles have the same size. We illustrate it in the following example.

Example 3. Let $P = \langle \mathcal{K}_1, \mathcal{K}_2, \mathcal{K}_3 \rangle$, and $P' = \langle \mathcal{K}_1, \mathcal{K}_2, \mathcal{K}_4 \rangle$ be two knowledge profiles, where $\mathcal{K}_1 = \{a, b\}, \mathcal{K}_2 = \{\neg a, b\}, \mathcal{K}_3 = \{\neg b\}$ and $\mathcal{K}_4 = \{\neg b, \neg a\}$. Note that the four knowledge bases are consistent, but are inconsistent with each other. For the D_η disagreement measure, we would have $D_\eta(P) = D_\eta(P') = 2/3$.

Note that, against $\mathcal{K}_1$ and $\mathcal{K}_2$, the knowledge base $\mathcal{K}_4$ presents more conflicts than $\mathcal{K}_3$. Thus, though both P and P' have the same size, P' is more conflicting than P. A disagreement measure should be able to distinguish this difference of conflicts. The D_η measure, however, does not distinguish these conflicts, since when knowledge bases are pairwise inconsistent, the measure considers only the size of the knowledge profile, which is rather simplistic.

On the other hand, disagreement measures on interpolants are able to distinguish this tenuous difference of conflicts. To illustrate this, consider the $\mathcal{D}_{\text{SI}}^\Sigma$ measure. For the knowledge bases above we would get $\mathcal{D}_{\text{SI}}(\mathcal{K}_1, \mathcal{K}_2) = 4, \mathcal{D}_{\text{SI}}(\mathcal{K}_1, \mathcal{K}_3) = 1, \mathcal{D}_{\text{SI}}(\mathcal{K}_1, \mathcal{K}_4) = 4, \mathcal{D}_{\text{SI}}(\mathcal{K}_2, \mathcal{K}_3) = 1$ and $\mathcal{D}_{\text{SI}}(\mathcal{K}_2, \mathcal{K}_4) = 4$. Thus, $\mathcal{D}_{\text{SI}}(P)^\Sigma = (4 + 1 + 1) * 2 = 12$ and $\mathcal{D}_{\text{SI}}^\Sigma(P') = (4 + 4 + 4) * 2 = 24$.

This shows that disagreement measures on interpolants, such as $\mathcal{D}_{\text{SI}}^\Sigma$, present ways of distinguishing sensible conflicts between knowledge bases. Towards this end, a deeper investigation of which rational behaviours interpolants yield for disagreement measures is a path worth to explore.

6   Computational Complexity

To conclude our analysis, we now investigate the computational complexity of problems related to our novel measures.

Let us first consider some general observations on interpolants, which seem—to the best of our knowledge—not have been explicitly mentioned in the literature thus far.

Theorem 3. *Let Φ and Φ' be sets of formulas and let ϕ be a formula. Deciding whether ϕ is an interpolant of Φ wrt. Φ' is* CONP-*complete.*

Proof. For CONP-*membership, consider the complement problem of deciding whether ϕ is not an interpolant of Φ wrt. Φ' and define the following non-deterministic algorithm. On instance (ϕ, Φ, Φ') we guess a triple (ω, ω', a) with interpretations ω, ω', and an atom a. Then ϕ is not an interpolant of Φ wrt. Φ' if either*

1. $\omega \models \Phi$ and $\omega \not\models \phi$,
2. $\omega' \models \Phi' \cup \{\phi\}$, or
3. $a \in \mathsf{At}(\phi) \setminus (\mathsf{At}(\Phi) \cap \mathsf{At}(\Phi'))$.

Observe that each of these checks correspond to disproving one item of Definition 2 and that they can be done in polynomial time. If follows CONP-*membership of the original problem.*

For CONP-*hardness, we reduce the problem* TAUT *to our problem. An instance to* TAUT *consists of a formula ψ and it has to be decided whether ψ is tautological. On input ψ we construct the instance (ϕ, Φ, Φ') for our problem (let a be a fresh atom not appearing in ψ) with*

$$\phi = a \qquad\qquad \Phi = \{\psi \rightarrow a\} \qquad\qquad \Phi' = \{\neg a\}$$

It remains to show that ψ is tautological if and only if ϕ is an interpolant of Φ wrt. Φ'. Assume that ψ is tautological, then it follows $\Phi \models a$. Obviously, $\Phi' \cup \{a\}$ is inconsistent and ϕ only contains atoms of the shared vocabulary. It follows that ϕ is an interpolant of Φ wrt. Φ'. The other direction is analogous. □

In [16] it is shown that, essentially, the size[3] of an interpolant $\phi \in \mathbb{SI}(\Phi, \Phi')$ is probably not bound polynomially in the size of both Φ and Φ' (very surprising results would follow in this case). This makes a characterisation of the complexity of various other computational problems hard. For example, a standard approach to decide whether some given formula ϕ would *not* be the strongest (weakest) interpolant, would be to guess another formula ϕ' (e.g., the actual strongest/weakest interpolant) and verify that ϕ' is an interpolant and $\phi' \models \phi$ ($\phi \models \phi'$). However, as ϕ' might be of exponential size this is not feasible to show membership in some class of the polynomial hierarchy or even PSPACE. We can therefore only provide a straightforward upper bound for the complexity of these and other problems of relevance to us.

Theorem 4. *Let Φ and Φ' be sets of formulas.*

1. For a formula ϕ, the problem of deciding whether $\phi = \mathsf{Strongest}(\Phi, \Phi')$ is in EXPSPACE.

---

[3] More specifically, the smallest size of a $\phi' \in [\phi]$.

2. For a formula ϕ, the problem of deciding whether $\phi = \mathsf{Weakest}(\Phi, \Phi')$ is in EXPSPACE.
3. The problem of determining $\mathsf{Strongest}(\Phi, \Phi')$ is in FEXPSPACE.
4. The problem of determining $\mathsf{Weakest}(\Phi, \Phi')$ is in FEXPSPACE.
5. The problem of counting $|\mathbb{SI}(\Phi, \Phi')|$ is in FEXPSPACE.

Proof. We first show 5. For that, we enumerate (by reusing space) every subset $M \subseteq \Omega(\mathsf{At})$ of interpretations (note that every interpretation is of polynomial size and there are exponentially many interpretations; therefore M is of exponential size). Each such set M characterises a formula ϕ_M and its equivalence class $[\phi_M]$ via $\mathsf{Mod}(\phi_M) = M$. We can then easily verify whether ϕ_M is an interpolant of Φ wrt. Φ, see Theorem 3 and the fact that CONP $\subseteq$ EXPSPACE. In the positive case we add 1 to some counter. After enumerating all possible M, the counter is exactly the number $|\mathbb{SI}(\Phi, \Phi')|$.

In order to prove 1–4, the above algorithm can easily be adapted. For 1., whenever we have verified a formula ϕ_M to be an interpolant, we can check whether $\phi_M \models \phi$ and $\phi \not\models \phi_M$. In that case, ϕ cannot be the strongest interpolant. Case 2 is analogous. Cases 3–4 can be realised by keeping track of the strongest (weakest) interpolant found so far and always comparing it newly discovered interpolants.   □

7 Summary and Conclusion

We investigated the problem of measuring disagreement in belief merging scenarios. For that, we made use of the concept of Craig interpolants and defined disagreement measures that consider the number of semantically equivalent interpolants between two knowledge bases and the information content in the strongest and weakest interpolants. We showed that our measures satisfy a number of desirable properties and we briefly discussed the computational complexity of related problems.

For future work, we will investigate the possibility of defining further measures based on interpolation and investigate their properties. Moreover, a deeper analysis of the differences of our measures with the measures proposed by Potyka [17] is needed. Precisely, our approach based on interpolants has a semantic perspective in assessing the culpability degree of the inconsistencies between two knowledge bases. We shall investigate what properties this semantic perspective brings upon these inconsistency measures. We will also explore algorithmic approaches to compute our measures and investigate applying our ideas to the area of inconsistency measurement [19].

Acknowledgements. The research reported here was partially supported by the Deutsche Forschungsgemeinschaft (grant DE1983/9-1).

References

1. Arrow, K.J.: Social Choice and Individual Values. Wiley, New York (1963)
2. Baral, C., Minker, J., Kraus, S.: Combining multiple knowledge bases. IEEE Trans. Knowl. Data Eng. **3**(2), 208–221 (1991)
3. Cholvy, L., Hunter, A.: Information fusion in logic: a brief overview. In: Gabbay, D.M., Kruse, R., Nonnengart, A., Ohlbach, H.J. (eds.) ECSQARU/FAPR -1997. LNCS, vol. 1244, pp. 86–95. Springer, Heidelberg (1997). https://doi.org/10.1007/BFb0035614
4. Corea, C., Delfmann, P.: Supporting business rule management with inconsistency analysis. In: Proceedings of the Dissertation Award, Demonstration, and Industrial Track at BPM 2018 co-located with 16th International Conference on Business Process Management (BPM 2018), Sydney, Australia, September 9–14, 2018, pp. 141–147 (2018)
5. Corea, C., Thimm, M.: Towards inconsistency measurement in business rule bases. In: Proceedings of the 24th European Conference on Artificial Intelligence (ECAI 2020), June 2020
6. Craig, W.: Linear reasoning. A new form of the Herbrand-Gentzen theorem. J. Symb. Log. **22**(3), 250–268 (1957)
7. Díaz, A.M., Pérez, R.P.: Impossibility in belief merging. Artif. Intell. **251**, 1–34 (2017)
8. Everaere, P., Konieczny, S., Marquis, P.: Belief merging versus judgment aggregation. In: Proceedings of the 14th International Conference on Autonomous Agents and Multiagent Systems (AAMAS 2015) (2015)
9. Grant, J.: Classifications for inconsistent theories. Notre Dame J. Formal Logic **19**(3), 435–444 (1978)
10. Grant, J., Hunter, A.: Measuring consistency gain and information loss in stepwise inconsistency resolution. In: Liu, W. (ed.) ECSQARU 2011. LNCS (LNAI), vol. 6717, pp. 362–373. Springer, Heidelberg (2011). https://doi.org/10.1007/978-3-642-22152-1_31
11. Grant, J., Hunter, A.: Measuring the good and the bad in inconsistent information. In: Proceedings of the 23rd International Joint Conference on Artificial Intelligence (IJCAI 2013) (2013)
12. Hunter, A., Konieczny, S.: Measuring inconsistency through minimal inconsistent sets. In: Brewka, G., Lang, J. (eds.) Proceedings of the Eleventh International Conference on Principles of Knowledge Representation and Reasoning (KR 2008), Sydney, Australia, pp. 358–366. AAAI Press, Menlo Park, September 2008
13. Knight, K.: Measuring inconsistency. J. Philos. Logic **31**(1), 77–98 (2002)
14. Konieczny, S., Pino Pérez, R.: On the logic of merging. In: Proceedings of the Sixth International Conference on Principles of Knowledge Representationa and Reasoning (KR 1998). Morgan Kaufmann (1998)
15. Kedian, M., Liu, W., Jin, Z.: A general framework for measuring inconsistency through minimal inconsistent sets. Knowl. Inf. Syst. **27**, 85–114 (2011)
16. Mundici, D.: Tautologies with a unique craig interpolant, uniform vs. nonuniform complexity. Ann. Pure Appl. Logic **27**(3), 265–273 (1984)
17. Potyka, N.: Measuring disagreement among knowledge bases. In: Proceedings of the 12th International Conference on Scalable Uncertainty Management (SUM 2018) (2018)
18. Thimm, M.: On the evaluation of inconsistency measures. In: Grant, J., Martinez, M.V. (eds.) Measuring Inconsistency in Information. Studies in Logic, vol. 73. College Publications, February 2018

19. Thimm, M.: Inconsistency measurement. In: Proceedings of the 13th International Conference on Scalable Uncertainty Management (SUM 2019), December 2019
20. Williamson, J.: Aggregating judgements by merging evidence. J. Logic Comput. **19**(3), 461–473 (2009)
21. Xiao, G., Ma, Y.: Inconsistency measurement based on variables in minimal unsatisfiable subsets. In: Proceedings of the 20th European Conference on Artificial Intelligence (ECAI 2012) (2012)

Inferring from an Imprecise Plackett–Luce Model: Application to Label Ranking

Loïc Adam, Arthur Van Camp$^{(\boxtimes)}$, Sébastien Destercke, and Benjamin Quost

UMR CNRS 7253 Heudiasyc, Sorbonne Université, Université de Technologie de Compiègne, CS 60319, 60203 Compiègne cedex, France
loic.adam@etu.utc.fr,
{arthur.van-camp,sebastien.destercke,benjamin.quost}@hds.utc.fr

Abstract. Learning ranking models is a difficult task, in which data may be scarce and cautious predictions desirable. To address such issues, we explore the extension of the popular parametric probabilistic Plackett–Luce model, often used to model rankings, to the imprecise setting where estimated parameters are set-valued. In particular, we study how to achieve cautious or conservative inference with it, and illustrate their application on label ranking problems, a specific supervised learning task.

Keywords: Preference learning · Cautious inference · Poor data

1 Introduction

Learning and estimating probabilistic models over rankings of objects has received attention for a long time: earlier works can be traced back at least to the 1920s [21]. Recently, this problem has known a revival, in particular due to the rising interest of machine learning in the issue [12]. Popular approaches range from associating a random utility to each object to be ranked, from which a distribution on rankings is derived [3], to directly defining a parametric distribution over the set of rankings [19].

Multiple reasons motivate making cautious inferences of ranking models. The information at hand may be scarce—this is typically the case in the cold-start problem of a recommender system, or partial—for instance because partial rankings are observed (e.g., pairwise comparisons, top-k items). In addition, since inferring a ranking model is difficult and therefore prone to uncertainty, it may be useful to output partial rankings as predictions, thus abstaining to predict when information is unreliable.

Imprecise probability theory is a mathematical framework where partial estimates are formalized in the form of sets of probability distributions. Therefore, it is well suited to making cautious inferences and address the aforementioned problems; yet, to our knowledge, it has not yet been applied to ranking models.

J. Davis and K. Tabia (Eds.): SUM 2020, LNAI 12322, pp. 98–112, 2020.
https://doi.org/10.1007/978-3-030-58449-8_7

In this paper, we use the imprecise probabilistic framework to infer a imprecise Plackett–Luce model, which is a specific parametric model over rankings, from data. We present the model in Sect. 2. We address its inference in Sect. 3, showing that for this specific parametric model, efficient methods can be developed to make cautious inferences based on sets of parameters. Section 4 will then present a direct application to label ranking, where we will use relative likelihoods [5] to proceed with imprecise model estimation.

2   Imprecise Plackett–Luce Models

In this paper, we consider the problem of estimating a probabilistic ranking model over a set of objects or labels $\Lambda = \{\lambda_1, \ldots, \lambda_n\}$. This model defines probabilities over *total orders on the labels*—that is, complete, transitive, and asymmetric relations $\succ$ on Λ. Any complete order $\succ$ over the labels can be identified with its induced permutation or *label ranking* τ, that is the unique permutation of Λ such that

$$\lambda_{\tau(1)} \succ \lambda_{\tau(2)} \succ \cdots \succ \lambda_{\tau(n)}.$$

We will use the terms "order on the labels", "ranking" and "permutation" interchangeably. We denote by $\mathcal{L}(\Lambda)$ all $n!$ permutations on Λ, and denote a generic permutation by τ.

We focus on the particular probability model $P \colon 2^{\mathcal{L}(\Lambda)} \to [0, 1]$ known as the Plackett–Luce (PL) model [6,13]. It is parametrised by n parameters or *strengths* $v_1, \ldots, v_n$ in $\mathbb{R}_{>0} := \{x \in \mathbb{R} : x > 0\}$.[1] The *strength vector* $v = (v_1, \ldots, v_n)$ completely specifies the PL model. For any such vector, an arbitrary ranking τ in $\mathcal{L}$ is assigned probability

$$P_v(\tau) := \prod_{k=1}^{n} \frac{v_{\tau(k)}}{\sum_{\ell=k}^{n} v_{\tau(\ell)}} = \frac{v_{\tau(1)}}{v_{\tau(1)} + \cdots + v_{\tau(n)}} \cdot \frac{v_{\tau(2)}}{v_{\tau(2)} + \cdots + v_{\tau(n)}} \cdots \frac{v_{\tau(n-1)}}{v_{\tau(n-1)} + v_{\tau(n)}}. \quad (1)$$

Clearly, the parameters $v_1, \ldots, v_n$ are defined up to a common positive multiplicative constant, so it is customary to assume that $\sum_{k=1}^{n} v_k = 1$. Therefore, the parameter $v = (v_1, \ldots, v_n)$ can be regarded as an element of the interior of the n-simplex $\Sigma := \{(x_1, \ldots, x_n) \in \mathbb{R}_{\geq 0}^n : \sum_{k=1}^{n} x_k = 1\}$, denoted $\mathrm{int}(\Sigma)$.

This model has the following nice interpretation: the larger a weight v_i is, the more preferred is the label λ_i. The probability that λ_i is ranked first is

$$\sum_{\substack{\tau \in \mathcal{L}(\Lambda) \\ \tau(1) = \lambda_i}} P_v(\tau) = v_i;$$

conditioning on λ_i being the first label, the probability that λ_j is ranked second (i.e. first among the remaining labels) is equal to $v_j / \sum_{k=1, k \neq i}^{n} v_k$. This reasoning can be repeated for each of the labels in a ranking. As a consequence, given a

---

[1] We also define the set of non-negative real numbers as $\mathbb{R}_{\geq 0} := \{x \in \mathbb{R} : x \geq 0\}$.

PL model defined by v, finding the "best" (most probable) ranking amounts to finding the permutation $\tau_v^\star$ which ranks the strengths in decreasing order:

$$\tau_v^\star \in \arg\max_{\tau \in \mathcal{L}(\Lambda)} P_v(\tau') \Leftrightarrow v_{\tau(1)} \geq v_{\tau(2)} \geq v_{\tau(3)} \cdots \geq v_{\tau(n-1)} \geq v_{\tau(n)}. \tag{2}$$

We obtain an *imprecise* Plackett–Luce (IPL) model by letting the strengths vary over a subset Θ of $\mathrm{int}(\Sigma)$.[2] Based on this subset of admissible strengths, we can compute the *lower* and *upper probabilities* of a ranking τ as

$$\underline{P}_\Theta(\tau) := \inf_{v \in \Theta} P_v(\tau) \quad \text{and} \quad \overline{P}_\Theta(\tau) := \sup_{v \in \Theta} P_v(\tau) \quad \text{for all } \tau \text{ in } \mathcal{L}(\Lambda).$$

The above notion of "best" ranking becomes ambiguous for an IPL model, since two vectors $v \neq u \in \Theta$ might be associated with different "best" rankings $\tau_v^\star \neq \tau_u^\star$.

Therefore, we consider two common ways to extend (2). The first one, *(Walley–Sen) maximality* [22,23], considers that τ_1 dominates τ_2 (noted $\tau_1 >_M \tau_2$) if it is more probable for any $v \in \Theta$:

$$\tau_1 >_M \tau_2 \Leftrightarrow (\forall v \in \Theta), P_v(\tau_1) > P_v(\tau_2). \tag{3}$$

The set $\mathcal{M}_\Theta$ of maximal rankings is composed of all such undominated rankings:

$$\mathcal{M}_\Theta := \{\tau \in \mathcal{L}(\Lambda) : \nexists \tau' \text{ s.t. } \tau' >_M \tau\}. \tag{4}$$

We may have $|\mathcal{M}_\Theta| > 1$ when Θ is imprecise.

The second one is *E-admissibility* [18]. A ranking τ is *E-admissible* if it is the "best", according to Eq. (2), for some $v \in \Theta$. The set $\mathcal{E}_\Theta$ of all E-admissible rankings is then

$$\mathcal{E}_\Theta := \bigcup_{v \in \Theta} \arg\max_{\tau \in \mathcal{L}(\Lambda)} P_v(\tau) = \left\{\tau : (\exists v \in \Theta) \text{ s. t. } (\forall \tau' \in \mathcal{L}(\Lambda)), P_v(\tau) \geq P_v(\tau')\right\}. \tag{5}$$

By comparing Eqs. (4) and (5), we immediately find that $\mathcal{E}_\Theta \subseteq \mathcal{M}_\Theta$.

3   Learning an Imprecise Plackett–Luce Model

We introduce here two methods for inferring an IPL model. The first one (Sect. 3.1), which does not make further assumptions about Θ, provides an outer approximation of the set of all maximal rankings. The second one (Sect. 3.2) computes the set of E-admissible rankings via an exact and efficient algorithm, provided that the set of strengths Θ has the form of *probability intervals*.

---

[2] Taking $\mathrm{int}(\Sigma)$ rather than Σ assures that all probabilities are positive and that Eq. (1) is well-defined.

3.1   General Case

Section 2 shows that the "best" ranking is found using Eq. (2). In the case of an IPL model, making robust and imprecise predictions requires to compare all possible ranks in a pairwise way, the complexity of which is $n!$—and thus generally infeasible in practice. However, checking maximality can be simplified. Notice that the numerator in Eq. (1) does not depend on τ (product terms can be arranged in any order). Hence, when comparing two permutations τ and τ' using Eq. (3), only denominators matter: indeed, $\tau \succ \tau'$ iff for all $v \in \Theta$,

$$\frac{P_v(\tau)}{P_v(\tau')} = \frac{v_{\tau'(1)} + \cdots + v_{\tau'(n)}}{v_{\tau(1)} + \cdots + v_{\tau(n)}} \cdot \frac{v_{\tau'(2)} + \cdots + v_{\tau'(n)}}{v_{\tau(2)} + \cdots + v_{\tau(n)}} \cdots \frac{v_{\tau'(n-1)} + v_{\tau'(n)}}{v_{\tau(n-1)} + v_{\tau(n)}} > 1. \qquad (6)$$

Assume for a moment that strengths are precisely known, and that τ and τ' only differ by a swapping of two elements: $\tau(k) = \tau'(k)$ for all $k \in \{1, \ldots, m\} \setminus \{i, j\}$ where $i \neq j$, and $\tau(j) = \tau'(i)$, $\tau(i) = \tau'(j)$. Assume, without loss of generality, that $i < j$. Then, the product terms in Eq. (6) only differ in the ratios involving rank j but not rank i; using furthermore $\tau(i) = \tau'(j)$, we get

$$\frac{P_v(\tau)}{P_v(\tau')} = \underbrace{\prod_{\substack{k=1 \\ k \notin \{i+1, \ldots, j\}}}^{n} \frac{\sum_{\ell=k}^{n} v_{\tau'(\ell)}}{\sum_{\ell=k}^{n} v_{\tau(\ell)}}}_{=1} \cdot \prod_{k=i+1}^{j} \frac{\sum_{\ell=k}^{n} v_{\tau'(\ell)}}{\sum_{\ell=k}^{n} v_{\tau(\ell)}} = \prod_{k=i+1}^{j} \frac{v_{\tau(i)} + \sum_{\ell=k, \ell \neq j}^{n} v_{\tau'(\ell)}}{v_{\tau(j)} + \sum_{\ell=k, \ell \neq j}^{n} v_{\tau(\ell)}}.$$

In this last ratio, we introduce now for any k in $\{i+1, \ldots, j\}$ the sums of strengths $C_k := \sum_{\ell=k, \ell \neq j}^{n} v_{\tau(\ell)} = \sum_{\ell=k, \ell \neq j}^{n} v_{\tau'(\ell)}$: these terms being positive, it follows that

$$\tau \succ \tau' \quad \Leftrightarrow \quad \frac{P_v(\tau)}{P_v(\tau')} > 1 \quad \Leftrightarrow \quad (\forall v \in \Theta),\, v_{\tau(i)} > v_{\tau(j)}.$$

In the case of imprecisely known strengths, the latter inequality will hold whenever the following (sufficient, but not necessary) condition is met:

$$\underline{v}_{\tau(i)} := \inf_{v \in \Theta} v_{\tau(i)} > \overline{v}_{\tau(j)} := \sup_{v \in \Theta} v_{\tau(j)}.$$

Now comes a crucial insight. Assume a ranking τ which prefers λ_ℓ to λ_k whereas $\underline{v}_k > \overline{v}_\ell$, for some $k \neq \ell$: then, we can find a "better" ranking τ' (i.e., which dominates τ according to Eq. (3)) by swapping labels λ_ℓ and λ_k. In other terms, as soon as $\underline{v}_k \geq \overline{v}_\ell$, all maximally admissible rankings satisfy $\lambda_k \succ \lambda_\ell$.

It follows that given an IPL model with strengths $\Theta \subseteq \mathrm{int}(\Sigma)$, we can deduce a partial ordering on objects from the pairwise comparisons of strength bounds: more particularly, we will infer that $\lambda_k \succ \lambda_\ell$ whenever $\underline{v}_k \geq \overline{v}_\ell$. This partial ordering can be obtained easily; it may contain solutions that are not optimal under the maximality criterion, but it is guaranteed to contain all maximal solutions.

3.2   Interval-Valued Case

We assume here that strengths are interval-valued: $v_k \in [\underline{v}_k, \overline{v}_k] \subseteq]0, 1[$; that is, the set Θ of possible strengths (called credal set hereafter) is defined by:

$$\Theta = \left(\underset{k=1}{\overset{n}{\bigtimes}} [\underline{v}_k, \overline{v}_k] \right) \cap \Sigma. \tag{7}$$

Note that we assume $\underline{v}_k > 0$ for each label λ_k: each object has a strictly positive lower probability of being ranked first. It follows that $\overline{v}_k < 1$, and thus $\Theta \subseteq \text{int}(\Sigma)$. Such interval-valued strengths fall within the category of *probability intervals on singletons* [1, Sect. 4.4], and are coherent (nonempty and convex) iff [10]:

$$(\forall k \in \{1, \dots, n\}), \ \left(\underline{v}_k + \sum_{\substack{i=1 \\ i \neq k}}^{n} \overline{v}_i \geq 1 \text{ and } \overline{v}_k + \sum_{\substack{i=1 \\ i \neq k}}^{n} \underline{v}_i \leq 1 \right). \tag{8}$$

From now on, we will assume this condition to hold, and thus that Θ is coherent.

We are interested in computing the set of E-admissible rankings, i.e. rankings τ such that there exists $v \in \Theta$ for which τ maximises P_v (see Sect. 2). Our approach relies on two propositions, the proofs of which will be omitted due to the lack of place.

Checking E-admissibility. We provide here an efficient way of checking whether a ranking τ is E-admissible. According to Eq. (2), it will be the case iff v is decreasingly ordered wrt to τ, i.e. $v_{\tau(1)} \geq v_{\tau(2)} \geq v_{\tau(3)} \geq \cdots$

Proposition 1. *Consider any interval-valued parametrisation of an IPL model such as defined by Eq. (7), and any ranking τ in $\mathcal{L}(\Lambda)$. Then, τ is E-admissible (i.e., $\tau \in \mathcal{E}_\Theta$) iff there exists an index $k \in \{1, \dots, n\}$ such that*

$$1 - \sum_{\ell=1}^{k-1} \min_{1 \leq j \leq \ell} \overline{v}_{\tau(j)} - \sum_{\ell=k+1}^{n} \max_{\ell \leq j \leq n} \underline{v}_{\tau(j)} \ \in \ [\max_{k \leq j \leq n} \underline{v}_{\tau(j)}, \min_{1 \leq j \leq k} \overline{v}_{\tau(j)}] \tag{9}$$

and

$$\begin{aligned} \underline{v}_{\tau(\ell)} &\leq \min\{\overline{v}_{\tau(1)}, \dots, \overline{v}_{\tau(\ell)}\} &&\text{for all } \ell \text{ in } \{1, \dots, k-1\}, \\ \overline{v}_{\tau(\ell)} &\geq \max\{\underline{v}_{\tau(\ell)}, \dots, \underline{v}_{\tau(n)}\} &&\text{for all } \ell \text{ in } \{k+1, \dots, n\}. \end{aligned} \tag{10}$$

Checking E-admissibility via Proposition 1 has a polynomial complexity in the number n of labels. Indeed, we need to check n different values of k: for each one, Eq. (9) requires to calculate a sum of $n - 1$ terms, and Eq. (10) to check $n - 1$ inequalities, which yields a complexity of $2n(n - 1)$.

Computing the Set of E-admissible Rankings. Although Eq. (9) opens the way to finding the set of E-admissible rankings, there are $n!$ many candidate rankings: checking all of them is intractable.

We propose to address this issue by considering a search tree, in which a node is associated with a specific sequence of labels. Each subsequent node adds a new element to this sequence: a leaf is reached when the sequence corresponds to a complete ranking. By navigating the tree top-down, we may progressively check whether a sequence corresponds to the beginning of an E-admissible ranking. Should it not, all completions of the sequence can be ignored.

This requires a way of checking whether a sequence $\kappa = (k_1, k_2, \ldots, k_m)$, by essence incomplete, may be completed into an E-admissible ranking—i.e., whether we can find $\tau \in \mathcal{E}_\Theta$ such that $\tau(1) = k_1, \tau(2) = k_2, \ldots, \tau(m) = k_m$. Proposition 2 provides a set of necessary and sufficient conditions to this end.

Proposition 2. *Consider any coherent parametrisation of an IPL model such as defined by Eq. (7), and a sequence of distinct labels $\kappa = (k_1, \ldots, k_m)$ of length $m \leq n-1$. Then, there exists an E-admissible ranking beginning with this initial sequence iff the following equations are satisfied for every j in $\{1, \ldots, m\}$:*

$$\sum_{\ell=1}^{j} \min\{\overline{v}_{k_1}, \ldots, \overline{v}_{k_\ell}\} + \sum_{\substack{i=1 \\ i \notin \{k_1, \ldots, k_j\}}}^{n} \min\{\overline{v}_{k_1}, \ldots, \overline{v}_{k_j}, \overline{v}_i\} \geq 1; \qquad (A_j)$$

$$\overline{v}_{k_j} \geq \max\{\underline{v}_i : i \in \{1, \ldots, n\} \setminus \kappa_j\}; \qquad (B_j)$$

$$\sum_{t=0}^{j-1} \max\{\underline{v}_i : i \in \{1, \ldots, n\} \setminus \kappa_t\} + \sum_{\substack{i=1 \\ i \notin \{k_1, \ldots, k_j\}}}^{n} \underline{v}_i \leq 1; \qquad (C_j)$$

here, κ_j ($j = 0, \ldots, m$) is the sub-sequence of the j first labels in κ (by convention, κ_0 is empty), and $\{1, \ldots, n\} \setminus \kappa_j$ is the set of labels not appearing in κ_j.

In the special case of $m = 1$, which is typically the case at depth one in the search tree, Eqs. (A_j), (B_j) and (C_j) reduce to:

$$\sum_{i=1}^{n} \min\{\overline{v}_{k_1}, \overline{v}_i\} \geq 1; \qquad (A_1)$$

$$\overline{v}_{k_1} \geq \max\{\underline{v}_i : i \in \{1, \ldots, n\}\}; \qquad (B_1)$$

$$\max\{\underline{v}_i : i \in \{1, \ldots, n\}\} + \sum_{\substack{i=1 \\ i \neq k_1}}^{n} \underline{v}_i \leq 1. \qquad (C_1)$$

Note that under the coherence requirement (8), Eq. (C_1) is a direct consequence of Eq. (B_1), but it is not the case for Eq. (C_j) when $j \geq 2$.

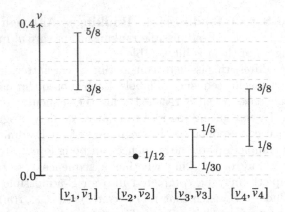

Fig. 1. Probability intervals for Example 1

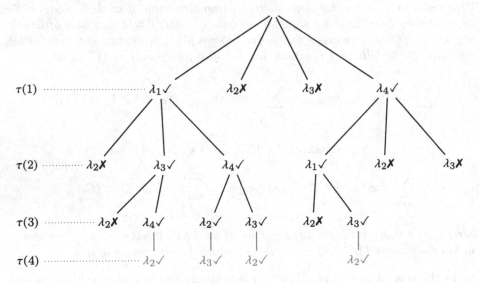

Fig. 2. Search tree for $n = 4$

Example 1. Consider an IPL model that is defined by strength intervals $[\underline{v}_1, \overline{v}_1] = [3/8, 5/8]$, $[\underline{v}_2, \overline{v}_2] = [1/12, 1/12]$, $[\underline{v}_3, \overline{v}_3] = [1/30, 1/5]$ and $[\underline{v}_4, \overline{v}_4] = [1/8, 3/8]$, displayed in Fig. 1 (the coherence of which can be checked using Eq. (8)).

Consider the tree in Fig. 2, which will help navigate the set of possible rankings with $n = 4$ labels. The left-most node at depth $m = 1$ corresponds to the sequence (λ_1); its left-most child (left-most node at depth $m = 2$) to the sequence (λ_1, λ_2). We can see that this sequence has been ruled out as a possible initial segment for an E-admissible ranking: no further completion (i.e., neither of the two rankings $(\lambda_1, \lambda_2, \lambda_3, \lambda_4)$ and $(\lambda_1, \lambda_2, \lambda_4, \lambda_3)$) will be checked.

The sequence $(\lambda_1, \lambda_3, \lambda_2)$ has been ruled out as well; however, the sequence $(\lambda_1, \lambda_3, \lambda_4)$ has been considered as valid, and can be straightforwardly completed into a valid E-admissible ranking (since only one possible label remains). Eventually, all E-admissible rankings $\tau = (\tau(1), \tau(2), \tau(3), \tau(4))$ corresponding to the IPL model are

$$\{(1, 3, 4, 2), (1, 4, 2, 3), (1, 4, 3, 2), (4, 1, 3, 2)\}.$$

A possible strength vector for which $\tau = (1, 3, 4, 2)$ dominates all others is given by $v = (5/8, 1/12, 1/6, 1/8)$: it can easily be checked that $v \in \Theta$ and that $v_{\tau(1)} = 5/8 \geq v_{\tau(2)} = 1/6 \geq v_{\tau(3)} = 1/8 \geq v_{\tau(4)} = 1/12$, i.e. τ is E-admissible according to Eq. (2). We provide below possible strength vectors for each of the E-admissible rankings associated with the IPL model considered:

Admissible strength vector $v \in \Theta$ $v = (v_1, v_2, v_3, v_4)$	Corresponding ranking $\tau \in \mathcal{E}_\Theta$ $\tau = (\tau(1), \tau(2), \tau(3), \tau(4))$
$(5/8, 1/12, 1/6, 1/8)$	$(1, 3, 4, 2)$
$(5/8, 1/12, 1/12, 5/24)$	$(1, 4, 2, 3)$
$(5/8, 1/12, 1/12, 5/24)$	$(1, 4, 3, 2)$
$(3/8, 1/12, 1/6, 3/8)$	$(4, 1, 3, 2)$

Let us show that there is no E-admissible ranking τ that starts for instance with $(1, 2)$. Assume *ex absurdo* that such an E-admissible ranking τ exists. This would imply that there exists $v \in \Theta$ such that $v_1 \geq v_2 \geq \max\{v_3, v_4\}$, which by Eq. (2) would imply that $1/12 = v_2 \geq v_4 \geq \underline{v}_4 = 1/8$, which is impossible. ◊

Algorithm. Eqs. (A_j), (B_j) and (C_j) used in Proposition 2 to check the E-admissibility of a ranking with a given initial sequence of labels can be turned into an efficient algorithm. We can indeed proceed recursively: checking whether there exists an E-admissible ranking starting with $(k_1, \ldots, k_m)$ basically requires to check whether it is the case for $(k_1, \ldots, k_{m-1})$ and then whether Eqs. (A_j), (B_j) and (C_j) still hold for $j = m$.

Algorithms 1 and 2 provide a pseudo-code version of this procedure. Note that as all branch-and-bound techniques, it does not reduce the worst-case complexity of building an E-admissible set. Indeed, if all the rankings are E-admissible—which typically happens when all probability intervals are wide, then no single branch can be pruned from the search tree. In that case, the algorithm navigates the complete tree, which clearly has a factorial complexity in the number of labels n. Then, even a simple enumeration of all E-admissible rankings has such a complexity.

However, in practice we can expect many branches of the tree to be quickly pruned: indeed, as soon as one of the Eqs. (A_j), (B_j) or (C_j) fail to hold, a branch can be pruned from the tree. We expect this to allow for efficient inferences in many circumstances.

Algorithm 1. Find the E-admissible rankings opt_n

Require: Probability intervals $[\underline{v}_k, \overline{v}_k]$ for $k \in \{1, \ldots, n\}$
Ensure: The set $\Theta = \{[\underline{v}_k, \overline{v}_k] : k \in \{1, \ldots, n\}\}$ is coherent
    $\text{opt}_n \leftarrow \emptyset$
    for all $k_1 \in \{1, \ldots, n\}$ **do**
        $\text{Recur}(1, (k_1))$
    end for

Algorithm 2. $\text{Recur}(j, (k_1, \ldots, k_j))$

---

if $j = n - 1$ **then**
    append the unique $k_n \in \{1, \ldots, n\} \setminus \{k_1, \ldots, k_{n-1}\}$ to the end of $(k_1, \ldots, k_{n-1})$
    add $(k_1, \ldots, k_n)$ to opt_n ▷we found a solution.
else

    for all $k_{j+1} \in \{1, \ldots, n\} \setminus \{k_1, \ldots, k_j\}$ **do**

        if Equations (A_{j+1}), (B_{j+1}) and (C_{j+1}) hold **then**
            append k_{j+1} to the end of $(k_1, \ldots, k_j)$
            $\text{Recur}(j + 1, (k_1, \ldots, k_{j+1}))$
        end if
    end for
end if

---

4 An Application to Label Ranking

In this section, we explore an application of the IPL model to supervised learning of label rankings. Usually, supervised learning consists in mapping any instance $\mathbf{x} \in \mathcal{X}$ to a single (preferred) label $\Lambda = \{\lambda_1, \ldots, \lambda_n\}$ representing its class. Here, we study a more complex issue called label ranking, which rather maps $\mathbf{x} \in \mathcal{X}$ to a predicted total order $\hat{y}$ on the labels in Λ—or a partial order, should we accept to make imprecise predictions for the sake of robustness.

For this purpose, we exploit a set of training instances associated with rankings $(\mathbf{x}_i, \tau_i)$, with $i \in \{1, \ldots, m\}$, in order to estimate the theoretical conditional probability measure $P_{\mathbf{x}} \colon 2^{\mathcal{L}(\Lambda)} \to [0, 1]$ associated to an instance $\mathbf{x} \in \mathcal{X}$. Ideally, observed outputs τ_i should be complete orders over Λ; however, this is seldom the case, total orders being more difficult to observe: training instances are therefore frequently associated with incomplete rankings τ_i (i.e., partial orders over Λ).

Here, we will apply the approach detailed in Sect. 3.1 to learning an IPL model from such training data, using the contour likelihood to get the parameter set corresponding to a specific instance $\mathbf{x}$.

4.1 Estimation and Prediction

Precise Predictions. In [7], it was proposed to use an instance-based approach: the predictions for any $\mathbf{x} \in \mathcal{X}$ are made locally using its nearest neighbours.

Let $\mathcal{N}_K(\mathbf{x})$ stand for the set of nearest neighbours of $\mathbf{x}$ in the training set, each neighbour $\mathbf{x}_i \in \mathcal{N}_K(\mathbf{x})$ being associated with a (possibly incomplete) ranking τ_i; and let M_i be the number of ranked labels in τ_i. Using the classical instance-based assumption that distributions are locally identical (i.e., in the neighborhood of $\mathbf{x}$), the probability of observing $\tau_1, \ldots, \tau_K$ given a parameter value v is:

$$P(\tau_1, \ldots, \tau_K | v) = \prod_{\mathbf{x}_i \in \mathcal{N}_K(\mathbf{x})} \prod_{m=1}^{M_i} \frac{v_{\tau_i(m)}}{\sum_{j=m}^{M_i} v_{\tau_i(j)}}. \tag{11}$$

We can then use maximum likelihood estimation (MLE) in order to determine v from $\tau_1, \ldots, \tau_K$, by maximizing (11)—or equivalently, its logarithm

$$l(v) = \sum_{i=1}^{K} \sum_{m=1}^{M_i} \left[\log(v_{\tau_i(m)}) - \log \sum_{j=m}^{M_i} v_{\tau_i(j)} \right].$$

Various ways to obtain this maximum have been investigated. We will use here the minorization-maximization (MM) algorithm [16], which aims, in each iteration, to maximize a function which minorizes the log-likelihood:

$$Q_k(v) = \sum_{i-1}^{K} \sum_{m=1}^{M_i} \left[\log(v_{\tau_i(m)}) - \frac{\log \sum_{j=m}^{M_i} v_{\tau_i(j)}}{\log \sum_{j=m}^{M_i} v_{\tau_i(j)}^{(k)}} \right]$$

where $v^{(k)}$ is the estimation of v in the k-th iteration. When the parameters are fixed, the maximization of Q_k can be solved analytically and the algorithm provably converges to the MLE estimate v^* of v. The best ranking τ^* is then

$$\tau^* \in \arg \max_{\tau \in \mathcal{L}(\Lambda)} P(\tau | v^*);$$

it is simply obtained by ordering the labels according to v^* (see Eq. (2)).

Imprecise Predictions. An IPL model is in one-to-one correspondence with an imprecise parameter estimate, which can be obtained here by extending the classical likelihood to the contour likelihood method [5]. Given a parameter space Σ and a positive likelihood function L, the contour likelihood function is:

$$L^*(v) = \frac{L(v)}{\max_{v \in \Sigma} L(v)};$$

by definition, L^* takes values in $]0, 1]$: the closer $L^*(v)$ is to 1, the more likely v is. One can then naturally obtain imprecise estimates by considering "cuts". Given β in $[0, 1]$, the β-cut of the contour likelihood, written B_β^*, is defined by

$$B_\beta^* = \{ v \in \Sigma : L^*(v) \geq \beta \}.$$

Once B_β^* is determined, for any test instance $\mathbf{x}$ to be processed, we can easily obtain an imprecise prediction $\hat{y}$ in the form of a partial ranking, using the results of Sect. 3.1: we will retrieve $\hat{y}$ such that $\lambda_i > \lambda_j$ for all $v_k \in B^*_\beta$. We stress here that the choice of β directly influences the precision (and thus the robustness) of the model: $B^*_1 = v^*$, which generally leads to a precise PL model; when β decreases, the IPL model is less and less precise, possibly leading to partial (and even empty) predictions.

In our experiments, the contour likelihood function is modelled by generating multiple strengths v according to a Dirichlet distribution with parameter $\beta = \gamma v^*$, where v^* is the ML estimate obtained with the best PL model (or equivalently, the best strength v) and $\gamma > 0$ is a coefficient which makes it possible to control the concentration of parameters generated around v^*.

4.2   Evaluation

When the observed and predicted rankings y and $\hat{y}$ are complete, various accuracy measures [15] have been proposed to measure how close they are to each other (0/1 accuracy, Spearman's rank, ...). Here, we retain Kendall's Tau:

$$A_\tau(y, \hat{y}) = \frac{C - D}{n(n-1)/2},\tag{12}$$

where C and D are respectively the number of concording and discording pairs in y and $\hat{y}$. In the case of imprecise predictions $\hat{y}$, the usual quality measures can be decomposed into two components [9]: correctness (CR), measuring the accuracy of the predicted comparisons, and completeness (CP):

$$CR(y, \hat{y}) = \frac{C - D}{C + D} \quad \text{and} \quad CP(y, \hat{y}) = \frac{C + D}{n(n-1)/2},\tag{13}$$

where C and D are the same as in Eq. (12). Should $\hat{y}$ be complete, $C + D = n(n-1)/2$, $CR(y, \hat{y}) = A_\tau(y, \hat{y})$ and $CP(y, \hat{y}) = 1$; while $CR(y, \hat{y}) = 1$ and $CP(y, \hat{y}) = 0$ if $\hat{y}$ is empty (since no comparison is done).

4.3   Results

We performed our experiments on several data sets, mostly adapted from the classification setting [7]; we report here those obtained on the Bodyfat, Housing and Wisconsin data sets. For each dataset, we tested several numbers of neighbours: $K \in \{5, 10, 15, 20\}$ (for the MLE estimate and using Eq. (12)), and chose the best by tenfold cross-validation. The sets of parameters B^*_β were obtained as explained above, by generating 200 strengths with $\gamma \in \{1, 10\}$, the best value being selected via tenfold cross validation repeated 3 times.

We also compared our approach to another proposal [8] based on a rejection threshold of pairwise preference probabilities, in three different configurations:

- using the original, unperturbed rankings;
- by deleting some labels in the original rankings with a probability $p \in [0, 1]$;

– by introducing some noise in the rankings, by randomly swapping adjacent labels with a probability $p \in [0, 1]$ (the labels being chosen at random).

Figure 3 displays the results of both methods for the Bodyfat data set (with $m = 252$ and $n = 7$) when rankings remain unperturbed, with a confidence interval of 95% (±2 standard deviation of measured correctness). Our approach based on the contour likelihood function is on par with the method based on abstention, which was the case with all tested data sets. Both methods see correctness increase once we allow for abstention. On the other data sets, the same behaviour can be seen: our approach seems to be on par with the one based on abstention, provided that the contour likelihood function has been correctly modelled (i.e., the generation of strengths is appropriate).

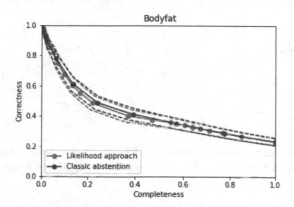

Fig. 3. Comparison of methods on Bodyfat with no perturbations

In order to be able to compare the two methods, we show underneath results on a specific range of the completeness. We only show the domain [0.6, 1]; however the behaviour is similar outside this range.

Figures 4 and 5 show that both methods are also on par on the Housing data set ($m = 506$, $n = 6$) even when the data sets are missing some labels. It can also be noticed that for a given completeness level, the correctness is lower than in the unperturbed case. On average, the greater the level of perturbation is, the lower the average correctness is. This also stands for the other data sets.

Figures 6 and 7 display that with a different method of perturbation (label swapping), our approach gives similar results on the Wisconsin data set ($m = 194$, $n = 16$). Moreover, the correctness is again lower in average for a given completeness level if the data set is perturbed. We observe the same behaviour for the label swapping perturbation method on the other data sets.

Such results are encouraging, as they show that we can at least achieve results similar to state-of-the-art approaches. We yet have to identify those cases where the two approaches significantly differ.

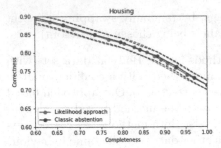

Fig. 4. Comparison of methods on Housing with no perturbations

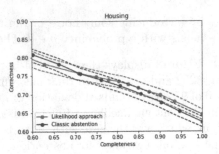

Fig. 5. Comparison of methods on Housing with 60% of missing label pairs

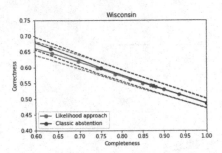

Fig. 6. Comparison of methods on Wisconsin with no perturbations

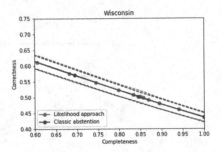

Fig. 7. Comparison of methods on Wisconsin with 60% of swapped label pairs

5   Conclusions

In this paper, we made a preliminary investigation into performing robust inference and making cautious predictions with the well known Plackett–Luce model, a popular ranking model in statistics. We have provided efficient methods to do so when the data at hand are poor, that is either of a low quality (noisy, partial) or scarce. We have demonstrated the interest of our approach in a label ranking problem, in presence of missing or noisy ranking information.

Possible future investigations may focus on the estimation problem, which may be improved, for example by extending Bayesian approaches [14] through the consideration of sets of prior distributions; or by developing a natively imprecise likelihood estimate, for instance by coupling recent estimation algorithms using stationary distribution of Markov chains [20] with recent works on imprecise Markov chains [17].

As suggested by an anonymous reviewer, it might be interesting to consider alternatives estimation methods such as epsilon contamination. There already exist non-parametric, decomposition-based approaches to label ranking with imprecise ranks; see [4,11]. However, the PL model, being tied to an order representation, may not be well-suited to such an approach. We intend to investigate this in the future.

Last, since the Plackett–Luce model is known to be strongly linked to particular RUM models [2,24], it may be interesting to investigate what becomes of this connection when the RUM model is imprecise (for instance, in our case, by considering Gumbel distributions specified with imprecise parameters).

Acknowledgements. This work benefited from the financial support of the projects PreServe ANR-18-CE23-0008 and LABEX MS2T ANR-11-IDEX-0004-02 of the French National Research Agency (ANR). We would like to thank three anonymous reviewers for their motivating comments.

References

1. Augustin, T., Coolen, F.P., De Cooman, G., Troffaes, M.C.: Introduction to Imprecise Probabilities. Wiley, Hoboken (2014)
2. Soufani, H.A., Parkes, D.C., Xia, L.: Random utility theory for social choice. In: Advances in Neural Information Processing Systems, pp. 126–134 (2012)
3. Baltas, G., Doyle, P.: Random utility models in marketing research: a survey. J. Bus. Res. **51**(2), 115–125 (2001)
4. Carranza-Alarcon, Y.-C., Messoudi, S., Destercke, S.: Cautious label-wise ranking with constraint satisfaction. In: Lesot, M.-J., et al. (eds.) IPMU 2020. CCIS, vol. 1238, pp. 96–111. Springer, Cham (2020). https://doi.org/10.1007/978-3-030-50143-3_8
5. Cattaneo, M.: Statistical Decisions Based Directly on the Likelihood Function. Ph.D. Thesis, ETH Zurich (2007)
6. Cheng, W., Dembczynski, K., Hüllermeier, E.: Label ranking methods based on the Plackett-Luce model. In: Proceedings of the 27th Annual International Conference on Machine Learning - ICML (2010)
7. Cheng, W., Hüllermeier, E., Dembczynski, K.J.: Label ranking methods based on the plackett-luce model. In: Proceedings of the 27th International Conference on Machine Learning, ICML-10, pp. 215–222 (2010)
8. Cheng, W., Hüllermeier, E., Waegeman, W., Welker, V.: Label ranking with partial abstention based on thresholded probabilistic models. In: Advances in Neural Information Processing Systems 25, NIPS-12, pp. 2510–2518 (2012)
9. Cheng, W., Rademaker, M., De Baets, B., Hüllermeier, E.: Predicting partial orders: ranking with abstention. In: Balcázar, J.L., Bonchi, F., Gionis, A., Sebag, M. (eds.) ECML PKDD 2010. LNCS (LNAI), vol. 6321, pp. 215–230. Springer, Heidelberg (2010). https://doi.org/10.1007/978-3-642-15880-3_20
10. de Campos, L., Huete, J., Moral, S.: Probability intervals: a tool for uncertain reasoning. Int. J. Uncertain. Fuzziness Knowl. Based Syst. **2**, 167–196 (1994)
11. Destercke, S., Masson, M.-H., Poss, M.: Cautious label ranking with label-wise decomposition. Eur. J. Oper. Res. **246**(3), 927–935 (2015)
12. Fürnkranz, J., Hüllermeier, E. (eds.): Preference Learning. Springer, Heidelberg (2011). https://doi.org/10.1007/978-3-642-14125-6
13. Gu, J., Yin, G.: Fast algorithm for generalized multinomial models with ranking data. In: Chaudhuri, K., Salakhutdinov, R. (eds.) Proceedings of the 36th International Conference on Machine Learning, volume 97 of Proceedings of Machine Learning Research, Long Beach, California, USA, 9–15 June 2019, pp. 2445–2453. PMLR (2019)

14. Guiver, J., Snelson, E.: Bayesian inference for Plackett-Luce ranking models. In: Proceedings of the 26th Annual International Conference on Machine Learning, pp. 377–384. ACM (2009)
15. Hüllermeier, E., Furnkranz, J., Cheng, W., Brinker, K.: Label ranking by learning pairwise preferences. Artif. Intell. **172**, 1897–1916 (2008)
16. Hunter, D.R., et al.: MM algorithms for generalized Bradley-Terry models. Ann. Stat. **32**(1), 384–406 (2004)
17. Krak, T., De Bock, J., Siebes, A.: Imprecise continuous-time Markov chains. Int. J. Approx. Reason. **88**, 452–528 (2017)
18. Levi, I.: The Enterprise of Knowledge. MIT Press, London (1980)
19. Marden, J.: Analyzing and Modeling Rank Data, vol. 64. Chapman & Hall/CRC, London (1996)
20. Maystre, L., Grossglauser, M.: Fast and accurate inference of Plackett-Luce models. In: Advances in Neural Information Processing systems, pp. 172–180 (2015)
21. Thurstone, L.: A law of comparative judgment. Psychol. Rev. **34**, 273–286 (1927)
22. Troffaes, M.: Generalising the conjunction rule for aggregating conflicting expert opinions. Int. J. Intell. Syst. **21**(3), 361–380 (2006)
23. Walley, P.: Statistical Reasoning with Imprecise Probabilities. Chapman and Hall, New York (1991)
24. Yellott Jr., J.I.: The relationship between Luce's Choice Axiom, Thurstone's Theory of Comparative Judgment, and the double exponential distribution. J. Math. Psychol. **15**(2), 109–144 (1977)

Inference with Choice Functions
Made Practical

Arne Decadt[✉], Jasper De Bock, and Gert de Cooman

FLip, ELIS, Ghent University, Ghent, Belgium
{arne.decadt,jasper.debock,gert.decooman}@ugent.be

Abstract. We study how to infer new choices from previous choices in a conservative manner. To make such inferences, we use the theory of choice functions: a unifying mathematical framework for conservative decision making that allows one to impose axioms directly on the represented decisions. We here adopt the coherence axioms of De Bock and De Cooman (2019). We show how to naturally extend any given choice assessment to such a coherent choice function, whenever possible, and use this natural extension to make new choices. We present a practical algorithm to compute this natural extension and provide several methods that can be used to improve its scalability.

Keywords: Choice functions · Natural extension · Algorithms · Sets of desirable option sets

1 Introduction

In classical probability theory, decisions are typically made by maximizing expected utility. This leads to a single optimal decision, or a set of optimal decisions all of which are equivalent. In imprecise probability theory, where probabilities are only partially specified, this decision rule can be generalized in multiple ways; Troffaes [9] provides a nice overview. A typical feature of the resulting imprecise decision rules is that they will not always yield a single optimal decision, as a decision that is optimal in one probability model may for example be sub-optimal in another.

We will not focus on one particular imprecise decision rule though. Instead, we will adopt the theory of choice functions: a mathematical framework for decision making that incorporates several (imprecise) decision rules as special cases, including the classical approach of maximizing expected utility [1,3,7]. An important feature of this framework of choice functions is that it allows one to impose axioms directly on the decisions that are represented by such a choice function [3,7,10]. We here adopt the coherence axioms that were put forward by De Bock and De Cooman [3].

As we will explain and demonstrate in this contribution, these coherence axioms can be used to infer new choices from previous choices. For any given assessment of previous choices that is compatible with coherence, we will achieve

© Springer Nature Switzerland AG 2020
J. Davis and K. Tabia (Eds.): SUM 2020, LNAI 12322, pp. 113–127, 2020.
https://doi.org/10.1007/978-3-030-58449-8_8

this by introducing the so-called natural extension of this assessment: the unique most conservative coherent choice function that is compatible with the assessment.

We start in Sect. 2 with an introduction to choice functions and coherence and then go on to define the natural extension of choice assessments in Sect. 3. From then on, we work towards a method to compute this natural extension. An important step towards this goal consists in translating our problem to the setting of sets of desirable option sets; we do this in Sect. 4. In this setting, as we show in Sect. 5, the problem takes on a more manageable form. Still, the complexity of the problem depends rather heavily on the size of the assessment that is provided. To address this issue, Sect. 6 presents several methods that can be used to replace an assessment by an equivalent yet smaller one. Section 7 then presents a practical algorithm that implements our approach. Section 8 concludes the paper and provides some suggestions for future work.

To adhere to the page-limit constraint, all proofs are omitted. They are available in an extended on-line version [4].

2   Choice Functions

A choice function C is a function that, when applied to a set of options, chooses one or more options from that set. Usually the options are actions that have a corresponding reward. This reward furthermore depends on the state of an unknown—uncertain—variable X that takes values in a set $\mathcal{X}$. We will assume that the rewards can be represented by real numbers, on a real-valued utility scale. In this setting, an option u is thus a function from states x in $\mathcal{X}$ to $\mathbb{R}$. We will denote the set of all possible options by $\mathscr{V}$. Evidently, $\mathscr{V}$ is a vector space with vector addition $u + v \colon x \mapsto u(x) + v(x)$ and scalar multiplication $\lambda u \colon x \mapsto \lambda u(x)$ for all $x \in \mathcal{X}$, $\lambda \in \mathbb{R}$ and $u, v \in \mathscr{V}$. Moreover, we endow $\mathscr{V}$ with the partial order $\leq$ defined by $u \leq v \Leftrightarrow (\forall x \in \mathcal{X}) u(x) \leq v(x)$ for all $u, v \in \mathscr{V}$ and a corresponding strict version $<$, where $u < v$ if both $u \neq v$ and $u \leq v$.[1] To make this more tangible, we consider the following toy problem as a running example.

Example 1.1. A farming company cultivates tomatoes and they have obtained a large order from a foreign client. However, due to government regulations they are not sure whether they can deliver this order. So, the state space $\mathcal{X}$ is {order can be delivered, order cannot be delivered}. The company now has multiple options to distribute their workforce. They can fully prepare the order, partially prepare the order or not prepare the order at all. Since $\mathcal{X}$ only has two elements, we can identify the options with vectors in $\mathbb{R}^2$. We will let the first component of these vectors correspond to the reward if the order can be delivered. For example, the option of fully preparing the order could correspond to the vector

---

[1] Our results in Sect. 2 and 3 also work for any ordered vector space over the real numbers; we restrict ourselves to the particular order $<$ for didactic purposes.

$v_1 := (5, -3)$. If the order goes through, then the company receives a payment—or utility—of 5 for that order. However, if the order does not go through, the company "receives" a negative reward -3, reflecting the large amount of resources that they spent on an order that could not be delivered in the end.   ◊

We will restrict ourselves to choices from finite sets of options. That is, the domain of our choice functions will be $\mathcal{Q} := \{A \subset \mathcal{V} : n \in \mathbb{N}, |A| = n\} \cup \{\emptyset\}$: the set of all finite subsets of $\mathcal{V}$, including the empty set. Formally, a choice function is then any function $C: \mathcal{Q} \to \mathcal{Q}$ such that $C(A) \subseteq A$ for all $A \in \mathcal{Q}$. We will also consider the corresponding rejection function $R_C: \mathcal{Q} \to \mathcal{Q}: A \mapsto A \setminus C(A)$.

Example 1.2. We will let the choice function C correspond to choices that the strategic advisor of the company makes or would make for a given set of options, where these choices can be multi-valued whenever he does not choose a single option. Suppose for example that he has rejected v_3 and v_4 from a set $B_1 := \{v_1, v_2, v_3, v_4\}$, with $v_1 := (5, -3)$, $v_2 := (3, -2)$, $v_3 := (1, -1)$, and $v_4 := (-2, 1)$, but remains undecided about whether to choose v_1 or v_2. This corresponds to the statement $C(B_1) = \{v_1, v_2\}$, or equivalently, $R_C(B_1) = \{v_3, v_4\}$.   ◊

We will give the following interpretation to these choice functions. For every set $A \in \mathcal{Q}$ and option $u \in A$, we take $u \in C(A)$—u is chosen—to mean that there is no other option in A that is preferred to u. Equivalently, $u \in R_C(A)$—u is rejected from A—if there is an option in A that is preferred to u. The preferences in this interpretation are furthermore taken to correspond to a strict partial vector order $\prec$ on $\mathcal{V}$ that extends the original strict order $<$. This implies that it should have the following properties: for all $u, v, w \in \mathcal{V}$ and $\lambda > 0$,

$O_1.$ $u \not\prec v$ or $v \not\prec u$,   (antisymmetry)

$O_2.$ if $u \prec v$ and $v \prec w$ then also $u \prec w$,   (transitivity)

$O_3.$ if $u \prec v$ then also $u + w \prec v + w$,   (translation invariance)

$O_4.$ if $u \prec v$ then also $\lambda u \prec \lambda v$.   (scaling invariance)

Crucially, however, the strict partial order $\prec$ need not be known. Instead, in its full generality, our interpretation allows for the use of a set of strict partial orders, only one of which is the true order $\prec$. As shown in reference [3], this interpretation can be completely characterized using five axioms for choice functions. Rather than simply state them, we will motivate them one by one starting form our interpretation.

The first axiom states that we should always choose at least one option, unless we are choosing from the empty set:

$C_0.$ $C(A) \neq \emptyset$ for all $A \in \mathcal{Q} \setminus \{\emptyset\}$.

This follows directly from our interpretation. Indeed, if every option in A would be rejected, then for every option v in A, there would be some other option in A that is preferred to v. Transitivity (O_2) would then imply that A contains an option that is preferred to itself, contradicting antisymmetry (O_1). To understand the second axiom, first observe that it follows from translation invariance (O_3) that $u \prec v$ if and only if $0 \prec v - u$. So if we let $A - u := \{v - u \colon v \in A\}$, then it follows from our interpretation that u is chosen from A if and only if 0 is chosen from $A - u$:

C_1. $u \in C(A) \Leftrightarrow 0 \in C(A - u)$ for all $A \in \mathscr{Q}$ and $u \in A$.

An important consequence of this axiom, is that knowing from which sets zero is chosen suffices to know the whole choice function. To formalize this, we introduce for any choice function C a corresponding set of option sets

$$K_C := \{A \in \mathscr{Q} \colon 0 \notin C(A \cup \{0\})\}. \tag{1}$$

For any option set $A \in K_C$, it follows from our interpretation for C that $A \cup \{0\}$ contains at least one option that is preferred to zero. Or equivalently, since zero is not preferred to itself because of O_1, that A contains at least one option that is preferred to zero. We call such an option set A a *desirable option set* and will therefore refer to K_C as the *set of desirable option sets* that corresponds to C. Since it follows from C_1 that

$$(\forall A \in \mathscr{Q})(\forall u \in A)\ u \notin C(A) \Leftrightarrow A - u \in K_C, \tag{2}$$

We see that the set of desirable option sets K_C fully characterizes the choice function C. Whenever convenient, we can therefore express axioms for C in terms of K_C as well. The next axiom is a first example where this is convenient. Since the preference order $\prec$ is taken to extend the order $<$, it follows that we must prefer all elements of $\mathscr{V}_{>0} := \{u \in \mathscr{V} \colon 0 < u\}$ to zero. Hence, $\{u\}$ is a desirable option set for all $u \in \mathscr{V}_{>0}$:

C_2. $\{u\} \in K_C$ for all $u \in \mathscr{V}_{>0}$.

Or to state it in yet another way: the set of positive singletons $\mathscr{V}_{>0}^s := \{\{u\} \colon u \in \mathscr{V}_{>0}\}$ should be a subset of K_C.

Another axiom that is easier to state in terms of K_C follows from the fact that we can take positive linear combinations of preferences. For example, if we have two non-negative real numbers λ_1, λ_2 and they are not both zero, and we know that $0 \prec u_1$ and $0 \prec u_2$, then it follows from O_2, O_3 and O_4 that also $0 \prec \lambda_1 u_1 + \lambda_2 u_2$. To state this more compactly, we let[2]

$$\mathbb{R}^{n,+} := \{\boldsymbol{\lambda} \in \mathbb{R}^n \colon (\forall j \in \{1, \ldots, n\})(\lambda_j \geqslant 0), \textstyle\sum_{j=1}^n \lambda_j > 0\}$$

for any positive integer n and introduce a product operation $\boldsymbol{\lambda}\mathbf{u} := \sum_{j=1}^n \lambda_j u_j$ for tuples $\boldsymbol{\lambda} = (\lambda_1, \ldots, \lambda_n) \in \mathbb{R}^n$ with tuples of vectors $\mathbf{u} = (u_1, \ldots, u_n) \in \mathscr{V}^n$. Then for any $\boldsymbol{\lambda} \in \mathbb{R}^{2,+}$ and any $\mathbf{u} \in \mathscr{V}^2$ such that $0 \prec u_1$ and $0 \prec u_2$, we have

---

[2] We will denote tuples in boldface and their elements roman with a positive integer index corresponding to their position.

that $0 \prec \boldsymbol{\lambda}\mathbf{u}$. Consider now two sets $A, B \in K_C$. Then as we explained after Eq. (1), each of them contains at least one option that is preferred to zero. Hence, by the reasoning above, there will be at least one pair of options $\mathbf{u} = (u_1, u_2)$ in $A \times B$ for which $0 \prec u_1$ and $0 \prec u_2$ and thus $0 \prec \boldsymbol{\lambda}\mathbf{u} = \lambda_1 u_1 + \lambda_2 u_2$ for $\boldsymbol{\lambda} \in \mathbb{R}^{2,+}$. Thus, the set of all such possible combinations—where $\boldsymbol{\lambda}$ can depend on $\mathbf{u}$—will contain at least one option preferred to zero en must therefore belong to K_C:

C_3. $\{\boldsymbol{\lambda}(\mathbf{u})\mathbf{u} \colon \mathbf{u} \in A \times B\} \in K_C$ for all $\boldsymbol{\lambda} \colon A \times B \to \mathbb{R}^{2,+}$ and $A, B \in K_C$.

The final axiom states that if an option u is rejected from an option set A, then it will also be rejected from any superset B:

C_4. $A \subseteq B \Rightarrow R_C(A) \subseteq R_C(B)$ for all $A, B \in \mathscr{Q}$.

Once more, this follows from our interpretation for C. If u is rejected from A, then this means that there is an element $v \in A$ that is preferred to u. Since v and u also belong to B, it thus follows that u is rejected from B as well. This axiom is also known as Sen's condition α [8].

Definition 1. *We call a choice function* $C \colon \mathscr{Q} \to \mathscr{Q}$ *coherent if it satisfies axioms* $C_0 - C_4$. *The set of all coherent choice functions is denoted by* $\mathcal{C}$.

A crucial point is that the axioms $C_0 - C_4$ are the same as the axioms $R_0 - R_4$ in [3], but tailored to our notation and in a different order; C_0 corresponds to R_1 whereas C_1 corresponds to R_0. Interestingly, as proven in reference [3], the axioms $C_0 - C_4$ are therefore not only necessary for a choice function to correspond to a set of strict partial orders, but sufficient as well.

3   Natural Extension

Fully specifying a coherent choice function is hard to do in practice, because this would amount to specifying a set-valued choice for every finite set of options, while at the same time taking into account the coherence axioms. Instead, a user will typically only be able to specify $C(A)$—or $R_C(A)$—for option sets A in some small—and often finite—subset $\mathcal{O}$ of $\mathscr{Q}$. For all other option sets $A \in \mathscr{Q} \setminus \mathcal{O}$, we can then set $C(A) = A$ because this adds no information to the choice function assessment. The resulting choice function may not be coherent though. To make this more concrete, let us go back to the example.

Example 1.3. Suppose that the strategic advisor of the farming company has previously rejected the options v_3 and v_4 from the option set B_1, as in Example 1.2, and has chosen v_6 from $B_2 := \{v_5, v_6\}$, where $v_5 := (3, 1)$ and $v_6 := (-4, 8)$. This corresponds to the choice function assessments $C(B_1) = \{v_1, v_2\}$ and $C(B_2) = \{v_6\}$. Suppose now that the company's strategic advisor has fallen ill and the company is faced with a new decision problem that amounts to choosing from the set $B_3 = \{(-3, 4), (0, 1), (4, -3)\}$. Since no such choice was made before, the conservative option is to make the completely uninformative statement $C(B_3) = B_3$. However, perhaps the company can make a more informative choice by taking into account the advisor's previous choices? $\Diamond$

In order to make new choices based on choices that have already been specified—as in the last question of the example—we can make use of coherence. Indeed, all the coherence axioms except C_0 allow one to infer new choices—or rejections—from other ones. In this way, we obtain a new choice function C' that is more informative than C in the sense that $C'(A) \subseteq C(A)$ for all $A \in \mathcal{Q}$. Any such choice function C' that is more informative than C, we call an extension of C. In order to adhere to the coherence axioms, we are looking for an extension C' of C that is coherent. We denote the set of all such extensions by $\mathcal{C}$.

Whenever at least one such coherent extension exists, we call C consistent. The least informative such coherent extension of C is then called its natural extension, as it is the only coherent extension of C that follows solely from C and the coherence axioms, without adding any additional unwarranted information. In order to formalize this notion, we let

$$\mathcal{C}_C := \{C' \in \mathcal{C} : C'(A) \subseteq C(A) \text{ for all } A \in \mathcal{Q}\}$$

and let $\text{Ex}(C)$ be defined by

$$\text{Ex}(C)(A) := \bigcup_{C' \in \mathcal{C}_C} C'(A) \text{ for all } A \in \mathcal{Q},$$

where, by convention, the union over an empty set is taken to be empty itself.

Definition 2. *For any choice function C, we call C consistent if $\mathcal{C}_C \neq \emptyset$ and then refer to $\text{Ex}(C)$ as the natural extension of C.*

Theorem 1. *For any choice function C that is consistent, $\text{Ex}(C)$ is the least informative coherent extension of C. That is, $\text{Ex}(C) \in \mathcal{C}_C$ and, for all $C' \in \mathcal{C}_C$, we have that $C'(A) \subseteq \text{Ex}(C)(A)$ for all $A \in \mathcal{Q}$. If on the other hand C is not consistent, then $\text{Ex}(C)$ is incoherent and $\text{Ex}(C)(A) = \emptyset$ for all $A \in \mathcal{Q}$.*

Given a choice function C that summarizes earlier choices and a new decision problem that consists in choosing from some set of options A, we now have a principled method to make this choice, using only coherence and the information present in C. First we should check if C is consistent. If it is not, then our earlier choices are not compatible with coherence, and would therefore better be reconsidered. If C is consistent, then the sensible choices to make are the options in $\text{Ex}(C)(A)$, since the other options in A can safely be rejected taking into account coherence and the information in C. If $\text{Ex}(C)(A)$ contains only one option, we arrive at a unique optimal choice. If not, then adding additional information is needed in order to be able to reject more options. The problem with this approach, however, is that it requires us to check consistency and evaluate $\text{Ex}(C)(A)$. The rest of this contribution is devoted to the development of practical methods for doing so. We start by translating these problems to the language of sets of desirable option sets.

4   Sets of Desirable Option Sets

As explained in Sect. 2, every coherent choice function C is completely determined by its corresponding set of desirable option sets K_C. Conversely, with any set of desirable option sets $K \subseteq \mathcal{Q}$, we can associate a choice function C_K, defined by

$$C_K(A) := \{u \in A : (A - u) \setminus \{0\} \notin K\} \text{ for all } A \in \mathcal{Q}. \tag{3}$$

In order for C_K to be coherent, it suffices for K to be coherent in the following sense.

Definition 3. *A set of desirable option sets $K \subseteq \mathcal{Q}$ is called* coherent *[2] if for all $A, B \in K$:*

K$_0$. $A \setminus \{0\} \in K$;

K$_1$. $\{0\} \notin K$;

K$_2$. $\mathcal{V}_{>0}^s \subseteq K$;

K$_3$. $\{\lambda(u)u : u \in A \times B\} \subseteq K$ for all $\lambda: A \times B \to \mathbb{R}^{2,+}$;

K$_4$. $A \cup Q \in K$ for all $Q \in \mathcal{Q}$.

We denote the set of all coherent K by $\bar{\mathbf{K}}$.

Proposition 2. *If a set of desirable option sets $K \subseteq \mathcal{Q}$ is coherent, then C_K is coherent too.*

Proposition 3. *If a choice function C is coherent, then K_C is coherent too.*

Theorem 4. *The map $\Phi: \mathcal{C} \to \bar{\mathbf{K}}: C \mapsto K_C$ is a bijection and has inverse $\Phi^{-1}(K) = C_K$ for all $K \in \bar{\mathbf{K}}$.*

In other words: every coherent choice function C corresponds uniquely to a coherent set of desirable option sets K, and vice versa.

The plan is now to use this connection to transform the problem of computing the natural extension of a choice function C to a similar problem for sets of desirable option sets. We start by transforming the choice function C into a set of option sets. One way to do this would be to consider the set of desirable option sets K_C. However, there is also a smarter way to approach this that yields a more compact representation of the information in C.

Definition 4. *An assessment is any subset of $\mathcal{Q}$. We denote the set of all assessments, the power set of $\mathcal{Q}$, by $\mathscr{A}$. In particular, with any choice function C, we associate the assessment*

$$\mathcal{A}_C := \{C(A) - u : A \in \mathcal{Q}, u \in R_C(A)\}.$$

Example 1.4. In our running example, the assessment that corresponds to C is $\mathcal{A}_C = \{A_1, A_2, A_3\}$, with $A_1 := \{w_1, w_2\}$, $A_2 := \{w_3, w_4\}$ and $A_3 := \{w_5\}$, where $w_1 := v_1 - v_3 = (4, -2)$ and similarly $w_2 := v_2 - v_3 = (2, -1)$, $w_3 := v_1 - v_4 = (7, -4)$, $w_4 := v_2 - v_4 = (5, -3)$ and $w_5 := v_6 - v_5 = (-7, 7)$. $\qquad \diamond$

An assessment such as $\mathcal{A}_C$ may not be coherent though. To extend it to a coherent set of desirable option sets, we will use the notions of consistency and natural extension that were developed for sets of desirable option sets by De Bock and De Cooman [2]. To that end, for any assessment $\mathcal{A} \in \mathscr{A}$, we consider the set $\bar{\mathbf{K}}(\mathcal{A}) := \{K \in \bar{\mathbf{K}}: \mathcal{A} \subseteq K\}$ of all coherent sets of desirable option sets that contain $\mathcal{A}$ and let

$$\mathrm{Ex}(\mathcal{A}) := \bigcap \bar{\mathbf{K}}(\mathcal{A}), \tag{4}$$

where we use the convention that $\bigcap \emptyset = \mathscr{Q}$.

Definition 5. *For any assessment $\mathcal{A} \in \mathscr{A}$, we say that $\mathcal{A}$ is* consistent *if $\bar{\mathbf{K}}(\mathcal{A}) \neq \emptyset$ and we then call $\mathrm{Ex}(\mathcal{A})$ the* natural extension *of $\mathcal{A}$ [2, Definition 9].*

As proven by De Bock and De Cooman [2], the consistency of $\mathcal{A}$ implies the coherence of $\mathrm{Ex}(\mathcal{A})$. Our next result establishes that the converse is true as well.

Theorem 5. *For any assessment $\mathcal{A} \in \mathscr{A}$ the following are equivalent:*

1. *$\mathcal{A}$ is consistent;*
2. *$\mathrm{Ex}(\mathcal{A})$ is coherent;*
3. *$\emptyset \notin \mathrm{Ex}(\mathcal{A})$.*

The connection with choice functions is established by the following result.

Theorem 6. *Let C be a choice function. Then C is consistent if and only if $\emptyset \notin \mathrm{Ex}(\mathcal{A}_C)$, and if it is, then $C_{\mathrm{Ex}(\mathcal{A}_C)}(A) = \mathrm{Ex}(C)(A)$ for all $A \in \mathscr{Q}$.*

By this theorem, we see that checking consistency and computing the natural extension of a choice function C amounts to being able to check for any option set whether it belongs to $\mathrm{Ex}(\mathcal{A}_C)$ or not. Indeed, given some choice function C, we can check its consistency by checking if $\emptyset \in \mathrm{Ex}(\mathcal{A}_C)$ and, taking into account Eq. (3), we can calculate $\mathrm{Ex}(C)(A)$ for a given set $A \in \mathscr{Q}$, by checking for any element $u \in A$ if $(A - u) \setminus \{0\} \in \mathrm{Ex}(\mathcal{A}_C)$.

5   Natural Extension and Consistency for Finite Assessments

In practice, we will typically be interested in computing the natural extension $\mathrm{Ex}(C)$ of choice functions C that are finitely generated, in the sense that $C(A) \neq A$ for only finitely many option sets $A \in \mathscr{Q}$. In that case, if we let $\mathcal{O}_C := \{A \in \mathscr{Q}: C(A) \neq A\}$, then

$$\mathcal{A}_C = \{C(A) - u: A \in \mathscr{Q}, u \in R_C(A)\} = \{C(A) - u: A \in \mathcal{O}_C, u \in R_C(A)\}.$$

Since $\mathcal{O}_C$ is finite and every $R_C(A) \subseteq A$ is finite because A is, it then follows that $\mathcal{A}_C$ is finite as well. Without loss of generality, for finitely generated choice functions C, $\mathcal{A}_C$ will therefore be of the form $\mathcal{A}_C = \{A_1, \ldots, A_n\}$, with $n \in \mathbb{N} \cup \{0\}$. The list of option sets $A_1, \ldots, A_n$ may contain duplicates in practice, for example if different $A \in \mathcal{O}_C$ and $u \in A$ yield the same option set $C(A) - u$. It is better to remove these duplicates, but our results will not require this. The only thing that we will assume is $\mathcal{A}_C = \{A_1, \ldots, A_n\}$.

The following theorem is the main result that will allow us to check in practice whether an option set belongs to $\mathrm{Ex}(\mathcal{A}_C)$ or not.

Theorem 7. *Let $\mathcal{A} = \{A_1, \ldots, A_n\}$, with $A_1, \ldots, A_n \in \mathscr{Q}$ and $n \in \mathbb{N} \cup \{0\}$. An option set $S \in \mathscr{Q}$ then belongs to $\mathrm{Ex}(\mathcal{A})$ if and only if either $S \cap \mathscr{V}_{>0} \neq \emptyset$ or, $n \neq 0$ and, for every $\mathbf{u} \in \times_{j=1}^n A_j$, there is some $s \in S \cup \{0\}$ and $\boldsymbol{\lambda} \in \mathbb{R}^{n,+}$ for which $\boldsymbol{\lambda}\mathbf{u} \leq s$.*

In combination with Theorem 6, this result enables us to check the consistency and compute the natural extension of finitely generated choice functions. Checking consistency is equivalent to checking if $\emptyset \notin \mathrm{Ex}(\mathcal{A}_C)$.

Example 1.5. We will now go ahead and test if the strategic advisor was at least consistent in his choices. Since $\emptyset \cap \mathscr{V}_{>0} = \emptyset$, we have to check the second condition in Theorem 7. In particular, we have to check if, for every tuple $\mathbf{u} \in A_1 \times A_2 \times A_3$, there is some $\boldsymbol{\lambda} \in \mathbb{R}^{3,+}$ such that $\boldsymbol{\lambda}\mathbf{u} \leq 0$. We will show that this is not the case for the particular tuple $\mathbf{u} = (w_2, w_4, w_5) \in A_1 \times A_2 \times A_3$. Assume that there is some $\boldsymbol{\lambda} = (\lambda_1, \lambda_2, \lambda_3) \in \mathbb{R}^{3,+}$ such that $\lambda_1 w_2 + \lambda_2 w_4 + \lambda_3 w_5 = \boldsymbol{\lambda}\mathbf{u} \leq 0$. Notice that $2w_4 \leq 5w_2$, so if we let $\mu_1 := \frac{2}{5}\lambda_1 + \lambda_2$ and $\mu_2 := 7\lambda_3$ then

$$\boldsymbol{\lambda}\mathbf{u} \geq \frac{2}{5}\lambda_1 w_4 + \lambda_2 w_4 + \lambda_3 w_5 = \mu_1 w_2 + \frac{1}{7}\mu_2 w_5 = (5\mu_1 - \mu_2, -3\mu_1 + \mu_2).$$

Since $\boldsymbol{\lambda}\mathbf{u} \leq 0$, this implies that $5\mu_1 \leq \mu_2 \leq 3\mu_1$ and thus $\mu_1 \leq 0$ and $\mu_2 \leq 0$. This is impossible though because $\boldsymbol{\lambda} \in \mathbb{R}^{3,+}$ implies that $\mu_1 > 0$ or $\mu_2 > 0$. Hence, $\emptyset \notin \mathrm{Ex}(\mathcal{A}_C)$, so $\mathcal{A}_C$ is consistent. We conclude that the decisions of the strategic advisor were consistent, enabling us to use natural extension to study their consequences.   ◊

If a finitely generated choice function C is consistent, then for any option set $\mathcal{A} \in \mathscr{Q}$, we can evaluate $\mathrm{Ex}(C)(A)$ by checking for every individual $u \in A$ if $u \in \mathrm{Ex}(C)(A)$. As we know from Theorem 6 and Eq. (3), this will be the case if and only if $(A - u) \setminus \{0\} \notin \mathrm{Ex}(\mathcal{A}_C)$.

Example 1.6. We can now finally tackle the problem at the end of Example 1.3: choosing from the set $B_3 = \{(-3, 4), (0, 1), (4, -3)\}$. This comes down to computing $\mathrm{Ex}(C)(B_3)$. Because of Theorem 6, we can check if $(4, -3)$ is rejected from B_3 by checking if $\{(-7, 7), (-4, 4)\} \in \mathrm{Ex}(\mathcal{A}_C)$. By Theorem 7, this requires us to check if for every $\mathbf{u} \in A_1 \times A_2 \times A_3$ we can find some $s \in \{(-7, 7), (-4, 4), (0, 0)\}$ and some $\boldsymbol{\lambda} \in \mathbb{R}^{3,+}$ such that $\boldsymbol{\lambda}\mathbf{u} \leq s$. Since $(-7, 7) = w_5 \in \{w_5\} = A_3$, $s = (-7, 7)$ and $\boldsymbol{\lambda} = (0, 0, 1)$ do the job for every

u. So we can reject $(4, -3)$. Checking if $(0, 1)$ is rejected is analogous: we have to check if $\{(-3, 3), (4, -4)\} \in \text{Ex}(\mathcal{A})$. In this case, we can use $s = (-3, 3)$ and $\lambda = (0, 0, \frac{3}{7})$ for every **u** in $A_1 \times A_2 \times A_3$ to conclude that $(0, 1)$ is rejected as well. Since $\text{Ex}(C)(B_3)$ must contain at least one option because of C_0 (which applies because Theorem 1 and the consistency of C imply that $\text{Ex}(C)$ is coherent) it follows that $\text{Ex}(C)(B_3) = \{(-3, 4)\}$. So based on the advisor's earlier decisions and the axioms of coherence, it follows that the company should choose $(-3, 4)$ from B_3.   ◇

In this simple toy example, the assessment $\mathcal{A}_C$ was small and the conditions in Theorem 7 could be checked manually. In realistic problems, this may not be the case though. To address this, we will provide in Sect. 7 an algorithm for checking the conditions in Theorem 7. But first, we provide methods for reducing the size of an assessment.

6 Simplifying Assessments

For any given assessment $\mathcal{A} = \{A_1, \ldots, A_n\}$, the number of conditions that we have to check to apply Theorem 7 is proportional to the size of $\times_{j=1}^{n} A_j$. Since Theorem 7 draws conclusions about $\text{Ex}(\mathcal{A})$ rather than $A_1, \ldots, A_n$, it can thus be useful to try to make $\times_{j=1}^{n} A_j$ smaller without altering $\text{Ex}(\mathcal{A})$, as this will reduce the number of conditions that we have to check. This is especially true when we want to apply Theorem 7 to several sets S, for example because we want to evaluate the natural extension of a choice function in multiple option sets.

To make $\times_{j=1}^{n} A_j$ smaller, a first straightforward step is to remove duplicates from $A_1, \ldots, A_n$; after all, it is only the set $\mathcal{A} = \{A_1, \ldots, A_n\}$ that matters, not the list $A_1, \ldots, A_n$ that generates it. To further reduce the size of $\times_{j=1}^{n} A_j$, we need to reduce the size of $\mathcal{A}$ and the option sets it consists of. To that end, we introduce a notion of equivalence for assessments.

Definition 6. *For any two assessments $\mathcal{A}_1, \mathcal{A}_2 \in \mathscr{A}$, we say that $\mathcal{A}_1$ and $\mathcal{A}_2$ are equivalent if $\bar{\mathbf{K}}(\mathcal{A}_1) = \bar{\mathbf{K}}(\mathcal{A}_2)$.*

It follows immediately from Definition 5 and Eq. (4) that replacing an assessment with an equivalent one does not alter its consistency, nor, if it is consistent, its natural extension.

The following result shows that it is not necessary to directly simplify a complete assessment; it suffices to focus on simplifying subsets of assessments.

Proposition 8. *If two assessments $\mathcal{A}_1, \mathcal{A}_2 \in \mathscr{A}$ are equivalent and $\mathcal{A}_1 \subseteq \mathcal{A} \in \mathscr{A}$, then $\mathcal{A}$ is equivalent to $(\mathcal{A} \setminus \mathcal{A}_1) \cup \mathcal{A}_2$.*

This result is important in practice because it means that we can build and simplify assessments gradually when new information arrives and that we can develop and use equivalence results that apply to small (subsets of) assessments.

A first simple such equivalence result is that we can always remove zero from any option set.

Proposition 9. *Consider an option set $A \in \mathcal{Q}$. Then the assessment $\{A\}$ is equivalent to $\{A \setminus \{0\}\}$.*

This result can be generalized so as to remove options for which there is a second option that can, by scaling, be made better than the first option.

Theorem 10. *Consider an option set $A \in \mathcal{Q}$ and two options $u, v \in A$ such that $u \neq v$ and $u \leq \mu v$ for some $\mu \geq 0$. Then the assessment $\{A\}$ is equivalent to $\{A \setminus \{u\}\}$.*

If $\mathcal{V} = \mathbb{R}^2$, this result—together with Proposition 8—guarantees that every option set in $\mathcal{A}$ can be replaced by an equivalent option set of size at most two.

Proposition 11. *Let $\mathcal{V} = \mathbb{R}^2$ and consider any option set $A \in \mathcal{Q}$. Then there is always an option set $B \in \mathcal{Q}$ with at most two options such that $\{A\}$ is equivalent to $\{B\}$, and this option set B can be found by repeated application of Theorem 10.*

In the case of our running example, all the option sets in $\mathcal{A}$ can even be reduced to singletons.

Example 1.7. In A_1 we see that $w_1 = 2w_2$ and in A_2 we see that $w_4 \leq \frac{5}{7} w_3$. So, by Proposition 8 and Theorem 10, we can simplify the assessment $\mathcal{A}_C$ of Example 1.4 to $\mathcal{A}'_C := \{\{w_2\}, \{w_3\}, \{w_5\}\}$. $\diamond$

Our equivalence results so far were all concerned with removing options from the option sets in $\mathcal{A}$. Our next result goes even further: it provides conditions for removing the option sets themselves.

Theorem 12. *Consider an assessments $\mathcal{A}$ and an option set $A \in \mathcal{A}$ such that $A \in \mathrm{Ex}(\mathcal{A} \setminus \{A\})$. Then $\mathcal{A}$ is equivalent to $\mathcal{A} \setminus \{A\}$.*

Example 1.8. Let us start from the assessment $\mathcal{A}'_C = \{\{w_2\}, \{w_3\}, \{w_5\}\}$ in Example 1.7. We will remove $\{w_2\}$ from this assessment using Theorem 12. To do that, we need to show that $\{w_2\} \in \mathrm{Ex}(\{\{w_3\}, \{w_5\}\})$. To that end, we apply Theorem 7 for $S = \{w_2\}$. Since $w_2 \not> 0$, it follows that $S \cap \mathcal{V}_{>0} = \emptyset$. We therefore check the second condition of Theorem 7. Since $\{w_3\} \times \{w_5\}$ is a singleton, we only need to check the condition for a single tuple $\mathbf{u} = (w_3, w_5)$. For $s = w_2$ and $\boldsymbol{\lambda} = (\frac{1}{4}, 0)$, we find that $\boldsymbol{\lambda}\mathbf{u} = \frac{1}{4}w_3 + 0w_5 = (\frac{7}{4}, -1) \leq (2, -1) = w_2 = s$. Hence, $\{w_2\} \in \mathrm{Ex}(\{\{w_3\}, \{w_5\}\})$ and we can therefore replace $\mathcal{A}'_C$ by the smaller yet equivalent assessment $\mathcal{A}''_C = \{\{w_3\}, \{w_5\}\}$. Taking into account our findings in Example 1.7, it follows that $\mathcal{A}_C$ is equivalent to $\mathcal{A}''_C$ and, therefore, that $\mathrm{Ex}(\mathcal{A}_C) = \mathrm{Ex}(\mathcal{A}''_C)$.

Obviously, this makes it easier to evaluate $\mathrm{Ex}(C)$. Suppose for example that we are asked to choose from $\{v_7, v_8\}$, with $v_7 := (5, -2)$ and $v_8 := (-4, 3)$. By Theorem 6 and Eq. (3), we can check if we can reject v_8 by checking if $\{v_7 - v_8\} = \{(9, -5)\} \in \mathrm{Ex}(\mathcal{A}''_C) = \mathrm{Ex}(\mathcal{A}_C)$. For $s = (9, -5)$, $\mathbf{u} = (w_3, w_5)$ and $\boldsymbol{\lambda} = (\frac{9}{7}, 0)$ we see that $\boldsymbol{\lambda}\mathbf{u} = (9, -\frac{36}{7}) \leq s$, so Theorem 7 tells us that, indeed, $\{(9, -5)\} \in \mathrm{Ex}(\mathcal{A}''_C)$. If we were to perform the same check directly for $\mathrm{Ex}(\mathcal{A}_C)$, then we would have to establish four inequalities—one for every $\mathbf{u} \in A_1 \times A_2 \times A_3$—while now we only had to establish one. $\diamond$

7   Practical Implementation

For large or complicated assessments, it will no longer be feasible to manually check the conditions in Theorem 7. In those cases, the algorithm that we are about to present can be used instead. According to Theorem 7, testing if an option set S belongs to $\mathrm{Ex}(\mathcal{A})$ for some assessment sequence $\mathcal{A} = \{A_1, \ldots, A_n\}$ requires us to check if $S \cap \mathcal{V}_{>0} \neq \emptyset$ and, in case $n \neq 0$, if there is for every $\mathbf{u} \in \times_{j=1}^n A_j$ some $s \in S \cup \{0\}$ and $\boldsymbol{\lambda} \in \mathbb{R}^{n,+}$ such that $\boldsymbol{\lambda}\mathbf{u} \leq s$. If one of these two conditions is satisfied, then S belongs to $\mathrm{Ex}(\mathcal{A})$. The first condition is not complicated, as we just have to check for every $s \in S$ if $s > 0$. For the second condition, the difficult part is how to verify, for any given $\mathbf{u} \in \times_{j=1}^n A_j$ and $s \in S \cup \{0\}$, whether there is some $\boldsymbol{\lambda} \in \mathbb{R}^{n,+}$ for which $\boldsymbol{\lambda}\mathbf{u} \leq s$. If one of these two conditions is satisfied then S belongs to $\mathrm{Ex}(\mathcal{A})$. Given the importance of this basic step, we introduce a boolean function ISFEASIBLE: $\mathcal{V}^n \times \mathcal{V} \to$ {True, False}. For every $\mathbf{u} \in \mathcal{V}^n$ and $s \in \mathcal{V}$, it returns True if $\boldsymbol{\lambda}\mathbf{u} \leq s$ for at least one $\boldsymbol{\lambda} \in \mathbb{R}^{n,+}$, and False otherwise.

So the only problem left is how to compute ISFEASIBLE$(\mathbf{u}, s)$. The tricky part is the constraint that $\boldsymbol{\lambda} = (\lambda_1, \ldots, \lambda_n) \in \mathbb{R}^{n,+}$. By definition of $\mathbb{R}^{n,+}$, this can be rewritten as $\lambda_j \geq 0$ for all $j \in \{1, \ldots, n\}$ and $\sum_{j=1}^n \lambda_j > 0$, which are all linear constraints. Since the condition $\boldsymbol{\lambda}\mathbf{u} \leq s$ is linear as well, we have a linear feasibility problem to solve. However, strict inequalities such as $\sum_{j=1}^n \lambda_j > 0$ are problematic for software implementations of linear feasibility problems. A quick fix is to choose some very small $\epsilon > 0$ and impose the inequality $\sum_{j=1}^n \lambda_j \geq \epsilon$ instead, but since this is an approximation, it does not guarantee that the result is correct. A better solution is to use the following alternative characterisation that, by introducing an extra free variable, avoids the need for strict inequalities.[3]

Proposition 13. *Consider any $s \in \mathcal{V}$ and any $\mathbf{u} = (u_1, \ldots, u_n) \in \mathcal{V}^n$. Then* ISFEASIBLE$(\mathbf{u}, s) = $ True *if and only if there is a* $\boldsymbol{\mu} = (\mu_1, \ldots, \mu_{n+1}) \in \mathbb{R}^{n+1}$ *such that* $\sum_{j=1}^n \mu_j u_j \leq \mu_{n+1} s$, $\mu_j \geq 0$ *for all* $j \in \{1, \ldots, n\}$, $\mu_{n+1} \geq 1$ *and* $\sum_{j=1}^n \mu_j \geq 1$.

Computing ISFEASIBLE$(\mathbf{u}, s)$ is therefore a matter of solving the following linear feasibility problem:

$$
\begin{aligned}
\text{find} \quad & \mu_1, \ldots, \mu_{n+1} \in \mathbb{R}, \\
\text{subject to} \quad & \mu_{n+1} s(x) - \sum_{j=1}^n \mu_j u_j(x) \geq 0 && \text{for all } x \in \mathcal{X}, \\
& \sum_{j=1}^n \mu_j \geq 1, \quad \mu_{n+1} \geq 1, \\
\text{and} \quad & \mu_j \geq 0 && \text{for all } j \in \{1, \ldots, n\}.
\end{aligned}
$$

For finite $\mathcal{X}$, such problems can be solved by standard linear programming methods; see for example [5].

---

[3] This result was inspired by a similar trick that Erik Quaegebeur employed in his CONEstrip algorithm [6].

Algorithm 1. Check if an option set $S \in \mathcal{Q}$ is in $\mathrm{Ex}(\mathcal{A})$ for an assessment $\mathcal{A} \in \mathscr{A}$

Precondition: $S, A_1, \ldots, A_n \in \mathcal{Q}$ with $n \in \mathbb{N} \cup \{0\}$ and $\mathcal{A} = \{A_1, \ldots, A_n\}$.

1: **function** ISINEXTENSION($S, A_1, \ldots, A_n$)   ▷ Check if S is in $\mathrm{Ex}(\mathcal{A})$.
2:    **for all** $s \in S$ **do**
3:      **if** $s > 0$ (so $s \geqslant 0$ and $s \neq 0$) **then**
4:        **return** True
5:    **if** $n = 0$ **then**
6:      **return** False
7:    **for all** $\mathbf{u} \in \times_{j=1}^{n} A_j$ **do**
8:      **for all** $s \in S \cup \{0\}$ **do**   ▷ Search an s and λ such that $\lambda\mathbf{u} \leq s$.
9:        res $\leftarrow$ ISFEASIBLE($\mathbf{u}, s$)
10:      **if** res **then**
11:        **break**   ▷ Stop the loop in s when a suitable s is found.
12:      **if** $\neg$ res **then**
13:        **return** False   ▷ If there is no such s.
14:    **return** True   ▷ When all $\mathbf{u}$ have been checked.

Together, Theorem 7 and Proposition 13 therefore provide a practical method to test if an option set S belongs to $\mathrm{Ex}(\mathcal{A})$, with $\mathcal{A} = \{A_1, \ldots, A_n\}$. Pseudocode for this method is given in Algorithm 1. First, in lines 2 to 4, we check if $S \cap \mathcal{V}_{>0} \neq \emptyset$. If it is, then $S \in \mathrm{Ex}(\mathcal{A})$ and we thus return True. If this is not the case, and the assessment is empty, i.e. $n = 0$, then we have to return False, as we do in lines 5 and 6. Next, in lines 7 to 11, we run through all $\mathbf{u} \in \times_{j=1}^{n} A_j$ and search an $s \in S \cup \{0\}$ for which ISFEASIBLE$(u, s) = $ True, i.e. for which there is some $\lambda \in \mathbb{R}^{n,+}$ such that $\lambda\mathbf{u} \leq s$. As soon as we have found such an s, we can break from the loop—halt the search—and go to the next $\mathbf{u}$. However, if we went through all of $S \cup \{0\}$ and we did not find such an s, then the second condition of Theorem 7 is false for the current $\mathbf{u}$ and thus S does not belong to $\mathrm{Ex}(\mathcal{A})$; we then return False, as in lines 12 and 13. On the other hand, if we went through all $\mathbf{u} \in \times_{j=1}^{n} A_j$ and for every one of them we have found an $s \in S \cup \{0\}$ such that ISFEASIBLE$(\mathbf{u}, s) = $ True, then we conclude that S is in $\mathrm{Ex}(\mathcal{A})$ by the second condition of Theorem 7, so we return True.

Example 2. Roger is an expert in the pro snooker scene. An important game is coming soon where two players will play two matches. Betting sites will offer bets on the following three possible outcomes: the first player wins 2-0, a 1-1 draw, or the second player wins 2-0. So a bet corresponds with an option in $\mathbb{R}^3$, the components of which are the rewards for each of the three outcomes. Before the possible bets are put online, we ask Roger to provide us with an assessment. He agrees to do so, but tells us that we should not contact him again when the bets are online. We ask him to choose from the sets $B_1 = \{(-4, 1, -1), (3, -5, -1), (-3, 1, -1), (4, 0, -4), (3, -5, 4)\}$, $B_2 = \{(-4, 2, 4), (-2, -4, 3), (0, -4, 2), (0, 3, -5), (2, 1, 3)\}$ and $B_3 = \{(-4, 1, 4), (-2, -2, 4), (-5, 3, 4)\}$. He provides us with the choice assessment $C(B_1) = \{(4, 0, -4), (3, -5, 4)\}$, $C(B_2) = \{(-4, 2, 4)\}$ and

$C(B_3) = \{(-4, 1, 4), (-2, -2, 4)\}$. Some time later, the betting site makes the following set of bets available:

$$A = \{(-1, -1, 2), (-4, -4, 6), (-2, -10, 6), (-1, 0, -2), (-2, 8, -6),$$
$$(2, -4, 4), (4, -6, 1), (-3, 8, 5), (2, 9, -9), (1, 7, -3)\}.$$

The question then is what bet to choose from A, based on what Roger has told us. The assessment $\mathcal{A}_C$ that corresponds to Roger's choice statements contains eight option sets. However, we can use Theorems 10 and 12 and Proposition 8 to reduce $\mathcal{A}_C$ to the equivalent assessment $\mathcal{A}^* = \{\{(3, -5, 0)\}, \{(-6, 1, 1), (-2, 2, -8)\}\}$. With Theorems 5 and 6 and Algorithm 1, we find that the assessment is consistent and with subsequent runs of Algorithm 1, and using Theorem 6 and Eq. (3), we find that $\text{Ex}(C)(A) = \{(-4, -4, 6), (-2, -10, 6), (-3, 8, 5)\}$. So we can greatly reduce the number of bets to choose from but we cannot, based on the available information, deduce entirely what Roger would have chosen.   ◇

8   Conclusion and Future Work

The main conclusion of this work is that choice functions provide a principled as well as practically feasible framework for inferring new decisions from previous ones. The two key concepts that we introduced to achieve this were consistency and natural extension. The former allows one to check if an assessment of choices is compatible with the axioms of coherence, while the latter allows one to use these axioms to infer new choices. From a practical point of view, our main contribution is an algorithm that is able to execute both tasks. The key technical result that led to this algorithm consisted in establishing a connection with the framework of sets of desirable option sets. This allowed us to transform an assessment of choices into an assessment of desirable option sets, then simplify this assessment, and finally execute the desired tasks directly in this setting.

Future work could add onto Sect. 6 by trying to obtain a 'simplest' representation for any given assessment, thereby further reducing the computational complexity of our algorithm. We would also like to conduct extensive experiments to empirically evaluate the time efficiency of our algorithm and the proposed simplifications, and how this efficiency scales with a number of important parameters. This includes the number of option sets in an assessment, the size of the option sets themselves, the dimension of the vector space $\mathcal{V}$ and the size of the option set A for which we want to evaluate the natural extension $\text{Ex}(C)(A)$. We also intend to consider alternative forms of assessments, such as bounds on probabilities, bounds on expectations and preference statements, and show how they can be made to fit in our framework. Finally, we would like to apply our methods to a real-life decision problem.

Acknowledgements. The work of Jasper De Bock was supported by his BOF Starting Grant "Rational decision making under uncertainty: a new paradigm based on choice functions", number 01N04819. We also thank the reviewers for their valuable feedback.

References

1. De Bock, J.: Archimedean choice functions: an axiomatic foundation for imprecise decision making. In: Lesot, M.-J., et al. (eds.) IPMU 2020. CCIS, vol. 1238, pp. 195–209. Springer, Cham (2020). https://doi.org/10.1007/978-3-030-50143-3_15
2. De Bock, J., de Cooman, G.: A desirability-based axiomatisation for coherent choice functions. In: Destercke, S., Denoeux, T., Gil, M.Á., Grzegorzewski, P., Hryniewicz, O. (eds.) SMPS 2018. AISC, vol. 832, pp. 46–53. Springer, Cham (2019). https://doi.org/10.1007/978-3-319-97547-4_7
3. De Bock, J., De Cooman, G.: Interpreting, axiomatising and representing coherent choice functions in terms of desirability. In: Proceedings of ISIPTA 2019, pp. 125–134. PMLR (2019)
4. Decadt, A., De Bock, J., de Cooman, G.: Inference with choice functions made practical (2020). https://arxiv.org/abs/2005.03098
5. Dantzig, G.B., Thapa, M.N.: Linear Programming 2: Theory and Extensions, 1st edn. Springer, New York (2003). https://doi.org/10.1007/b97283
6. Quaeghebeur, E.: The CONEstrip algorithm. In: Kruse, R., Berthold, M., Moewes, C., Gil, M., Grzegorzewski, P., Hryniewicz, O. (eds.) Synergies of Soft Computing and Statistics for Intelligent Data Analysis. Advances in Intelligent Systems and Computing, vol. 190, pp. 45–54. Springer, Heidelberg (2013). https://doi.org/10.1007/978-3-642-33042-1_6
7. Seidenfeld, T., Schervish, M J., Kadane, J.B.: Coherent choice functions under uncertainty. Synthese 172(1), 157 (2010)
8. Sen, A.K.: Choice functions and revealed preference. Rev. Econ. Stud. 38(3), 307–317 (1971)
9. Troffaes, M.C.: Decision making under uncertainty using imprecise probabilities. Int. J. Approx. Reason. 45(1), 17–29 (2007)
10. Van Camp, A., Miranda, E., de Cooman, G.: Natural extension of choice functions. In: Medina, J., et al. (eds.) IPMU 2018. CCIS, vol. 854, pp. 201–213. Springer, Cham (2018). https://doi.org/10.1007/978-3-319-91476-3_17

A Formal Learning Theory for Three-Way Clustering

Andrea Campagner[(✉)] and Davide Ciucci[iD]

Università degli Studi di Milano-Bicocca, Milan, Italy
a.campagner@campus.unimib.it, davide.ciucci@unimib.it

Abstract. In this work, we study the theoretical properties, from the perspective of learning theory, of three-way clustering and related formalisms, such as rough clustering or interval-valued clustering. In particular, we generalize to this setting recent axiomatic characterization results that have been discussed for classical hard clustering. After proposing an axiom system for three-way clustering, which we argue is a compatible weakening of the traditional hard clustering one, we provide a constructive proof of an existence theorem, that is, we show an algorithm which satisfies the proposed axioms. We also propose an axiomatic characterization of the three-way k-means algorithm family and draw comparisons between the two approaches.

Keywords: Three-way clustering · Rough clustering · Interval-set clustering · Learning theory

1 Introduction

Clustering, that is the unsupervised task of grouping objects into groups by account of their similarity [30], is a popular and important task in data analysis and related fields. Several clustering approaches have been proposed, such as hierarchical and partitive [24], density-based [16] or subspace-based [27], and have been successfully applied to different domains [9].

Compared to other areas in Machine Learning, however, the study of the formal properties of clustering, from a learning theory perspective [26], have been lacking, convergence or soundness results for specific algorithms aside [25].

In the recent years, starting from the seminal work of Kleinberg [10], there has been an increasing interest toward the study of the learnability of clustering, focusing on formal characterizations based on an *axiomatic* perspective: that is, studying systems of axioms that clustering methods should satisfy and then prove either impossibility theorems or characterization results.

A major limitation of these works consists in the fact that they apply only to *hard clustering* methods, that is methods in which each object is definitely assigned to one and only cluster. Today, however, many *soft clustering* [23] approaches have been developed and shown to be effective in practical applications: probabilistic clustering methods [17], fuzzy clustering [3], possibilistic

J. Davis and K. Tabia (Eds.): SUM 2020, LNAI 12322, pp. 128–140, 2020.
https://doi.org/10.1007/978-3-030-58449-8_9

clustering [12], credal clustering [8], three-way clustering [32] and related formalisms [13,31]. Contrary to traditional hard clustering approaches, soft clustering methods allow clusters to overlap or, either, the relation of containment of objects into cluster to be only partially or imprecisely defined.

In this article, we start to address this gap by extending the available results to soft clustering, in particular we will study a formal characterization of three-way clustering and related approaches (e.g. rough clustering, interval clustering). Specifically, in Sect. 2, we present the necessary background about three-way clustering and the learning theory of clustering; in Sect. 3, we present the main content of the paper by generalizing the learning theory of hard clustering to three-way clustering; finally, in Sect. 4, we discuss the proposed approach and describe some possible future work.

2   Background

2.1   Formal Theory of Clustering

Let X be a set of objects and $d : X \times X \mapsto \mathbb{R}$ be a *distance* function, i.e. a function s.t.:

$$d(x,y) \geq 0 \wedge d(x,y) = 0 \text{ iff } x = y \tag{1}$$

$$d(x,y) = d(y,x) \tag{2}$$

Remark. *We notice that, formally, a distance d should also satisfy the triangle inequality $\forall x, y, z \in X \ d(x,y) \leq d(x,z) + d(z,y)$. Functions not satisfying the triangle inequality are more usually denoted as* semi-distances. *However, since [10], in the literature on formal clustering theory it is customary to not make such a distinction.*

Let $\mathcal{D}_X$ be the collection of all distance functions over X and $\Pi(X)$ be the collection of partitions over X. A partition π is *trivial* if $\pi = antidiscr(X) = \{X\}$ or $\pi = discr(X) = \{\{x\} : x \in X\}$ and denote as $\hat{\Pi}(X)$ the collection of non-trivial partitions.

Definition 1. *A clustering algorithm is a computable function $c : \mathcal{D}_X \mapsto \Pi(X)$.*

Given d, then $c(d) = \pi = \{\pi_1, ..., \pi_n\}$ where each π_i is a *cluster*. We denote the case in which two objects x, y belong to the same cluster π_i as $x \sim_\pi y$ The formal study of clustering algorithms, after Kleinberg [10], starts from the definition of characterization axioms:

Axiom 1 (Scale Invariance). *A clustering algorithm c is scale invariant if, for any $d \in \mathcal{D}_X$ and $\alpha > 0$, $c(d) = c(\alpha \cdot d)$.*

Axiom 2 (Richness). *A clustering algorithm c is rich if $Range(c) = \Pi(X)$: that is, for each $\pi \in \Pi(X)$, $\exists d \in \mathcal{D}_X$ s.t. $c(d) = \pi$.*

Axiom 3 (Consistency). *Let $d, d' \in \mathcal{D}_X$ and $\pi \in \Pi(X)$. Then, d' is a π-transformation of d if*

$$\forall x \sim_\pi y.d'(x,y) \leq d(x,y) \tag{3}$$

$$\forall x \not\sim_\pi y.d'(x,y) \geq d(x,y) \tag{4}$$

A clustering algorithm c is consistent *if, given d s.t. $c(d) = \pi$, for any d' $\pi(X)$-transformation of d it holds that $c(d') = c(d)$.*

The following impossibility theorem represents a seminal result in the formal learning theory of clustering:

Theorem 1 ([10]). *If $|X| \geq 2$ then no clustering algorithm satisfies Axioms 1, 2, 3.*

Corollary 1 ([10]). *For each pair of Axioms among 1, 2, 3 there exists a clustering algorithm that satisfies it.*

Remark. *We note that the axioms, and the proofs of Theorems 1 and Corollary 1, allow one to arbitrarily choose the distance function, irrespective of the nature and topological structures of the instances in X. While this assumption may seem overly general, it is to note that in the definitions the instances of X are completely abstract, and the topological space is entirely determined by the function d. In this respect, letting d vary arbitrarily may be seen as requiring that, no matter the nature of the distance function chosen for the given application, a clustering should respect some properties w.r.t. the chosen distance. This is in analogy with the distribution-independence assumption in the definition of PAC learnability for supervised learning theory [26].*

It is to note that these results can be interpreted similarly to the No Free Lunch theorem for supervised learning: that is, there is no clustering algorithm that (under the requirement of allowing to return every possible clustering) satisfies two intuitively appealing criteria. Indeed, Zadeh et al. [35] have shown that the most problematic constraint is due to the Richness axiom and proposed an alternative formalization based on the concept of k-clustering algorithms (i.e. clustering algorithms which require an additional input $k \in \mathbb{N}^+$) and the *k-Richness* axiom:

Axiom. 2' (k-Richness). *A k-clustering algorithm $c_k : \mathcal{D}_X \mapsto \Pi_k(X)$ is k-rich if $\forall k, Range(c_k) = \Pi_k(X)$, where $\Pi_k(X)$ is the collection of k-partitions on X.*

The authors also showed that, considering k-Richness in place of Richness, provides a consistent set of axioms:

Theorem 2 ([35]). *There exists a k-clustering algorithm that satisfies Axioms 1, 2', 3.*

Lastly, we recall the work of Ben-David et al. [2] on *clustering quality measures* (CQM), i.e. functions $q : \Pi(X) \times \mathcal{D}_X \mapsto \mathbb{R}$, which showed that the following set of axioms represents a consistent formalization of these measures:

Axiom 1_q (Scale Invariance). *A CQM q is scale invariant if $\forall \alpha > 0, \pi \in \Pi(X)$, $q(\pi, d) = q(\pi, \alpha \cdot d)$.*

Axiom 2_q (Richness). *A CQM q is rich if $\forall \pi \in \hat{\Pi}(X)$ exists $d \in \mathcal{D}_X$ s.t. $\pi = argmax_{\pi' \in \hat{\Pi}(X)} \{q(\pi', d)\}$.*

Axiom 3_q (Consistency). *A CQM q is consistent if given $d \in \mathcal{D}_x$, $\pi \in \Pi(x)$, for any $\pi(X)$-transformation d' of d it holds that $q(\pi, d') \geq q(\pi, d)$.*

Theorem 3 ([2]). *There exists a CQM that satisfies Axioms 1_q, 2_q, 3_q.*

We can note that, even though Axioms 1_q through 3_q are defined for *clustering quality measures*, they also implicitly define a *clustering algorithm* by $c(d; q) = argmax_{\pi' \in \hat{\Pi}(X)} \{q(\pi', d)\}$

2.2   Three-Way Clustering and Related Formalisms

An orthopair on a universe X is defined as $O = \langle P, N \rangle$, where $P \cap N = \emptyset$. From P and N a third set, can be defined as $Bnd = X \setminus (P \cup N)$. In the setting of clustering an orthopair can be understood as an uncertain or imprecisely known cluster: the objects in P are those that surely belong to the cluster (P — Positive), those in N are the ones that surely do not belong to the cluster (N = negative), and the objects in Bnd are those that may possibly belong to the cluster (Bnd = Boundary). In the setting of three-way clustering P is also called the *Core* region of the cluster, and Bnd as the *Fringe* region. In the following, we will denote a cluster as the orthopair $O_i = (Core_i, Fringe_i)$.

Different clustering frameworks have been proposed based on the idea of employing orthopairs as a representation of clusters, namely rough clustering [15], interval-set clustering [31], three-way clustering [32] and shadowed set clustering [18]. In these frameworks, a variety of different clustering algorithms have been proposed: rough k-means [15,19,21] and variations based on evolutionary computing [14] or the principle of indifference [22] for the optimal selection of the thresholds that define the *Core* and *Fringe* regions, three-way c-means [28,36], different three-way clustering algorithms that automatically determine the appropriate thresholds or number of clusters such as gravitational search-based three-way clustering [33], variance-based three-way clustering [1], three-way clustering based on mathematical morphology [29] or density-based [34] and hierarchical [5] three-way clustering, and many others.

In the context of this paper we will not consider specific three-way clustering algorithms, as we will be primarily interested in the general formalism behind these clustering frameworks that we now recall. As highlighted previously different frameworks have been proposed, based on similar but different axiom requirements: rough clustering, interval-set clustering, three-way clustering.

A rough clustering is defined as a collection of $\mathcal{O} = \{O_1, ..., O_n\}$ of orthopairs satisfying:

(R1) $\forall i \neq j$, $Core_i \cap Core_j = Core_i \cap Fringe_j = Core_j \cap Fringe_i = \emptyset$

(R2) $\forall x \in X, \nexists i$ s.t. $x \in Core_i \to \exists i \neq j$ s.t. $x \in Fringe_i, Fringe_j$.

On the other hand, both interval-set clustering and three-way clustering are defined as collections $\mathcal{O} = \{O_1, ..., O_n\}$ of orthopairs s.t.:

(T1) $\forall i, Core_i \neq \emptyset$
(T2) $\bigcup_i (Core_i \cup Fringe_i) = X$
(T3) $\forall i \neq j, Core_i \cap Core_j = Core_i \cap Fringe_j = Core_j \cap Fringe_i = \emptyset$

Finally, shadowed set clustering [18] adopts a framework which is instead based on fuzzy clustering, where the degree of membership of an object $x \in X$ to a cluster $C \in \pi$ is given by a membership function $C : X \mapsto [0,1]$. Compared with standard fuzzy clustering, in shadowed set clustering the membership function for each cluster C are then discretized into three regions, which are then equivalent to the three regions in three-way clustering (i.e., $Core_C, Fringe_C$ and $Ext_C = (Core_C \cup Fringe_C)^c)$ [20].

Thus, while the different clustering frameworks are based on the same mathematical representation (i.e., orthopairs), there are some differences: rough clustering allows the core regions to be empty (in this case, the object is required to belong to at least two fringe regions); interval–set (and three–way) clustering require the core regions to be non–empty and allows objects to be in only one fringe region.

Recently, the notion of an *orthopartition* [4] has been proposed as a unified representation for clustering based on orthopairs. Formally, an orthopartition is defined as a collection $\mathcal{O}$ of orthopairs s.t.:

(O1) $\forall i \neq j \; Core_i \cap Core_j = Core_i \cap Fringe_j = Core_j \cap Fringe_i = \emptyset$
(O2) $\bigcup_i (Core_i \cup Fringe_i) = X$
(O3) $\forall x \in U \; (\exists i \text{ s.t. } x \in Fringe_i) \to (\exists j \neq i \text{ s.t. } x \in Fringe_j)$

It can easily be seen that the axioms for orthopartitions more closely follow the ones for rough clustering (Axiom O3 does not hold for three-way clustering). However, in [4], it has been shown that every three-way clustering can easily be transformed in an orthopartition by isolating in an ad-hoc cluster the elements not satisfying (O3). As all the different representations can be transformed into each other, in the following, we will thus refer generally to three-way clustering as a general term for clustering based on orthopairs.

Since a three-way clustering represents an incomplete or uncertain state of knowledge about a clustering (i.e. about which specific clusters do the objects belong), we can also represent a three-way clustering as a collection of *consistent clusterings*, that is given a three-way clustering $\mathcal{O}$:

$$\Sigma(\mathcal{O}) = \{\pi \in \Pi(X) : \pi \text{ is consistent with } \mathcal{O}\} \tag{5}$$

where π is consistent with $\mathcal{O}$ iff $\forall O_i \in \mathcal{O}, \exists \pi_j \in \pi$ s.t. $\pi_j \subseteq Core_i \cup Fringe_i$.

We notice that, in general, a collection of clusterings $\mathcal{C}$ does not necessarily represents the collection of consistent clusterings for any given three-way clustering $\mathcal{O}$. However, it can also be easily seen that each collection of clusterings

$\mathcal{C}$ can be extended to a collection of consistent clusterings (for a given three-way clustering $\mathcal{O}$). Thus, when we refer to a collection of clusterings, we will implicitly refer to its extension that we denote as $tw(\Sigma)$, where Σ is a collection of clusterings. The vacuous three-way clustering is defined as $\mathcal{O}_v$ s.t. $\Sigma(\mathcal{O}_v) = \Pi(X)$. Let $\mathcal{O}$ be a three-way clustering, we denote by $Core(\mathcal{O}) \subset X$ the collection of objects in the core regions of the clusters of $\mathcal{O}$.

If we denote as $\mathbb{O}(X)$ the set of three-way clusterings over X, then a *three-way clustering algorithm* is a computable function $c_{tw} : \mathcal{D}_X \mapsto \mathbb{O}(X)$.

3 Formal Theory of Three-Way Clustering

Our aim is to study the learnability properties and formal characterization of three-way clustering. One aspect that should be considered, in this respect, is the increased flexibility derived from adopting the three-way formalism, which is due not only to the increased model complexity, but also to the fact that it allows to conceive *weakenings* of the axioms proposed for hard clustering, as long as they retain compatibility with the standard case. In particular, since as previously argued, the Richness axiom represents the most problematic constraint, we will study possible weakenings of it which are meaningful in the three-way clustering setting. This, however, should be done with care, e.g.. the following naive consistent weakening of the Richness axiom:

$$\bigcup_{\mathcal{O} \in Range(c_{tw})} \Sigma(\mathcal{O}) = \Pi(X)$$

would clearly be too permissive, as it would admit always returning the *vacuous* three-way clustering as output. Similarly, requiring that $Range(c_{tw}) = \mathbb{O}(X)$ would be too strong a requirement: as a consequence of Theorem 1 it would result in an unsatisfiable constraint.

The following axiom, which is intermediate in strength between Axiom 2 and Axiom 2_q (as shown previously, any clustering quality measure implicitly defines a clustering algorithm), represents a weakening of the Richness axiom which is coherent with the three-way clustering setting:

Axiom 2_{tw} (Almost Richness). *A three-way clustering algorithm c_{tw} is almost rich if*

$$\bigcup_{\mathcal{O} \in Range(c_{tw})} \Sigma(\mathcal{O}) = \Pi(X) \tag{6}$$

and

$$\forall \pi \in \hat{\Pi}(X), \exists d \in \mathcal{D}_X \ s.t. \ c_{tw}(d) = \pi \tag{7}$$

First, we notice that, obviously, when restricted to hard clustering algorithms Almost Richness reduces to Richness.

Proposition 1. *If c is a clustering algorithm, then it satisfies Axiom 2 iff it satisfies Axiom 2_{tw}.*

Proof. For a clustering algorithm c its output $c(d)$ is always a single partition π. Thus, $\Sigma(\{\pi\}) = \pi$ and thus, if c is almost rich it is also rich. Equation (7) becomes redundant in this particular case. The converse (richness implies almost richness) is evident.

Second, we note that in Axiom 2_{tw}, we restrict the range to $\hat{\Pi}(X)$ rather than $\Pi(X)$, in analogy with Axiom 2_q; in this sense, as stated above, the proposed Axiom is intermediate in strength between Axioms 2 and 2_q.

On the other hand, as regards Axioms 1 and 3, we simply require that they hold for each possible three-way clustering $\mathcal{O}$ (thus, we do not weaken these two axioms).

In order to prove that Axiom 2_{tw}, together with Axioms 1 and 3, characterizes three-way clustering, we first introduce the notion of a CQM s.t. the resulting c_{tw}^q is almost rich.

For a pair of clusters π_i, π_j let $s(\pi_i) = \frac{1}{2|\pi_i|} \sum_{x \neq y \in \pi_i} d(x, y)$ be the mean distance of the elements inside cluster π_i and $d(\pi_i, \pi_j) = \frac{1}{|\pi_i||\pi_j|} \sum_{x \in \pi_i} \sum_{y \in \pi_j} d(x, y)$ the mean distance between elements belonging to two different clusters π_i, π_j. Given a partition π and a distance d, we define a CQM $q_{tw} : \Pi(X) \times \mathcal{D}_X \mapsto \mathbb{R}^2$ as

$$q_{tw}(\pi, d) = \langle q_{intra}, q_{inter} \rangle \tag{8}$$

where

$$q_{intra}(\pi, d) = \frac{1}{|\pi|} \sum_{\pi_i \in \pi} s(\pi_i) - min_{x \neq y \in X} d(x, y) \tag{9}$$

$$q_{inter}(\pi, d) = \frac{1}{|\pi|^2} \sum_{\pi_i \neq \pi_j \in \pi} d(\pi_i, \pi_j) - min_{x \neq y \in X} d(x, y) \tag{10}$$

Remark. *We notice that, strictly speaking, the introduced quality measure q_{tw} is not a CQM, as a CQM is defined as a function $q : \Pi(X) \times \mathcal{D}_X \mapsto \mathbb{R}$ while $q_{tw} : \Pi(X) \times \mathcal{D}_X \mapsto \mathbb{R}^2$.*

Definition 2. Given two clustering π^1, π^2 and a distance function, we say that $q_{tw}(\pi^1, d) < q_{tw}(\pi^2, d)$ if both:

$$q_{intra}(\pi^1, d) \geq q_{intra}(\pi^2, d) \tag{11}$$

$$q_{inter}(\pi^1, d) \leq q_{inter}(\pi^2, d) \tag{12}$$

and at least one of the two is strict. Then, we say that $\pi^1 <_q \pi^2$ if $q_{tw}(\pi^1, d) < q_{tw}(\pi^2, d)$.

The idea is that if $\pi^1 <_q \pi^2$ then instances in π_1 have greater intra-cluster distance and smaller inter-cluster distance.

The following result shows that, indeed, the three Axioms provide a characterization for three-way clustering.

Algorithm 1. Three-way Clustering based on q_{tw}

Require: d distance function
  $\Sigma = \emptyset$
  for $\pi \in \Pi(X)$ **do**
    $check := \top$
    for $\pi' \neq \pi \in \Pi(X)$ **do**
      if $\pi' >_q \pi$ **then**
        $check := \bot$
        break
      end if
    end for
    if $check$ **then**
      $\Sigma.append(\pi)$
    end if
  end for
  Return $tw(\Sigma)$

Theorem 4. *There exists a three-way clustering algorithm that satisfies Axioms 1, 2_{tw} and 3.*

Proof. The Theorem can be proven based on the previously defined CQM q_{tw}.

Indeed, from q_{tw}, we can define a three-way clustering algorithm c_{tw}^q as shown in Algorithm 1. It can be verified that, $\forall d \in \mathcal{D}_X$, $c_{tw}^q(d) = \{\pi \in \Pi(X) : \nexists \pi' \in \Pi(X), \pi' >_q \pi\}$. It is easily shown that c_{tw}^q satisfies Scale-Invariance and Consistency (when restricted to pairs of objects x, y in the core regions of $tw(\Sigma)$, where Σ is the result of c_{tw}^q). We thus show only the proof for Almost Richness.

Let π be a given non-trivial clustering and let d be the distance function defined $\forall x, y$ as

$$d(x,y) = \begin{cases} \epsilon & \text{if } x \sim_\pi y \\ \alpha & \text{if } x \not\sim_\pi y \end{cases}$$

with $\alpha \gg \epsilon$. Then, $q_{tw}(\pi, d) = \langle 0, \alpha - \epsilon \rangle$ and evidently, for any other π', $q_{tw}(\pi', d) < q_{tw}(\pi, d)$ (hence, $\pi' <_q \pi$). This satisfies the second condition of the Axiom.

As regards the first condition, let d be s.t. $\forall x, y$, $d(x, y) = \epsilon$. Then, for any $\pi \in \Pi(X)$ $q_{tw}(\pi, d) = \langle 0, 0 \rangle$. The condition, and hence the result, follows. $\square$

The proof of Theorem 4 is constructive and directly provides a three-way clustering algorithm satisfying Axioms 1, 2_{tw}, 3. Further, evidently $\hat{\Pi}(X) \subset Range(c_{tw}^q) \subset \mathbb{O}(X)$ but future work should study how to provide a more precise specification of the range of c_{tw}^q.

As a limitation of this result, it is easy to observe that an exact implementation of this algorithm is not practical from a time complexity perspective: indeed, as the algorithm requires to compute the value of q for all possible clusterings its complexity is evidently exponential in $|X|$. A possible solution to this problem would be to define heuristic or randomized algorithms for implementing approximation of the c^q_{tw} three-way clustering algorithm (possibly, with proven quality bounds).

A different approach, instead, consists in studying other, more efficient three-way clustering algorithms and providing their axiomatic characterization, in order to understand their flexibility when compared with c^q_{tw}. We will provide such a characterization for the three-way k-means algorithm [28, 36] (and related ones, such as rough k-means [15, 21]). In order to prove this result, we will focus on a simplified single-step version of the algorithm, as defined in Algorithm 2.

Algorithm 2. Single-Step Three-way K-Means

Require: d distance function
Require: $x_1, ..., x_k \in X$ cluster centroids
Require: $\delta, \Delta \in [0, 1]$ parameters
    Let $\hat{d}(x, y) := \frac{d(x, y)}{max_{a, b \in X} d(a, b)}$
    for $x \in X$ **do**
        $I := \{i : \hat{d}(x, x_i) \leq \Delta\}$
        for $i \in I$ **do**
            if $(|I| = 1) \wedge (\hat{d}(x, x_i) \leq \delta)$ **then**
                $x \in Core_i$
            else
                $x \in Fringe_i$
            end if
        end for
    end for

Evidently, Algorithm 2 is Scale-Invariant (as it only uses the normalized distance) but it can easily be shown that is neither Consistent, nor Almost Rich, but it is k-Rich.

Example 1. For Consistency, consider a distance function d, let $a, b \in X$ s.t. $d(a, b) = max_{x, y \in X} d(x, y)$ and suppose that the result of Algorithm 2, denoted as $\mathcal{O}$, assigns a, b to two different clusters. Further, let d' s.t. $\forall x, y \in X \setminus \{a, b\} d'(x, y) = d(x, y)$, while $d(a, b) \ll d'(a, b)$. Then, evidently, d' is a $\mathcal{O}$-transformation of d, but Consistency is violated. For Almost Richness, it easily follows from the fact the result of Algorithm 2 contains exactly k clusters.

We can characterize this Algorithm (and similar algorithms such as three-way k-means [28] or rough k-means [13]) via the following two Axioms (together with Scale-Invariance):

Axiom 2_{twk} (Three-way k-Richness). *A three-way k-clustering algorithm* $c^k_{tw} : \mathcal{D}_X \mapsto \mathbb{O}_k(X)$ *is k-rich if* $\forall k, Range(c^k_{tw}) = \mathbb{O}_k(X)$ *where* $\mathbb{O}_k(X) = \{\mathcal{O} \in \mathbb{O}(X) : |\mathcal{O}| = k\}$.

That is, a three-way clustering algorithm is k-rich if as possible outputs (by changing the distance function) we can obtain all the orthopartions made exactly of k orthopairs. This requirement is a natural generalization of k-Richness to the setting of three-way clustering.

Axiom 3_{twk} ((δ, Δ)-Consistency). *Let $d, d' \in \mathcal{D}_X$ and $\delta < \Delta \in [0, 1]$. Then, d' is a (δ, Δ)-transformation of d if:*

$$sign(\frac{d(x, y)}{max_{a,b} d(a, b)} - \delta) = sign(\frac{d'(x, y)}{max_{a,b} d'(a, b)} - \delta) \tag{13}$$

$$sign(\frac{d(x, y)}{max_{a,b} d(a, b)} - \Delta) = sign(\frac{d'(x, y)}{max_{a,b} d'(a, b)} - \Delta) \tag{14}$$

A three-way clustering algorithm c_{tw} is (δ, Δ)-consistent if, given d s.t. $c_{tw}(d) = \mathcal{O}$, for any d' (δ, Δ)-transformation of d it holds that $c_{tw}(d') = \mathcal{O}$.

So, (δ, Δ)-consistency means that small changes in the distance function do not alter the clustering result. The notion of (δ, Δ)-consistency can be seen as a restricted form of Consistency, determined by two thresholds that are used to describe three different regions (a natural requirement in the setting of three-way clustering): the objects whose normalized distance is lower than δ; those for which the normalized distance is between Δ and δ; and those for which the normalized distance is greater than Δ).

Theorem 5. *Algorithm 2 satisfies Axioms 1, 2_{twk}, 3_{twk}.*

Proof. Evidently, Algorithm 2 is Scale-Invariant. Further, by construction, it is also (δ, Δ)-consistent w.r.t. its input parameters δ, Δ. Thus, we only need to show that it is Three-way k-Rich.

Let $\mathcal{O} \in \mathbb{O}_k(X)$ be the target three-way clustering and $\delta < \Delta$ the input parameters. For each cluster $O_i \in \mathcal{O}$ select one element $x_i \in O_i$. Then, for each x, if $x \in Core_i$ set $d(x, x_i) < \delta$ and if $x \in Fringe_i$ set $\delta < d(x, x_i) < \Delta$. For any two x, y s.t. they belong to different clusters set $d(x, y) = 1$. Then the output of Algorithm 2 in this case is exactly $\mathcal{O}$.

We can thus compare the two algorithms, c_{tw}^q and Three-way K-Means, through their characterization. Indeed, we can observe that the two algorithms can be seen as offering a trade-off between representational flexibility and computational efficiency. Indeed, c_{tw}^q is more flexible (as it does not require to set, a-priori, the number of clusters k) and satisfies a stricter notion of consistency. Thus, its result is more well-behaved w.r.t. coherent modifications to the distance function. However, it has exponential complexity in the size of X, as it requires an enumeration of all $\pi \in \Pi(X)$. On the other hand, having fixed both k, the number of clusters, and the cluster centroids (i.e., $x_1, ..., x_k$ in Algorithm 2), the complexity of Three-way K-Means is linear in $|X|$. However, the Three-way K-Means family of algorithms satisfies a weaker form of consistency (i.e., (δ, Δ)-consistency) and, further, requires to set both the number of clusters (which in practice is selected heuristically using criteria such as the Silhouette or cross-validation [11]) and the cluster centroids: this usually involves an iterative approach which, however, only guarantees convergence to a local optimum (as, even for traditional k-Means, the problem of finding the optimal k-clustering is NP-hard [6]).

4 Conclusion

In this article, we set out the foundations for the study of the theoretical properties of three-way clustering and related formalisms, from the perspective of computational learning theory. We provided an axiomatic characterization of three-way clustering and proved that, contrary to the case of traditional clustering, these requirements are consistent, i.e., there exists a three-way clustering algorithm satisfying them, which however has exponential time complexity. We then studied an axiomatic characterization of the popular Three-way k-Means family of clustering algorithm, showing that it provides a trade-off, favoring better time complexity against reduced flexibility. Our results represent a first step towards a formal study of three-way clustering and, as such, we think that the following open problems may be important to understand the formal properties of this increasingly popular clustering framework:

- What is the exact characterization of $Range(c_{tw}^q)$? As we previously argued, it can easily be shown that $\hat{\Pi}(X) \subsetneq Range(c_{tw}^q) \subsetneq \mathbb{O}(X)$, but it is not clear which proper three-way clusterings can be represented by c_{tw}^q;
- Is there a three-way clustering algorithm satisfying the following generalized Almost Richness axiom (together with Consistency and Scale-Invariance):

$$\forall \mathcal{O} \in \hat{\mathbb{O}}(X) \exists d \in \mathcal{D}_X \text{ s.t } c_{tw}(d) = \mathcal{O} \qquad (15)$$

  where $\hat{\mathbb{O}}(X)$ is the set of non-trivial three-way clusterings? Otherwise, what is the greatest subset of $\mathbb{O}(X)$ which admits a consistent and scale-invariant three-way clustering algorithm?
- While the time complexity of c_{tw}^q is exponential in $|X|$, can we find an approximation or randomization scheme for c_{tw}^q with provable error bounds?
- What is the learning-theoretic axiomatic characterization of other three-way clustering algorithms, such as three-way density-based clustering [34] or rough-set hierarchical clustering [5]?

More generally, and observing that rough k-means can be seen as a particular case of both evidential clustering [7] and possibilistic clustering [12], we can think to extend the learning-theoretic axiomatic characterization to these other soft clustering approaches.

References

1. Afridi, M.K., Azam, N., Yao, J.: Variance based three-way clustering approaches for handling overlapping clustering. Int. J. Approximate Reasoning **118**, 47–63 (2020)
2. Ben-David, S., Ackerman, M.: Measures of clustering quality: A working set of axioms for clustering. Proc. NIPS **2009**, 121–128 (2009)
3. Bezdek, J.C., Ehrlich, R., Full, W.: Fcm: The fuzzy c-means clustering algorithm. Comput. Geosci. **10**(2), 191–203 (1984)

4. Campagner, A., Ciucci, D.: Orthopartitions and soft clustering: Soft mutual information measures for clustering validation. Knowl. Based Syst. **180**, 51–61 (2019)
5. Chen, D., Cui, D.W., Wang, C.X., Wang, Z.R.: A rough set-based hierarchical clustering algorithm for categorical data. Int. J. Inf. Technol. **12**(3), 149–159 (2006)
6. Dasgupta, S.: The hardness of k-means clustering. Department of Computer Science and Engineering, University of California, San Diego, Technical report (2008)
7. Denœux, T., Kanjanatarakul, O.: Beyond fuzzy, possibilistic and rough: An investigation of belief functions in clustering. In: Ferraro, M.B., Giordani, P., Vantaggi, B., Gagolewski, M., Gil, M.Á., Grzegorzewski, P., Hryniewicz, O. (eds.) Soft Methods for Data Science. AISC, vol. 456, pp. 157–164. Springer, Cham (2017). https://doi.org/10.1007/978-3-319-42972-4_20
8. Denœux, T., Masson, M.H.: Evclus: Evidential clustering of proximity data. IEEE Trans. Syst. Man Cybern. **34**(1), 95–109 (2004)
9. Kameshwaran, K., Malarvizhi, K.: Survey on clustering techniques in data mining. IJCSIT **5**(2), 2272–2276 (2014)
10. Kleinberg, J.M.: An impossibility theorem for clustering. Pro. NIPS **2003**, 463–470 (2003)
11. Kodinariya, T.M., Makwana, P.R.: Review on determining number of cluster in k-means clustering. Int. J. Adv. Res. Comput. Sci. Manag. Stud. **1**(6), 90–95 (2013)
12. Krishnapuram, R., Keller, J.M.: A possibilistic approach to clustering. IEEE Trans. Syst. **1**(2), 98–110 (1993)
13. Lingras, P., Peters, G.: Rough clustering. WIREs Data Min. Knowl. Discov. **1**, 65–72 (2011)
14. Lingras, P.: Evolutionary rough K-means clustering. In: Wen, P., Li, Y., Polkowski, L., Yao, Y., Tsumoto, S., Wang, G. (eds.) RSKT 2009. LNCS (LNAI), vol. 5589, pp. 68–75. Springer, Heidelberg (2009). https://doi.org/10.1007/978-3-642-02962-2_9
15. Lingras, P., West, C.: Interval set clustering of web users with rough k-means. J. Intell. Inf. Syst. **23**(1), 5–16 (2004)
16. Loh, W.K., Park, Y.H.: A survey on density-based clustering algorithms. In: Ubiquitous Information Technologies and Applications, pp. 775–780. Springer, Berlin (2014)
17. MacKay, D.J.C.: Information Theory, Inference and Learning Algorithms. Cambridge University Press, New York (2002)
18. Mitra, S., Pedrycz, W., Barman, B.: Shadowed c-means: Integrating fuzzy and rough clustering. Pattern Recogn. **43**(4), 1282–1291 (2010)
19. Murugesan, V.P., Murugesan, P.: A new initialization and performance measure for the rough k-means clustering. Soft Comput. 1–15 (2020)
20. Pedrycz, W.: Interpretation of clusters in the framework of shadowed sets. Pattern Recogn. Lett. **26**(15), 2439–2449 (2005)
21. Peters, G.: Some refinements of rough k-means clustering. Pattern Recogn. **39**(8), 1481–1491 (2006)
22. Peters, G.: Rough clustering utilizing the principle of indifference. Inf. Sci. **277**, 358–374 (2014)
23. Peters, G., Crespo, F., Lingras, P., Weber, R.: Soft clustering-fuzzy and rough approaches and their extensions and derivatives. Int. J. Approximate Reason. **54**(2), 307–322 (2013)
24. Reddy, C.K., Vinzamuri, B.: A survey of partitional and hierarchical clustering algorithms. In: Data Clustering, pp. 87–110. Chapman and Hall/CRC (2018)

25. Selim, S.Z., Ismail, M.A.: K-means-type algorithms: A generalized convergence theorem and characterization of local optimality. IEEE Trans. Pattern Anal. Mach. Intell. **1**, 81–87 (1984)
26. Shalev-Shwartz, S., Ben-David, S.: Understanding Machine Learning: From Theory to Algorithms. Cambridge University Press, Cambridge (2014)
27. Vijendra, S.: Efficient clustering for high dimensional data: Subspace based clustering and density based clustering. Inf. Technol. J. **10**(6), 1092–1105 (2011)
28. Wang, P., Shi, H., Yang, X., Mi, J.: Three-way k-means: Integrating k-means and three-way decision. Int. J. Mach. Learn. Cybern. **10**(10), 2767–2777 (2019). https://doi.org/10.1007/s13042-018-0901-y
29. Wang, P., Yao, Y.: Ce3: A three-way clustering method based on mathematical morphology. Knowl. Based Syst. **155**, 54–65 (2018)
30. Xu, R., Wunsch, D.: Clustering, vol. 10. Wiley, Hoboken (2008)
31. Yao, Y., Lingras, P., Wang, R., Miao, D.: Interval set cluster analysis: a reformulation. In: Sakai, H., Chakraborty, M.K., Hassanien, A.E., Ślęzak, D., Zhu, W. (eds.) RSFDGrC 2009. LNCS (LNAI), vol. 5908, pp. 398–405. Springer, Heidelberg (2009). https://doi.org/10.1007/978-3-642-10646-0_48
32. Yu, H.: A framework of three-way cluster analysis. In: Polkowski, L., Yao, Y., Artiemjew, P., Ciucci, D., Liu, D., Ślęzak, D., Zielosko, B. (eds.) IJCRS 2017. LNCS (LNAI), vol. 10314, pp. 300–312. Springer, Cham (2017). https://doi.org/10.1007/978-3-319-60840-2_22
33. Yu, H., Chang, Z., Wang, G., Chen, X.: An efficient three-way clustering algorithm based on gravitational search. Int. J. Mach. Learn. Cybern. **11**(5), 1003–1016 (2019). https://doi.org/10.1007/s13042-019-00988-5
34. Yu, H., Chen, L., Yao, J., Wang, X.: A three-way clustering method based on an improved dbscan algorithm. Physica A Stat. Mech. Appl. **535**, 122289 (2019)
35. Zadeh, R.B., Ben-David, S.: A uniqueness theorem for clustering. In: Proceedings of the Twenty-Fifth Conference on Uncertainty in Artificial Intelligence, pp. 639–646 (2009)
36. Zhang, K.: A three-way c-means algorithm. Appl. Soft Comput. **82**, 105536 (2019)

Belief Functions for Safety Arguments Confidence Estimation: A Comparative Study

Yassir Idmessaoud[1](✉), Didier Dubois[2], and Jérémie Guiochet[1]

[1] LAAS-CNRS, University of Toulouse, Toulouse, France
{yassir.id-messaoud,jeremie.guiochet}@laas.fr
[2] IRIT, University of Toulouse, Toulouse, France
didier.dubois@irit.fr

Abstract. Structured safety arguments are widely applied in critical systems to demonstrate their safety and other attributes. Graphical formalisms such as Goal Structuring Notation (GSN) are used to represent these argument structures. However, they do not take into account the uncertainty that may exist in parts of these arguments. To address this issue, several frameworks for confidence assessment have been proposed. In this paper, a comparative study is carried out on three approaches based on Dempster-Shafer theory. We extract and compare the implicit logic at work in these works, and show that, to some extent, these current approaches fail to provide a consistent relationship between the informal statement of arguments, their logical model and the use of belief functions. We also propose recommendations to improve this consistency.

Keywords: Confidence assessment · Goal Structuring Notation (GSN) · Dempster-Shafer theory (DST) · Evidence fusion · Safety cases

1 Introduction

The deployment of autonomous systems in the highly uncertain human environment raises the issue of safety. Argument structures are widely used to evaluate and prove the safety of these systems. They are a clearly represented collection of rational pieces of evidence like test or simulation results, expert judgments, analysis reports, etc. They aim to demonstrate that a certain property of the system is satisfied. Many studies and standards define safety arguments as "Safety cases" (e.g, in the automotive [17] or railway [12] industries), but it is now extended to more general domains like dependability, assurance or trust cases. These cases are presented in the form of texts, tables or, more interestingly, graphically. Using graphical tools to represent arguments structures is more relevant because graphs are simpler to review, offer a clear overlook, help to understand the connection between pieces of evidence, and moreover they are easier to use and manage. Formalisms such as Goal Structuring Notation (GSN) [19] and Claims-Arguments-Evidence (CAE) [2] are commonly used in this field.

© Springer Nature Switzerland AG 2020
J. Davis and K. Tabia (Eds.): SUM 2020, LNAI 12322, pp. 141–155, 2020.
https://doi.org/10.1007/978-3-030-58449-8_10

However, even with all these benefits, these tools do not take into account the uncertainties pervading this sort of arguments. Especially since autonomous systems, and critical systems in general, are becoming much more complex, they are affected by many sources of uncertainty like any decision support system, AI based system and the like. As a response to this issue many research projects are conducted to find a solution.

Several works have proposed methods based on Bayesian Networks (BN) to model uncertainty in a safety argument [6,14,15]. Nevertheless, these approaches have a major impediment, which is the need for *data*. As a matter of fact, using BN requires statistical information that is often not available. Moreover, the use of subjective probabilities is questionable in the presence of partial ignorance. Dempster-Shafer theory (DST) (aka Theory of Evidence) was developed to address the issue of imprecise evidence [21]. It represents a form of generalized probability theory where probability masses are assigned to sets of possible values, instead of singletons. In some works on safety argumentation, aggregation rules stemming from DST are used to merge confidence degrees in pieces of evidence (represented by mass functions) and calculate an overall mass function, in order to estimate an overall confidence in the top statement of a safety argument (e.g., "the system is acceptably safe"). In our problem, the connection between pieces of evidence is represented by various types of arguments. In the literature, they strongly influence the choice of an aggregation rule. However, no real consensus emerges in current research works to relate argument types, their logical modeling, and aggregation rules based on DST.

In this paper, three approaches to uncertain safety cases using DST are compared as to the different definitions of types of arguments they propose and we review and discuss the aggregation methods they use. Section 2 presents the background on safety cases and introduces the existing selected approaches. Section 3 extracts the formal definitions of arguments from the selected articles [1,4,24]. Then, we compare and analyze the aggregation rules used to compute belief degrees of the top statement of an argument. Section 4 suggests the existence of two basic types of arguments and proposes a rigorous methodology.

2   Baseline and Related Work

This section introduces a safety case formalism (GSN), some works on confidence quantification, and basic concepts of DST.

2.1   Background

Safety arguments or safety cases can be defined in multiple ways. In fact, the definition may vary slightly according to the field where it is used. For instance, in the automotive industry [17], it is defined as : *argument that functional safety is achieved for items, or elements, and satisfied by evidence compiled from work products of activities during development.* This concept has been generalized

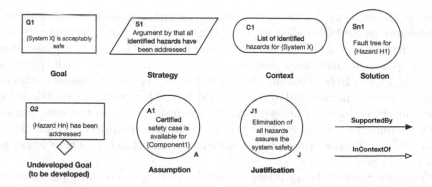

Fig. 1. GSN main components.

in the OMG (Object Management Group) standardized Assurance Case Meta-model [5] and an instance of it is the goal structuring notation (GSN), which is commonly used to represent safety cases [19]. As presented in Fig. 1, it includes nine main elements. It breaks down the conclusion called a *Goal* (following a given *Context* and in accordance with a specific *Strategy*) into *Sub-goals* and supports each of them by evidence items called *Solutions*. The choice of strategies and sub-goals is supported by the use of so-called *Justifications*. Figure 2 presents an example of a GSN pattern.

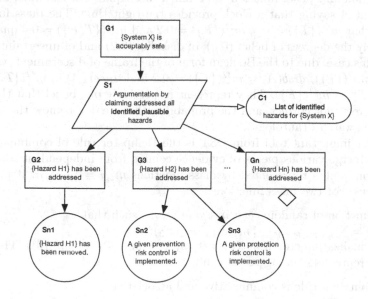

Fig. 2. GSN example adapted from Hazard Avoidance Pattern [20].

GSN is categorised as a qualitative method to justify safety. However, in this example, many uncertainties may exist. For instance, what is the uncertainty linked to the element "C1: List of identified hazards" or what is the confidence in the solutions Sn (also called pieces of evidence). In order to estimate the confidence in the top goal $G1$, all these uncertainties should be assessed. Some quantitative approaches have been proposed to assess uncertainties in such arguments [13]. In [6,14,15], Belief Bayesian Networks (BBN) are used to assess confidence in safety case structure and pieces of evidence. They measure confidence by computing probabilities from evidence to conclusion. Due to the huge amount of data required to apply BBN, other works based on subjective logic [18,29] or on DST are proposed to define and estimate uncertainties. These works are presented in Sect. 2.2.

Dempster-Shafer Theory offers a powerful setting to combine pieces of evidence. A mass function, or basic belief assignment (BBA), assigns probabilities over the power set of the universe of possibilities (Ω), known as the *frame of discernment*. Formally, a mass function $m^\Omega : 2^\Omega \rightarrow [0, 1]$ is such that $\sum_{E \subseteq \Omega} m(E) = 1$, and $m(\emptyset) = 0$. Any subset E of Ω such as $m(E) > 0$ is called a focal set of m. Mass assignment induces the concept of belief function ($bel : 2^\Omega \rightarrow [0, 1]$). It represents the summation of all the masses supporting the same statement and is defined by : $bel(A) = \sum_{E \subseteq \Omega, E \neq \emptyset} m(E)$. Belief in the denial or uncertainty of the statement A are respectively represented by : $disb(A) = bel(\neg A)$ and $uncer(A) = 1 - bel(A) - disb(A)$.

In our case, we use propositional variables for which the frame of discernment has two states: $\Omega = \{True(T), False(F)\}$. As a consequence, in such frames, mass function and belief function for T and F are equal. For example, consider a statement A saying that a clock provides the right time. The mass function m^Ω such that $m^\Omega(\{T\}) = 0.5$, $m^\Omega(\{F\}) = 0.2$ and $m^\Omega(\{T, F\}) = 0.3$ quantifies respectively the degrees of belief (0.5), of disbelief (0.2) and of uncertainty (0.3) in A. In this case, due to the Boolean form of the frame of discernment, we have $bel(A) = m^\Omega(\{T\})$, $disb(A) = m^\Omega(\{F\}) = 0.2$ and $uncer(A) = m^\Omega(\{T, F\}) = 1 - m^\Omega(\{T\}) - m^\Omega(\{F\})$. They represent respectively our belief that the time read is correct, not correct, and the probability that we don't know the time by reading the watch (Tautology).

Another important tool from DST is the Dempster rule of combination. It is used to merge various pieces of evidence coming from independent sources of information, and is represented by mass functions $m_i, i = 1, \ldots, n$. It proceeds in two steps. For two mass functions:

1. a conjunction of random sets: $m_\cap = m_1 \otimes m_2$ such that
$m_\cap(A) = \sum_{E_1, E_2 : E_1 \cap E_2 = A} m_1(E_1) \cdot m_2(E_2)$;
2. a renormalization step if $m_\cap(\emptyset) > 0$: $m(A) = m_\cap(A)/(1 - m_\cap(\emptyset))$. The value $m_\cap(\emptyset)$ represents the degree of conflict between m_1 and m_2.

This combination rule is commutative and associative.

2.2   Some DST-Based Approaches to Safety Cases

In this subsection, we discuss three uncertainty management methods in safety cases proposed in the literature. All such methods are DST-based. Two of them use Goal Structuring Notation (GSN) for structuring arguments. Our objective is to extract, in each paper, definitions of argument types, and to evaluate their consistency with the proposed aggregation rule that computes degrees of belief of conclusions.

Cyra and Gorski [4] present an argument model called VAA inspired by Toulmin [23] to graphically represent all pieces of evidence that support a conclusion (e.g., "The system is safe"). It proposes a method that, in a first step, transforms qualitative expert opinions expressed in natural language, about pieces of evidence (forming premises), into belief and plausibility functions, using [18]. In a second step, the authors define five argument patterns and associate to each of them an appropriate belief aggregation rule. These rules use as inputs the values obtained at the first step, to calculate the overall confidence in the conclusion.

Anaheed et al. [1] define four basic types of arguments. Each argument type is composed of at least two premises that support a conclusion. Premises are assessed by two parameters (sufficiency and insufficiency) and every argument type is associated to an aggregation rule. In this paper, they propose an algorithm based on a bottom-up approach that computes the degree of confidence in each premise to calculate the overall confidence in the system.

Wang et al. [24–27] propose a confidence assessment method by converting qualitative expert opinions, on their confidence in pieces of evidence appearing in a GSN, into mass functions. These values are then merged by Dempster rule of combination to obtain the overall confidence in the studied system. The paper also defines two parameters to assess confidence: *Trustworthiness* to quantify confidence in the evidence; *Appropriateness* to quantify the confidence in the claim that the evidence supports the conclusion.

3   Comparative Study

In this section, we introduce a framework for confidence assessment. Then, we compare argument types given in the studied papers and the propagation rules used to calculate the overall confidence.

3.1   General Framework for BF-Based Confidence Estimation

In order to estimate the overall confidence of the argument structure, our main issue is how to propagate the quantitative values coming from the confidence in premises in accordance with the characteristics of its structure. In this regard, it is important to propose a general method to compute degrees of belief in conclusions of safety cases. Most works omit to provide this general method putting together logic and belief functions. Such a methodology was described more than 30 years ago [3] and is recalled here.

The first step is to define the nature of the relationship between premises in their support of the conclusion, known as *argument types*. These types should be firstly expressed informally in a natural language (e.g, if premises P_1 and P_2 are true, then the conclusion (C) is true) because it is more understandable for the human expert. Then this verbal relation should be transformed into a formal logical sentence (e.g., $P_1 \wedge P_2 \Rightarrow C$). The importance of these definitions lies in the fact that they significantly affect the logical expression of the links between pieces of evidence and the conclusion. The second step is mass assignment. This task consists in defining masses assigned to the focal sets deduced from the logical expressions obtained from argument definitions.

Consider a set of well-formed formulas $K = \{\phi_1, \dots \phi_n\}$ in propositional logic, and a formula C such that $K \vdash C$. Assume each formula ϕ_i is a piece of evidence that comes from a specific source independent of other ones. Uncertainty about the validity of each formula can be represented by a mass function m_i assigning some probabilities to $\phi_i, \neg\phi_i$ and the tautology $\top$. Take for example the case of a simple premise P and a rule in the form of an equivalence $P \equiv C$. One mass function will be assigned to the premise P in the form of three values $m_1(P)$, $m_1(\neg P)$ and $m_1(\top)$ summing to 1, and another will be assigned to the rule $(m_2(P \equiv C) + m_2(\top) = 1)$.

The third step is to choose the appropriate aggregation rule. This rule will be used to calculate the belief in the top goal (conclusion) based on beliefs about premises and rules. Extending classical logic inference to this uncertain environment can be done by means of Dempster rule of combination [3], first computing an overall mass function, $m = m_1 \otimes \cdots \otimes m_n$ and then computing the degree of belief in the conclusion C as $Bel(C) = \sum_{\phi_i \vdash C} m(\phi_i)$. There are also several variants of this combination rule that could be used in evidence fusion. However, each method obeys certain assumptions and describes some kind of situation. That is why it is needed to make sure that every definition resulting from the first step verifies the assumptions and each fusion rule fits with the given situation. Here, pieces of evidence and rules are supposed to come from independent sources. If this assumption is not satisfied, idempotent combination rules can be used as discussed in [7,8].

The complete process includes an additional preliminary step, which consists in transforming expert opinions (qualitative values) expressed in natural language (*safe, little safe, uncertain*, etc.) into a numerical format that can be computed with (i.e. mass, belief or plausibility functions). This could be done in [4] using the triangle method of Josang [18]. This is needed to compute the belief in the conclusion. The choice of this transformation has a profound impact on the results. However this aspect of the evaluation process will not be addressed in this paper.

3.2   Definition of Argument Types

The concept of *argument type* pertains to the logical relationship between the premises and the conclusion. In other words, it answers the question : In which format do the premises support the conclusion ? The terminology is not uniform. For instance, in [4] this relation is named a *warrant*, in [29] it is called an *affection factor* and in [24] it is named *appropriateness*. Moreover, most papers only give an informal definition.

Table 1. Formal definitions of arguments. Note that argument types 4 and 5 are logically equivalent $(\wedge_{i=1}^{n}[p_i \Rightarrow C] \equiv [\vee_{i=1}^{n}p_i] \Rightarrow C)$.

	Formal definition	Terminology for argument types
Type 1	$(\wedge_{i=1}^{n}p_i) \equiv C$	NSC-Arg [4], Consensus [29], Full complementary [24]
Type 2	$\wedge_{i=1}^{n}(p_i \equiv C)$	Disparate [24]
Type 3	$(\wedge_{i=1}^{n}p_i) \Rightarrow C$	SC-Arg, C-Arg [4], Conjunctive argument [29]
Type 4	$\wedge_{i=1}^{n}(p_i \Rightarrow C)$	A-Arg [4], Alternative argument [1]
Type 5	$(\vee_{i=1}^{n}p_i) \Rightarrow C$	Disjunctive argument [29], Full redundant [24]
Type 6	$\begin{cases} \wedge_{i=1}^{n}(p_i \equiv C) \\ (\wedge_{i=1}^{n}p_i) \equiv C \end{cases}$	AC-Arg [4], Complementary [24], Containment, Overlapping arguments [1]
Type 7	$\begin{cases} \wedge_{i=1}^{n}(p_i \equiv C) \\ (\vee_{i=1}^{n}p_i) \equiv C \end{cases}$	R-Arg [24]

Table 1 presents formal definitions of argument types that we infer from the reviewed papers. We notice from the formal definitions given in Table 1 that the premises are related to the conclusion by either an equivalence or an implication connective. The choice could be justified by the intuitive perception of the relation between premises and conclusion (e.g. "The system is safe" is supported by "Hazard H1 has been addressed").

As already noticed, there are two types of arguments, one using implication, the other using equivalence. Using equivalence assumes that there is a symmetry between the conclusion and premises. Consider a small safety case where the statement "The system is safe" is supported by the premise "All tests are conclusive". Using equivalence means, on the one hand, that the system is safe because we are confident in our tests; on the other hand, that the actual safety of the system can only be ensured by the success of the tests, which is not necessary true. In addition to this, the use of equivalence generates cases for the denial of the conclusion (disbelief) which appears in the calculation as a conjunction of one or several premises with the negation of the conclusion ($\neg C$). For instance, consider the case of a conclusion (C) supported by a single premise (p) with an equivalence relation between them (Type 1 or 2). Combining the masses of (p) and the rule ($p \equiv C$) with DS combination rule reveals two cases where the conclusion is not satisfied ($\neg p \wedge \neg C$) and ($p \wedge \neg C$). Using implication can only indicates that, due to the tests, the system is safe; it cannot prove that it is faulty. Choosing between equivalence or implication could also be justified by the purpose of the safety case. Generally speaking, a safety case is used to demonstrate that a system is acceptably safe. Its purpose is to provide a structured argument in order to certify a critical system, and not to present statements that it could be faulty, i.e. that there is disbelief about safety greater than 0. This is actually guaranteed when using only implication. In contrast, if the goal is to use the safety case at the debugging phase, i.e. to consider that disbelief in the conclusion may be not null, the equivalence may be an appropriate choice between premises and conclusion. Since we are interested, in this study, in the certification aspect of safety cases, the remainder of the paper will be focusing on argument types modelled by implication from tests to a statement of safety.

We also notice from Table 1 that the premises are linked with each other by AND, OR logic gates or by a combination of the two. It depends on whether, for instance, tests justifying a conclusion are alternative or complementary. For example, type 3 represents the situation where the conjunction of all premises is needed to support the conclusion. In the contrary, type 4 represents the case of separate rules where each premise alone can support the whole conclusion.

3.3   Mass Assignment

As seen in previous sections, masses are allocated to propositions of interest. Apart from the assignment of masses to logical expressions resulting from the definition types arguments (called *appropriateness* in [24]), masses are also assigned to premises to assess their degree of confidence. This evaluation is used under the name *trustworthiness* in [24] and *affection factor* in [29]. Normally, mass functions assigned to premises (P_i) have two parameters : belief (i.e. $m_p(P_i)$), disbelief (i.e. $m_p(\neg P_i)$) and the remainder is their uncertainty (i.e $m_p(\top) = 1 - m_p(P_i) - m_p(\neg P_i)$). In cases when the argument is an implication, not an equivalence, the disbelief in the premises will not affect the conclusion, and need not be taken into account in the uncertainty propagation. This remark reduces the number of useful focal sets and simplifies the calculation.

The choice of mass functions is a very important step in the assessment process. It has a huge impact on the form of the final result. We can either define several mass functions, one for each logical expression, to emphasize the fact that there are multiple independent sources of information. Or, one mass function is distributed over all the logical expressions to represent the situation where a single source supplies these pieces of information.

3.4   Belief Aggregation

As we saw in the previous Sect. 3.2 the informal definition of argument types is important for the belief assessment process. Since masses are also assigned to the logical formulas resulting from these definitions, many authors confuse logical and numerical aspects. But as explained in Sect. 3.1, the definition of argument type, especially the informal ones, conditions the choice of focal sets, on the one hand. On the other hand, it also affects the choice of the aggregation rule. For example, one may think of using the disjunctive consensus rule [10], if disjunction is expressed in the definition. In this section, we are interested in choosing aggregation rules based on Dempster-Shafer Theory and observing the effect of mass functions assignment on the degree belief of the conclusion.

DST offers many aggregation rules (Dempster rule, disjunctive consensus, Yager's rule, etc. see [21,22] for surveys). However, we are going to focus on the methods used in the studied papers listed in Table 2. In general, Dempster combination rule computes the intersection of focal sets. If some focal sets from one source are inconsistent with some from another source, a renormalization must take place. It also assumes that sources are independent and reliable. Other combination rules that express a conjunction exist (e.g. Yager's [28] and Inagaki's rules [16]). For instance, Yager's rule uses a renormalization scheme different from the one of Dempster, reallocating mass of the empty set to the whole frame. The disjunctive rule of combination is a union of random sets and does not need renormalization. Finally, the weighted average rule [21] is used to make a trade-off between conjunctive and disjunctive methods.

So long as the focal propositions involved in a safety case are not conflicting, there is no need to renormalize the resulting mass function, nor to use the disjunctive rule. By taking into account the definition of a safety case and the underying assumptions, applying Dempster rule (without normalization) to aggregate evidences is well adapted.

Table 2. Consistency between argument types and combination rules.

Authors	Argument types	Combination rules	Consistency
Cyra and Gorski [4]	NSC-Arg	DS rule	Yes
	SC-Arg	DS rule	Yes
	A-Arg	Yager's rule	No
	C-Arg	Weighted average	No
	AC-Arg	-	-
Anaheed et al. [1]	Alternative	DS rule	No
	Disjoint	Weighted average	No
	Containment	DS + Weighted average	No
	Overlapping		No
Wang et al. [24]	Disparate	DS rule	Yes
	Complementary		Yes
	Full complementary		Yes
	Redundant		Yes
	Full redundant		Yes

As can be seen from Table 2, several combination rules are proposed to obtain the overall confidence in the conclusion. But, none of the reviewed papers clearly justifies its choice of the applied method, nor does it lay bare the underlying independence assumptions. Also, some of the proposed expressions are not consistent with the verbal definition. For instance, in [1], the type *Alternative argument* is used when several independent premises support the conclusion. The formal definition induced is : $\wedge_{i=1}^{n}(p_i \Rightarrow C)$ (Table 1). However, the authors only consider the confidence in the premises (also called trustworthiness, in [24]), but they did not consider the confidence in the relation between them and the conclusion (the rule). A possible formula that takes into consideration this definition could be Eq. (1).

Consider the example of a conclusion (C) separately supported by two premises p_1 and p_2, which refers to an argument of type 4 in Table 1, i.e., $(p_1 \Rightarrow C) \wedge (p_2 \Rightarrow C)$. We develop this example below. For other argument types, the calculation follows the same method. The confidence assessment process is measured through two parameters. The confidence in premises is modelled by mass functions m_{p_i} and the confidence in the support of the the conclusion by each premise (the rules) is modelled by the mass function m_{r_i}. Then, we apply Dempster combination rule, presented earlier, to merge each premise with its appropriate rule (see Table 3), and secondly merge the two resulting mass functions m_i (see Table 4). Notice that we could also merge premises and rules separately first, then fuse partial conclusions together. The result will be the same because Dempster rule is associative and commutative.

Table 3. Combination of a premise with its rule

$m_1 = m_{p_1} \otimes m_{r_1}$	$m_{r_1}(p_1 \Rightarrow C)$	$m_{r_1}(\top)$
$m_{p_1}(p_1)$	$p_1 \wedge C$	p_1
$m_{p_1}(\top)$	$p_1 \Rightarrow C$	$\top$

In the example given in Table 3, the focal formula $p_1 \wedge C$ results from the conjunction between formulas p_1 and $p_1 \Rightarrow C$. Its mass is calculated by multiplying the masses values in the corresponding line and column. Since the frame of discernment (Ω) of elementary mass functions has two states, masses and belief functions of non-tautological inputs are equal. The calculation of the remaining masses follows the same logic. An example is given here, where, $bel_{\Rightarrow}^i(p_i \Rightarrow C)$ represents the degree of belief that the ith premise supports the conclusion and $bel_p^i(p_i)$ represents the belief degree in the ith premise. For instance, $m_1(p_1 \wedge C) = m_{p_1}(p_1) \times m_{r_1}(p_1 \Rightarrow C) = bel_p^1(p_1) \times bel_{\Rightarrow}^1(p_1 \Rightarrow C)$. Likewise the combination of m_1 and m_2, yields mass function m_{12} using Table 4.

Table 4. Combination of confidence in type 4 : $\wedge_{i=1}^n (p_i \Rightarrow C)$

$m_{12} = m_1 \otimes m_2$	$m_2(P_2 \wedge C)$	$m_2(P_2)$	$m_2(P2 \Rightarrow C)$	$m_2(\top)$
$m_1(P_1 \wedge C)$	$P_1 \wedge P_2 \wedge C$	$P_1 \wedge P_2 \wedge C$	$P_1 \wedge C$	$P_1 \wedge C$
$m_1(P_1)$	$P_1 \wedge P_2 \wedge C$	$P_1 \wedge P_2$	$P_1 \wedge (P_2 \Rightarrow C)$	P_1
$m_1(P_1 \Rightarrow C)$	$P_2 \wedge C$	$P_1 \wedge (P_2 \Rightarrow C)$	$(P_1 \vee P_2) \Rightarrow C$	$P_1 \Rightarrow C$
$m_1(\top)$	$P_2 \wedge C$	P_2	$P_2 \Rightarrow C$	$\top$

The calculation of the degree of belief $bel_c(C)$ in the conclusion for type 4 arguments is as follows (in Table 4 cells in gray identify ϕ's that imply C):

$$bel_c^4(C) = \sum_{\phi:\phi\ implies\ C} m_{12}(\phi) = m_{12}(P_1 \wedge P_2 \wedge C) + m_{12}(P_1 \wedge C) + m_{12}(P_2 \wedge C)$$

$$= m_1(P_1 \wedge C)\sum_{\phi_2} m_2(\phi_2) + m_2(P_2 \wedge C)\sum_{\phi_1} m_1(\phi_1) - m_1(P_1 \wedge C)m_2(P_2 \wedge C)$$

$$= m_1(P_1 \wedge C) + m_2(P_2 \wedge C) - m_1(P_1 \wedge C)m_2(P_2 \wedge C)$$

$$= bel_p^1(P_1)bel_\Rightarrow^1(P_1 \Rightarrow C) + bel_p^2(P_2)bel_\Rightarrow^2(P_2 \Rightarrow C)$$

$$- bel_p^1(P_1)bel_\Rightarrow^1(P_1 \Rightarrow C)bel_p^2(P_2)bel_\Rightarrow^2(P_2 \Rightarrow C)$$

$$= 1 - [1 - bel_p^1(P_1)bel_\Rightarrow^1(P_1 \Rightarrow C)][1 - bel_p^2(P_2)bel_\Rightarrow^2(P_2 \Rightarrow C)]$$

In general, with n premises, the formula for type 4 arguments is as follows.

$$bel_c^4(C) = 1 - \prod_{i=1}^{n}[1 - bel_p^i(p_i)bel_\Rightarrow^i(p_i \Rightarrow C)] \tag{1}$$

Letting $bel_c^i(C) = bel_p^i(p_i)bel_\Rightarrow^i(p_i \Rightarrow C)$ be the degree of belief in C due to premise p_i, the expression in Eq. (1) is a many-valued disjunction connective aggregating the weights $bel_c^i(C)$. So it is enough that $bel_c^i(C) = 1$ for some p_i to get $bel_c(C) = 1$, which is in agreement with the argument type.

It is important to mention that in Eq. (1), a mass function $m_\Rightarrow$ was assigned to each rule $p_i \Rightarrow C$, assuming independence between them, according to type 4 in Table 1. In type 5 argument, we assign a single mass function $m_\Rightarrow$ to the complete rule with a disjunction of premises ($[\vee_{i=1}^n p_i] \Rightarrow C$). The formula resulting from this new mass assignment is given in (2). In general, the belief in the conclusion for type 5 arguments is as follows, using Dempster rule of combination:

$$bel_c^5(C) = bel_\Rightarrow([\vee_{i=1}^n p_i] \Rightarrow C)[1 - \prod_{i=1}^{n}(1 - bel_p^i(p_i))] \tag{2}$$

where $bel_\Rightarrow([\vee_{i=1}^n p_i] \Rightarrow C)$ is the belief that the disjunction of all premises support the conclusion. We stress again that in Eq. (1), we assign one mass function to each simple rule, while in (2), we assign a single mass to a composite rule. So, even though types 4 and 5 are logically equivalent ($\wedge_{i=1}^n[p_i \Rightarrow C]) \equiv ([\vee_{i=1}^n p_i] \Rightarrow C$), because the assignment of masses is different in types 4 and 5, they produce different results for the belief calculation. The same combination pattern applies to arguments of type 3 in Table 1. It requires all premises be true to justify the conclusion, and a simple support mass is assigned to the implication $[\wedge_{i=1}^n p_i] \Rightarrow C$. It yields for type 3 arguments:

$$bel_c^3(C) = bel_\Rightarrow([\wedge_{i=1}^u p_i] \Rightarrow C) \prod_{i=1}^{n} bel_p^i(p_i) \tag{3}$$

In (2) a multivalued disjunction connective is applied to the degrees of belief in the premises while in (3), it is a multivalued conjunction.

In contrast, equation (4) below for type 4 arguments supposes a single mass function $m_\Rightarrow$ with masses distributed over the elementary rules $p_i \Rightarrow C$, assuming $\sum_{i=1}^{n} m_\Rightarrow(p_i \Rightarrow C) + m_\Rightarrow(\top) = 1$. The resulting belief in the conclusion is then a weighted sum of the degrees of belief in the premises:

$$bel'_c(C) = \sum_{i=1}^{n} bel_\Rightarrow(p_i \Rightarrow C) bel_p^i(p_i) \tag{4}$$

In front of these three expressions (Eqs. (1), (2) and (4)) which result from the same logical form of the argument, a question arises. Which of them is the most appropriate for this argument type? The choices of the assignment of mass function in each equation do not model the same situation. Comparing the first and second formulas, on the one hand (1) suggests that each argument is based on one piece of evidence that could support the whole conclusion and is independent from the other ones. On the other hand, 2 supposes that all such arguments are provided at once by a single source. If we also compare Eqs. (1) and (4), (1) represents the situation when the elementary arguments are independent, so that a mass is allocated each implication independently of the others. On the contrary, in (4) the belief mass assigned to in one implication affects those assigned to other ones, because, due to the use of a single mass function, the sum of all such masses must be one.

It is important to use each formula giving the belief in the conclusion in the appropriate situation, laying bare the underlying assumptions. For instance, the A-Arg presented in [4] is formally defined by $\wedge_{i=1}^{n}(p_i \Rightarrow C)$, and uses Yager's combination rule to calculate the overall confidence in the conclusion. However, Yager's rule was developed to deal with highly conflicting sources in place of Dempster rule. But the authors of [4] do not explain the presence of a conflict between pieces of evidence. Conflicts occur in cases when the intersection between focal sets is empty, which could be the case if masses were assigned to expressions supporting the negation of the implication (e.g., $p_i \wedge \neg C$), or in argument types involving equivalence. In the argument types discussed above, the focal sets resulting from handling argument of types 3, 4, 5 inspired by the selected articles do not generate such conflicts. In particular, the degree of disbelief in the conclusion is always 0 with these argument types.

4   Lessons Learned

As shown in the previous sections, defining an "Argument Type" is a very delicate process. It depends on the assessor's understanding of the argument. Four important issues emerge from this paper:

Formal representation of the argument: The assessor should faithfully translate the informal definition of each argument into a formal one, by choosing the proper logical connectives relating premises to one another (conjunction, disjunction) or between them and the conclusion (equivalence, implication). In order to do so, it is necessary to avoid vagueness and imprecision in the formulation of (informal) verbal definitions and to describe the characteristics of each argument type as accurately as possible. In this paper, two basic types have been laid bare : Those using conjunction of premises (Type 3) and those using a disjunction (Type 4 and 5). Indeed, it is important to know if for instance the truth in one premise is sufficient to ensure the conclusion, or if all premises are necessary to ensure the conclusion.

Using equivalence vs. implication connectives: Implication is the commonly used logical operator for representing arguments supporting a conclusion. However, we have seen that some verbal descriptions used in the literature on safety arguments can be formally translated into equivalences. The equivalence operator should be used carefully because it involves several situations that we may not intentionally want to encounter. Consider for example a conclusion (C) supported by one premise (P). Using the rule $(P \equiv C)$ implies that $(P \Rightarrow C)$ and that $(C \Rightarrow P)$. We saw in Sect. 3.4 that the second implication is not necessary true. In addition to this, equivalence is also expressed as $(\neg P \equiv \neg C)$, in particular, $\neg P \Rightarrow \neg C)$; this is why using equivalence, disbelief in the conclusion may be different from zero.

Assigning mass functions: It should be clear that the formal definition of the argument types is not enough determine the degree of belief in the conclusion. For instance, when we changed the mass functions definition in the 4th and 5th logically equivalent types in Table 1 while using the same combination rule (i.e., Dempster rule), we obtained three distinct formulas (1, 2, 4). It is necessary to be sure that the choice of mass assignment reflects as well as possible the situation described in the arguments.

Choosing a combination rule: Changing the combination rule obviously affects the result of uncertainty propagation. So, it is important to choose the right one. However, as we saw in Sect. 3.4, applying Dempster combination rule is well adapted to computing the overall confidence, because no conflict is met during the combination step using arguments modelled by implication. In that case it is equivalent to Yager's rule. A possible use of this rule could be justified to cope with conflicts between the involved pieces of evidence, in place of Dempster rule. On the other hand, the disjunctive rule is too weak to be applied if one wants to jointly exploit the pieces of evidence in the safety case. A trade-off could be the rule in [11] which combines focal sets conjunctively when they are consistent and disjunctively when they conflict.

5   Conclusion

In this paper, we propose a comparative study between some confidence propagation methods in safety cases. We highlight four important elements to be considered in the development of a safety case. First, arguments should be expressed in formal logic. Second, we advocate the use of the implication connective, rather than equivalence, to describe the relationship between premises and conclusion.

Then, we propose a simplified framework to define mass functions attached to premises and arguments. Finally, we argue that Dempster rule of combination should be preferred when the focal sets issued from independent mass functions to be combined do not conflict.

In future works, we plan to experiment an approach that exploits this methodology in an application pervaded with high uncertainties, such as autonomous vehicles. Another issue is the improvement of methods proposed for translating expert opinions into usable numerical values, such as those proposed in [18] and applied in [4,24]. In this regard, it would also be interesting to develop non-quantitative approaches using qualitative counterparts of belief functions as suggested in [9].

References

1. Ayoub, A., Chang, J., Sokolsky, O., Lee, I.: Assessing the overall sufficiency of safety arguments. In: 21st Safety-critical Systems Symposium (SSS'13), Bristol, United Kingdom (2013)
2. Bloomfield, R., Netkachova, K.: Building blocks for assurance cases. In: 2014 IEEE International Symposium on Software Reliability Engineering Workshops, pp. 186–191. IEEE (2014)
3. Chatalic, P., Dubois, D., Prade, H.: An approach to approximate reasoning based on Dempster rule of combination. Int. J. Expert Syst. Res. Appl. 1, 67–85 (1987)
4. Cyra, L., Górski, J.: Support for argument structures review and assessment. Reliab. Eng. Syst. Saf. 96(1), 26–37 (2011)
5. De La Vara, J.L., Génova, G., Álvarez-Rodríguez, J.M., Llorens, J.: An analysis of safety evidence management with the structured assurance case metamodel. Comput. Stan. Interfaces 50, 179–198 (2017)
6. Denney, E., Pai, G., Habli, I.: Towards measurement of confidence in safety cases. In: 2011 International Symposium on Empirical Software Engineering and Measurement, pp. 380–383. IEEE (2011)
7. Denoeux, T.: Conjunctive and disjunctive combination of belief functions induced by nondistinct bodies of evidence. Artif. Intell. 172(2–3), 234–264 (2008)
8. Destercke, S., Dubois, D.: Idempotent conjunctive combination of belief functions: Extending the minimum rule of possibility theory. Inf. Sci. 181(18), 3925–3945 (2011)
9. Dubois, D., Faux, F., Prade, H., Rico, A.: A possibilistic counterpart to Shafer evidence theory. In: IEEE International Conference on Fuzzy Systems (FUZZ-IEEE), New Orleans, LA, USA, June 23–26, pp. 1–6. IEEE (2019)
10. Dubois, D., Prade, H.: A set-theoretic view of belief functions. Logical operations and approximation by fuzzy sets. Int. J. General Syst. 12(3), 193–226 (1986)
11. Dubois, D., Prade, H.: Representation and combination of uncertainty with belief functions and possibility measures. Comput. Intell. 4, 244–264 (1988)
12. EN50129: Railway applications - Safety related electronic systems for signaling (2003), CENELEC, European Committee for Electrotechnical Standardization

13. Graydon, P.J., Holloway, C.M.: An investigation of proposed techniques for quantifying confidence in assurance arguments. Saf. Sci. **92**, 53–65 (2017)
14. Guiochet, J., Do Hoang, Q.A., Kaaniche, M.: A model for safety case confidence assessment. In: Koornneef, F., van Gulijk, C. (eds.) SAFECOMP 2015. LNCS, vol. 9337, pp. 313–327. Springer, Cham (2015). https://doi.org/10.1007/978-3-319-24255-2_23
15. Hobbs, C., Lloyd, M.: The application of Bayesian belief networks to assurance case preparation. In: Achieving Systems Safety, pp. 159–176. Springer (2012)
16. Inagaki, T.: Interdependence between safety-control policy and multiple-sensor schemes via Dempster-Shafer theory. IEEE Trans. Reliab. **40**(2), 182–188 (1991)
17. ISO 26262: Software considerations in airborne systems and equipment certification. In: International Organization for Standardization (ISO) (2011)
18. Jøsang, A.: Subjective Logic. Springer (2016)
19. Kelly, T.: Arguing Safety - A Systematic Approach to Safety Case Management. Ph.D. thesis, Department of Computer Science, University of York, UK (1998)
20. Kelly, T.P., McDermid, J.A.: Safety case construction and reuse using patterns. In: International Conference on Computer Safety, Reliability, and Security (Safecomp) 97, pp. 55–69. Springer (1997)
21. Sentz, K., Ferson, S., et al.: Combination of evidence in Dempster-Shafer theory. Technical Report 0835, Sandia National Laboratories, Albuquerque, NM, USA (2002)
22. Smets, P.: Analyzing the combination of conflicting belief functions. Inf. Fusion **8**(4), 387–412 (2007)
23. Toulmin, S.E.: The Uses of Argument. Cambridge Univ. Press, Cambridge (1958)
24. Wang, R.: Confidence in safety argument-An assessment framework based on belief function theory. Ph.D. thesis, Institut National des Sciences Appliquées de Toulouse, France (2018)
25. Wang, R., Guiochet, J., Motet, G., Schön, W.: D-S theory for argument confidence assessment. In: Vejnarová, J., Kratochvíl, V. (eds.) BELIEF 2016. LNCS (LNAI), vol. 9861, pp. 190–200. Springer, Cham (2016). https://doi.org/10.1007/978-3-319-45559-4_20
26. Wang, R., Guiochet, J., Motet, G., Schön, W.: Modelling confidence in railway safety case. Saf. Sci. **110**(part B), 286–299 (2018)
27. Wang, R., Guiochet, J., Motet, G., Schön, W.: Safety case confidence propagation based on Dempster-Shafer theory. Int. J. Approximate Reason. **107**, 46–64 (2019)
28. Yager, R.R.: On the Dempster-Shafer framework and new combination rules. Inf. Sci. **41**(2), 93–137 (1987)
29. Yuan, C., Wu, J., Liu, C., Yang, H.: A subjective logic-based approach for assessing confidence in assurance case. Int. J. Perform. Eng. **13**(6), 807–822 (2017)

Incremental Elicitation of Capacities for the Sugeno Integral with a Maximin Approach

Agnès Rico[1] and Paolo Viappiani[2(✉)]

[1] ERIC, Université Claude Bernard Lyon 1, Villeurbanne, France
`agnes.rico@univ-lyon1.fr`
[2] Sorbonne Université, UMR7606 CNRS, LIP6, 4 pl. Jussieu, 75005 Paris, France
`paolo.viappiani@lip6.fr`

Abstract. We propose an interactive elicitation protocol for the Sugeno integral. Our approach at each step asks the decision maker whether the overall evaluation of a given alternative attains at least a certain level. This information is encoded in terms of constraints on the capacity and a lattice of feasible capacities is identified. The procedure continues until a necessary winner is identified. The efficiency of our methodology is evaluated in numerical experiments.

Keywords: Sugeno integral · Capacity · Preference elicitation

1 Introduction

The Sugeno integral is used in multicriteria decision making as a tool for guiding decision support [7,12]. Sugeno integrals are qualitative aggregation operators that take as input some local evaluation of alternatives and output a global evaluation. In this paper we provide an elicitation protocol for the Sugeno integral.

The problem of eliciting some Sugeno integrals agreeing with a dataset has received some attention [6,10–12] both from theoretical and practical point of view. The theoretical results concern the elicitation of a unique family of Sugeno integrals expected to be consistent with a set of data. Inconsistency is usually to be avoided; in [8] the data is partitioned in classes and a fuzzy integral is calculated for each one. In [10,11] the aim is to identify the bounds of the set of the family of Sugeno integrals consistent with the data. If the dataset is not fully consistent with only one family of Sugeno integrals, they consider several ones; this point of view is motivated by the fact that the dataset may contain many classes of profiles.

Sugeno integrals are defined with capacities used in multicriteria decision making to represent the weights of subsets of criteria. Sugeno integrals are used in [10] to extract knowledge from experimental data in multi-factorial evaluation; a capacity associated to a Sugeno integral consistent with the dataset is calculated and then a set of rules corresponding to the capacity are derived. Some other work on eliciting or learning a capacity from examples are based on linear programming methods that minimize the total error [1].

© Springer Nature Switzerland AG 2020
J. Davis and K. Tabia (Eds.): SUM 2020, LNAI 12322, pp. 156–171, 2020.
https://doi.org/10.1007/978-3-030-58449-8_11

The approach presented in this paper is different. We are proposing incremental elicitation and winner determination processes in which preference queries are selected one at a time. The elicitation is targeted towards the determination of the best choice among a set of alternatives.

The aim of this paper is to introduce an adaptive elicitation procedure in the context of the Sugeno integral for the fast determination of a necessary winner and to evaluate the practical efficiency of this procedure.

The elicitation continues until we have enough information about the capacity to identify the alternative associated with the highest Sugeno value; to do this we adopt the maximin criterion as proposed in a previous work in multi attribute decision making [15]. The proposed method bears strong similarity to the regret-based approach for eliciting a capacity for the Choquet integral [2]; the main difference lies in the fully ordinal setting under consideration in this paper, that makes regret a meaningless concept in our context.

The paper is organized as follows. In Sect. 2 we provide some background on the Sugeno integral and its use in multiple criteria decision making. We then describe our elicitation method in Sect. 3. In Sect. 4 we provide numerical tests to evaluate our approach; we conclude with final remarks in Sect. 5.

2   Background and Notation

Let X be a finite set of alternatives or objects that need to be compared in order to make a decision. An object $x \in X$ is evaluated with respect to a set of n criteria $C = \{1, \cdots, n\}$. An object is represented by a vector $(x_1, \cdots, x_n)$ where x_i represents the evaluation of x according to the criterion i. The criteria are evaluated on a common (finite) evaluation scale L. The global evaluation is also given on L. We assume that L is a bounded totally ordered finite set with a bottom denoted by 0_L, a top denoted by 1_L. Moreover L is equipped with an involutive negation denoted by 1_L- which is an order reverse function.

2.1   Lattice of Capacities

A *capacity* (or fuzzy measure) v is a set function, defined over subsets of C, that is monotone with respect to set inclusion, i.e., $v : 2^C \to L$ such that if $A \subseteq B \subseteq C$ then $v(A) \leq v(B)$, $v(\emptyset) = 0$ and $v(C) = 1$.

We denote by V_C the set of all capacities on C, and we drop the subscript C when it is clear from the context. A partial order $\leq$ between capacities is established as follows:

$$v^1 \leq v^2 \text{ whenever } v^1(G) \leq v^2(G) \text{ for all } G \in 2^C.$$

The pair $(V_C, \leq)$ is a bounded lattice. It can also be identified by the tuple $(V_C, \wedge, \vee, \perp, \top)$ where the binary operators $\vee$ (join) and $\wedge$ (meet), and the elements $\perp$ and $\top$ are established as follows:

- given $v^1, v^2 \in V_C$, the capacity $v^1 \wedge v^2$ is such that $(v^1 \wedge v^2)(G) = \min(v^1(G), v^2(G))$ for all $G \in 2^C$;

- given $v^1, v^2 \in V_C$, the capacity $v^1 \vee v^2$ is such that $(v^1 \vee v^2)(G) = \max(v^1(G), v^2(G))$ for all $G \in 2^C$;
- $\perp$ gives 0 to all proper subsets of C, $\perp(G) = 0$ for all $G \subset C$;
- the $\top$ element is the capacity that associates 1 to every non-empty subset, $\top(G) = 1$ for all $G \subseteq C$.

Considering two capacities $\check{v}, \hat{v}$, an interval $[\check{v}, \hat{v}]_{V_C}$ is the subset $\{v \in V_C | \check{v} \leq v \leq \hat{v}\}$; the interval is nonempty if and only if $\check{v} \leq \hat{v}$. Note that a nonempty interval is a sublattice [5] i.e., the interval is closed with the infimum $\wedge$ and the supremum $\vee$.

Since the intersection of two sublattices is a sublattice, then the intersection of two intervals which is an interval is also a sublattice. The intersection of n intervals $[\check{v}^i, \hat{v}^i]_V$, with $i = 1, \ldots, n$ is given by

$$\bigcap_{i=1,\ldots,n} [\check{v}^i, \hat{v}^i]_V = \left[\bigvee \check{v}^i, \bigwedge \hat{v}^i \right]_V.$$

It follows that the intersection of n intervals is not empty if and only if $\bigvee \check{v}^i \leq \bigwedge \hat{v}^i$ i.e for all $i, j \in \{1, \cdots, n\}$, we have $\check{v}^i \leq \hat{v}^j$.

We provide a direct proof, using lattice theory, of a statement mentioned in the proof of Proposition 6 in the paper by Prade et al. [10].

Proposition 1. *Assume n intervals $[\check{v}^i, \hat{v}^i]_V$, with $i = 1, \ldots, n$ whose pairwise intersections $[\check{v}^i, \hat{v}^i]_V \cap [\check{v}^j, \hat{v}^j]_V$ for all $i, j \in \{1, \ldots, n\}$, are not empty. Then the intersection $\bigcap_{i=1,\ldots,n} [\check{v}^i, \hat{v}^i]_V$ is not empty.*

Proof. The fact the pairwise intersections are not empty means that

$$[\check{v}^i, \hat{v}^i]_V \cap [\check{v}^j, \hat{v}^j]_V \neq \emptyset \ \forall i, j \iff \check{v}^i \vee \check{v}^j \leq \hat{v}^i \wedge \hat{v}^j \ \forall i, j$$

Since, by definition of $\vee$ and $\wedge$, it holds $\check{v}^i \leq \check{v}^i \vee \check{v}^j$ and $\hat{v}^i \wedge \hat{v}^j \leq \hat{v}^j$, it also follows that $\check{v}^i \leq \hat{v}^j \ \forall i, j$. It follows that $\bigvee_i \check{v}^i \leq \bigwedge_i \hat{v}^i$ that exactly means $\bigcap_{i=1,\ldots,n} [\check{v}^i, \hat{v}^i]_V$ is not empty.

2.2 Discrete Sugeno Integral

We now review the definition of the Sugeno integral, as used in Multi Criteria Decision Analysis (MCDA) to aggregate into a single score the evaluation of an object with respect to several criteria.

Let σ be a permutation on C such that $x_{\sigma(1)} \leq \ldots \leq x_{\sigma(n)}$. The Sugeno integral [13] of an alternative x with respect to capacity v can be defined by means of several equivalent expressions:

$$S_v(x) = \max_{A \subseteq C} \min(v(A), \min_{i \in A} x_i) = \min_{A \subseteq C} \max(v(\overline{A}), \max_{i \in A} x_i), \tag{1}$$

where $\overline{A}$ is the complement of A. These expressions can be simplified as follows:

$$S_v(x) = \max_{\alpha \in L} \min(v(\{i : x_i \geq \alpha\}), \alpha) = \min_{a \in L} \max(v(\{i : x_i > \alpha\}), \alpha). \tag{2}$$

A basic property of the Sugeno integral is that the result of the aggregation is between the minimum and the maximum component.

$$\min_{i=1,\ldots,n} x_i \leq S_v(x_1,\ldots,x_n) \leq \max_{i=1,\ldots,n} x_i$$

A direct consequence is that $S_v(c,\ldots,c) = c$ for any capacity v (idempotency or unanimity). Note also that the value of the Sugeno integral of an alternative x is monotone with respect to the order between capacities:

$$\text{If } v^1 \leq v^2 \text{ then } S_{v^1}(x) \leq S_{v^2}(x) \quad \forall x \in X \tag{3}$$

2.3   The Set of Capacities Consistent with Preference Data

We now summarize the results presented in [12] about the identification of the family of Sugeno integrals consistent with a dataset of statements comparing alternatives to a global evaluation level. More precisely, we consider preference statements of the type "the global evaluation of x is higher or equal than a level α" or "the global evaluation of y is lower or equal than a level λ", and we want to derive the set of capacities consistent with such statements.

For a pair $(x, \alpha) \in X \times L$, we define the capacities $\check{v}_{x,\alpha}$ and $\hat{v}_{x,\alpha}$ as follows.

Definition 1. *Given $x \in X$ and $\alpha \in L$, the capacities $\check{v}_{x,\alpha}$ and $\hat{v}_{x,\alpha}$ are defined as:*

$$\check{v}_{x,\alpha}(A) = \begin{cases} 1_L & \text{if } A = C \\ \alpha & \text{if } \{i \in C | x_i \geq \alpha\} \subseteq A \\ 0_L & \text{otherwise} \end{cases} \quad \text{and} \quad \hat{v}_{x,\alpha}(A) = \begin{cases} 0 & \text{if } A = \emptyset \\ \alpha & \text{if } A \subseteq \{i \in C | x_i > \alpha\} \\ 1_L & \text{otherwise.} \end{cases}$$

Note that we always have $\check{v}_{x,\alpha} \leq \hat{v}_{x,\alpha}$. Using $\check{v}_{x,\alpha}$ and $\hat{v}_{x,\alpha}$ we can determine the set of capacities (that is a sub-interval of the lattice of capacities) consistent with a statement of the type $S_V(x) \geq \alpha$, $S_v(x) \leq \alpha$, or $S_v(x) = \alpha$.

Proposition 2. *The set of capacities satisfying the equation $S_v(x) \geq \alpha$ is:*

$$\{v \in V | S_v(x) \geq \alpha\} = \{v \in V | \check{v}_{x,\alpha} \leq v \leq \top_V\} = [\check{v}_{x,\alpha}, \top_V]_V \tag{4}$$

while the set of capacities satisfying the equation $S_v(x) \leq \alpha$ is:

$$\{v \in V | S_v(x) \leq \lambda\} = \{v \in V | \bot_V \leq v \leq \hat{v}_{x,\lambda}\} = [\bot_V, \hat{v}_{x,\alpha}]_V. \tag{5}$$

Therefore, the set of capacities satisfying $S_v(x) = \alpha$ is:

$$\{v \in V | S_v(x) = \alpha\} = \{v \in V | \check{v}_{x,\alpha} \leq v \leq \hat{v}_{x,\alpha}\} = [\check{v}_{x,\alpha}, \hat{v}_{x,\alpha}]_V.$$

In [12] the authors focus in considering a set $\mathcal{P}$ of assignments of alternatives to global evaluations, that is the constraints $S_v(x^k) = \alpha_k$ for $k = 1,\ldots,m$. The set of the capacities compatible with all assignments in $\mathcal{P}$ is

$$V^{\mathcal{P}} = \left\{ v \in V \middle| \bigvee_{k=1}^{m} \check{v}_{x^k,\alpha_k} \leq v \leq \bigwedge_{k=1}^{m} \hat{v}_{x^k,\alpha_k} \right\} = \left[\bigvee_{k=1}^{m} \check{v}_{x^k,\alpha_k}, \bigwedge_{k=1}^{m} \hat{v}_{x^k,\alpha_k} \right]_V.$$

In order to know if the set of capacities $V^{\mathcal{P}}$ consistent with the preferences $\mathcal{P}$, is empty it is not necessary to compare the capacities $\vee_{k=1}^{m} \check{v}_{x^k, \alpha_k}$ and $\wedge_{k=1}^{m} \hat{v}_{x^k, \alpha_k}$ for all subsets of criteria A since it is is proved in [12] the following property (that makes use of Proposition 1).

Proposition 3. *The set of capacities* $V_{\mathcal{P}} = [\vee_{k=1}^{m} \check{v}_{x^k \alpha_k}, \wedge_{k=1}^{m} \hat{v}_{x^k, \alpha_k}]_V$ *is not empty if and only if for all* $\alpha_k < \alpha_l$ *we have* $\{i | x_i^l \geq \alpha_l\} \not\subseteq \{i | x_i^k > \alpha_k\}$.

In this work we do not assume that all input statements are assignments, but we collect, using an interactive process, statements of the type $S_v(x) \geq \alpha$ or $S_v(x) \leq \alpha$. Let $\mathcal{P}$ be divided in two parts $(x^k, \alpha_k)_{k=1,...,m_1}$ and $(y^k, \lambda_k)_{k=1,...,m_2}$ such that the global evaluation of x^k is bigger than α_k and the global evaluation of y^k is lower than λ_k. Hence the set of consistent capacities $V^{\mathcal{P}}$ is:

$$V^{\mathcal{P}} = \left\{ v \in V \,\Big|\, \bigvee_k \check{v}_{x^k, \alpha_k} \leq v \right\} \cap \left\{ v \in V \,|\, v \leq \bigwedge_k \hat{v}_{y^k, \lambda_k} \right\} \tag{6}$$

$$= \left[\bigvee_{k=1}^{m_1} \check{v}_{x^k, \alpha_k}, \bigwedge_{k=1}^{m_2} \hat{v}_{y^k, \lambda_k} \right]_V. \tag{7}$$

Note that this intersection can be empty. We will see, in the next section, that this intersection is always non empty with the proposed algorithm.

We conclude this section with a remark concerning the focal sets of a capacity. The qualitative Moebius transform of a capacity v is the set function $v_\#$ defined as follows:

$$v_\#(A) = \begin{cases} v(A) \text{ if } v(B) < v(A) \ \forall B \subset A \\ 0 \text{ otherwise} \end{cases}$$

The sets A such that $v_\#(A) > 0$ are call the focal sets of v. The qualitative Moebius transform contains all the information to compute v since for all A, $v(A) = \vee_{B \subseteq A} v_\#(B)$, and the qualitative Moebius transform is sufficient to calculate the Sugeno integral:

$$S_v(x) = \max_{A \subseteq \mathcal{C}} \min(v_\#(A), \min_{i \in A} x_i) \tag{8}$$

This means that we just need to identify the focal sets in order to calculate the Sugeno integral. The preferences, described above, may be just given for objects with local evaluations equal to 0_L or 1_L. Nevertheless, in practice these theoretical objects could be inappropriate. For instance, imagine the situation of caregivers assessing the overall health of given patients: it would be difficult for them to assess abstract patients without referring to real cases. This difficulty of reasoning with abstract items is the reason why we decide to not use focal sets in the method proposed in this paper.

3   Incremental Elicitation Protocol

We provide an interactive elicitation method based on the maximin decision criteria. The goal of the elicitation is to determine a necessary winner.

Our method bears similarity to methods, relying on minimax regret, for the incremental elicitation of a capacity for the Choquet integral [2]. We note that, in our qualitative framework, minimax regret is not applicable since the difference of two Sugeno value is meaningless in decision context.

First of all, in Sect. 3.1, we introduce some concept of decision-making under uncertainty to be used to identify the most promising alternative when the capacity is not known precisely. We focus in the case where the capacity lies in an interval between a lower and a upper capacity.

Then in Sect. 3.2 we use these concepts to design our interactive elicitation protocol.

3.1   Reasoning with an Uncertain Capacity

Suppose now that a set $\mathcal{P}$ of statements have been collected and that the set of capacities $V^{\mathcal{P}} \subseteq V$ consistent with $\mathcal{P}$ has been identified; each $v \in V^{\mathcal{P}}$ is such that v satisfies all preferences in $\mathcal{P}$.

First of all, we observe that, in some cases, the set $\mathcal{P}$ is enough to identify the best alternative in X. An alternative is a *necessary winner* if it is the optimal alternative with respect to all capacities in $V^{\mathcal{P}}$.

Definition 2. *A necessary winner with respect to $\mathcal{P}$ is an alternative $x \in \mathcal{X}$ such that*

$$x \in \arg\max_{x \in X} S_v(x) \qquad \forall v \in V^{\mathcal{P}}.$$

If a necessary winner exists, it is not necessary to elicit further information from the decision maker in order to make a decision, since the current available information is enough to identify the best choice (or one of the best choices, in case of ties) among the set of alternatives.

In most cases, however, a necessary winner does not exists. When it is necessary to make a choice with the only knowledge that the capacity lies in $V^{\mathcal{P}}$, we can recommend the alternative(s) ensuring the highest Sugeno value in the worst-case. We therefore adopt the *maximin* criterion (similarly to a previous work in multiattribute decision making [15]), that is particularly apt to the ordinal settings where the Sugeno integral is typically used.

Given a set of capacities $V^{\mathcal{P}}$ and an alternative $x \in X$, the minimum (or pessimistic) value according to Sugeno is $s^{\downarrow}_{\mathcal{P}}(x) = \min_{v \in V^{\mathcal{P}}} S_v(x)$ while its maximum (or optimistic) value is $s^{\uparrow}_{\mathcal{P}}(y) = \max_{v \in V^{\mathcal{P}}} S_v(x)$. We now define the *maximin Sugeno value* $s^*_{\mathcal{P}}$ as

$$s^*_{\mathcal{P}} = \max_{x \in X} s^{\downarrow}_{\mathcal{P}}(x) = \max_{x \in X} \min_{v \in V^{\mathcal{P}}} S_v(x)$$

and a *maximin recommendation* $x^*_{\mathcal{P}}$ is such that:

$$x^*_{\mathcal{P}} \in \arg\max_{x \in X} s^{\downarrow}_{\mathcal{P}}(x) = \arg\max_{x \in X} \min_{v \in V^{\mathcal{P}}} S_v(x)$$

and $x_{\mathcal{P}}^*$ is said to be maximin optimal ($x_{\mathcal{P}}^*$ has the highest "pessimistic" value). The value $s_{\mathcal{P}}^{\downarrow}(x)$ is the worst-case "utility" associated with recommending alternative x; any choice that is not maximin optimal has strictly lower Sugeno value than x^* for some capacity $v \in V^{\mathcal{P}}$.

We further assume that all preference statements are of the kind $S(x) \geq \alpha$ or $S(x) \leq \alpha$. Then, as we have seen previously in Sect. 2.3, the set of capacities consistent with $\mathcal{P}$ can be written as the intersection of intervals. We will see that, using the proposed algorithm, this set of capacities is a lattice interval; i.e.,

$$V^{\mathcal{P}} = [\check{v}, \hat{v}]_V.$$

Given an interval of capacities, the *maximin* alternative (the choice that maximizes the worst-case Sugeno value) is easily found: using the property described in Eq. 3 and the fact that the set of the valid capacities is a lattice, $s_{\mathcal{P}}^{\downarrow}(x)$, the minimum Sugeno value of an alternative x, is just $S_{\check{v}}(x)$, the Sugeno integral of x computed with the "bottom" capacity $\check{v}$.

Proposition 4. *Assuming that the set of feasible capacities $V^{\mathcal{P}}$ is an interval (sublattice) of V, then we have:*

$$s_{\mathcal{P}}^{\downarrow}(x) = S_{\check{v}}(x), \; s_{\mathcal{P}}^* = \max_{x \in X} S_{\check{v}}(x) \; and \; x_{\mathcal{P}}^* = \arg\max_{x \in X} S_{\check{v}}(x) \qquad (9)$$

We need a measure of how "uncertain" we are with respect to our recommendation. Now, we consider the most optimistic Sugeno value, i.e. the *maximax* value, that can be attained by any alternative y different from the recommendation $x_{\mathcal{P}}^*$.

$$s_{\mathcal{P}}^{\circ} = \max_{y \neq x_{\mathcal{P}}^*} s_{\mathcal{P}}^{\uparrow}(y) = \max_{y \neq x_{\mathcal{P}}^*} \max_{v \in V^{\mathcal{P}}} S_v(y) = \max_{y \neq x_{\mathcal{P}}^*} S_{\hat{v}}(y)$$

$$y_{\mathcal{P}}^{\circ} \in \arg\max_{y \neq x_{\mathcal{P}}^*} s_{\mathcal{P}}^{\uparrow}(y) = \arg\max_{y \neq x_{\mathcal{P}}^*} \max_{v \in V^{\mathcal{P}}} S_v(y) = \arg\max_{y \neq x_{\mathcal{P}}^*} S_{\hat{v}}(y)$$

We dub $y_{\mathcal{P}}^{\circ}$ as the "adversary", since it is the alternative that may have the highest value. Recall that $s_{\mathcal{P}}^*$ is the value of the maximin optimal recommendations. By comparing $s_{\mathcal{P}}^*$ and $s_{\mathcal{P}}^{\circ}$ we determine whether there is any residual uncertainty about which is the optimal alternative. We notice that if $s_{\mathcal{P}}^* \geq s_{\mathcal{P}}^{\circ}$ then it means that the current maximin recommendation $x_{\mathcal{P}}^*$ is surely an optimal recommendation. This observation is formally stated in the following proposition, whose proof is very straightforward.

Proposition 5. *If $s_{\mathcal{P}}^* \geq s_{\mathcal{P}}^{\circ}$ then $x_{\mathcal{P}}^*$ is a necessary winner.*

Proof. For all $v \in V^{\mathcal{P}}$, $S_v(x_{\mathcal{P}}^*) \geq s_{\mathcal{P}}^*$, and $s_{\mathcal{P}}^{\circ} \geq s_{\mathcal{P}}^{\uparrow}(y) \geq S_v(y)$, for all $y \neq x_{\mathcal{P}}^*$. Therefore, if $s_{\mathcal{P}}^* \geq s_{\mathcal{P}}^{\circ}$, then by transitivity we have $S_v(x_{\mathcal{P}}^*) \geq S_v(y)$ for all $v \in V^{\mathcal{P}}$ and for all $y \neq x_{\mathcal{P}}^*$.

Example 1. Suppose that the available alternatives are $X = \{a, b, c, d\}$ whose performances are given in the following table of criteria.

Alternative	Criteria		
	1	2	3
a	0.2	0.4	0.5
b	0.7	0.2	0.4
c	0.1	1	0.7
d	0	0.5	0

The scale is $L = \{0, 0.1, 0.2, \ldots, 0.9, 1\}$. Assume that we know that alternative a is deemed to have value higher or equal than 0.4, that alternative b has Sugeno value at least 0.5, and that alternative c has Sugeno value at most 0.8; that is $S_v(a) \geq 0.4$, $S_v(b) \geq 0.5$, and $S_v(c) \leq 0.8$. We now inspect the lower bound $\check{v}$ and the upper bound $\hat{v}$ capacities, based on combining Eqs. 4 and 5.

Subset	$\emptyset$	$\{1\}$	$\{2\}$	$\{3\}$	$\{1,2\}$	$\{1,3\}$	$\{2,3\}$	$\{1,2,3\}$
$\check{v}(\cdot)$	0	0.5	0	0	0.5	0.5	0.4	1
$\hat{v}(\cdot)$	0	1	0.8	1	1	1	1	1

We determine the optimistic $s_{\mathcal{P}}^{\uparrow}$ and the pessimistic value $s_{\mathcal{P}}^{\downarrow}$ of each alternative by computing the Sugeno integral of a, b, c with respect to $\check{v}, \hat{v}$.

Alternative	a	b	c	d
$s_{\mathcal{P}}^{\downarrow}(\cdot)$	0.4	0.5	0.4	0
$s_{\mathcal{P}}^{\uparrow}(\cdot)$	0.5	0.7	0.8	0.5

We can determine that b attains the maximin optimal value $s_{\mathcal{P}}^{*} = s_{\mathcal{P}}^{\downarrow}(b) = 0.5$, while the adversary is c that can obtain up to $s_{\mathcal{P}}^{\circ} = s_{\mathcal{P}}^{\uparrow}(c) = 0.8$ in the optimistic case.

We conclude this part by observing that the condition in the proposition above gives us a *sufficient* condition for detecting a necessary winner, but not a *necessary* one[1]: it is possible that a necessary winner exists even when such condition is not satisfied (see example below). This means that, in some cases, the interactive approach that we present next may pose some questions that could be avoided with a more precise check for determining a necessary winner. However, by proceeding in this way we keep the algorithm rather simple and efficient.

---

[1] In future works, we will consider optimization techniques for identifying necessary winners.

Example 1 (continued). Now consider the set of alternatives to be restricted to c and d. Note that c dominates d, that is, the former has a strictly higher performance than the latter with respect to all three criteria; it follows that Sugeno of c is higher than the value of d. Alternative c is a necessary winner in $X' = \{c, d\}$. However, we have $s_{\mathcal{P}}^{\downarrow}(c) = 0.4 < 0.5 = s_{\mathcal{P}}^{\uparrow}(d)$ and the condition of Proposition 5 is not met.

Algorithm 1. Interactive Elicitation and Winner Determination

---

1: **procedure** INTERACTIVEELIC$(X, \mathcal{P})$
2:     $(\check{v}, \hat{v}) = \text{INIT}(\mathcal{P})$ ▷ Initialization
3:     **repeat**
4:         $x_{\mathcal{P}}^* = \arg\max_{x \in X} S_{\check{v}}(x)$ ▷ Determine maximin recommendation
5:         $s_{\mathcal{P}}^* = \max_{x \in X} S_{\check{v}}(x)$
6:         $y_{\mathcal{P}}^{\circ} = \arg\max_{y \neq x_{\mathcal{P}}^*} S_{\hat{v}}(y)$
7:         $s_{\mathcal{P}}^{\circ} = \max_{y \neq x_{\mathcal{P}}^*} S_{\hat{v}}(y)$
8:         $q \leftarrow \text{SELECTQUERY}(\check{v}, \hat{v}, x_{\mathcal{P}}^*, y_{\mathcal{P}}^{\circ})$ ▷ Use a query strategy
9:         $p \leftarrow \text{ASKQUERY}(q)$ ▷ Ask query q to user
10:        $\mathcal{P} = \mathcal{P} \cup p$
11:        $(\check{v}, \hat{v}) = \text{UPDATE}(\check{v}, \hat{v}, p)$ ▷ Update lattice of capacities
12:    **until** $s_{\mathcal{P}}^* \geq s_{\mathcal{P}}^{\circ}$ ▷ Termination condition
13:    **return** $x_{\mathcal{P}}^*$ ▷ We return the necessary winner

---

3.2 An Interactive Elicitation Scheme for Determining a Necessary Winner

This section proposes an interactive elicitation process based on the concepts introduced above; Algorithm 1 depicts the pseudocode of our procedure. The input parameters are X, the dataset, and the preference statements $\mathcal{P}$. During the course of the process, we maintain an explicit representation of the set $V^{\mathcal{P}}$ of feasible capacities.

The pair $(\check{v}, \hat{v})$ is initialized depending on $\mathcal{P}$:

- In the case that we start from an empty set of statements ($\mathcal{P} = \emptyset$) we initialize the pair $(\check{v}, \hat{v}) = (\bot, \top)$. These capacities entails particular cases for Sugeno integral: $S_{\bot}(x) = \min_{i=1}^{n} x_i$ and $S_{\top}(x) = \max_{i=1}^{n} x_i$.
- If $\mathcal{P}$ is not empty, for each $p \in \mathcal{P}$, we use Eq. 7 in order to initialise $(\check{v}, \hat{v})$.

At each step of the elicitation, a query is asked and a new statement is acquired. Based on this the lattice is updated. We then we compute the new maximin optimal alternative ensuring the highest value $s_{\mathcal{P}}^{\downarrow}$.

Questions are chosen considering the value $s_{\mathcal{P}}^*$ of the maximin alternative, given preferences $\mathcal{P}$, and the value of $y_{\mathcal{P}}^{\circ}$ that is the alternative, different than x, that have the highest Sugeno value. Proposition 5 gives us a termination condition for ending the elicitation process.

The function UPDATE updates the lower bound $\check{v}$ and the upper bound $\hat{v}$ of the lattice of capacities consistent with the current information. That is,

$$\check{v} := \check{v} \vee \check{v}_{x,\text{succ}(\alpha)} \qquad\qquad \text{if } p \text{ of type } S(x) > \alpha$$
$$\hat{v} := \hat{v} \wedge \hat{v}_{x,\alpha} \qquad\qquad \text{if } p \text{ of type } S(x) \leq \alpha$$

where $\text{succ}(\alpha)$ is the level in L right above α; see Definition 1 for how $\check{v}_{x,\alpha}$ and $\hat{v}_{x,\alpha}$ are defined.

A question is identified by a pair (x, α): the alternative x that we are asking about, and the level α. The space of possible queries at a given step of the elicitation is

$$Q(\check{v}, \hat{v}) = \{(x, \alpha) | x \in X, S_{\check{v}(x)} \leq \alpha \leq S_{\hat{v}(x)}\}.$$

We now formally state a property ensuring that the algorithm cannot lead to an empty set of feasible capacities.

Proposition 6. *During all the steps of procedure* INTERACTIVEELIC *we have* $\check{v} \leq \hat{v}$ *(that means the sublattice of capacities that they represent is not empty).*

Proof. The property is true when we start the procedure. The proof is based on showing that the property is still satisfied after updating the lattice of capacities.

Suppose to have $\check{v}$ and $\hat{v}$ with $\check{v} \leq \hat{v}$. Consider any pair $(x, \alpha) \in Q(\check{v}, \hat{v})$, i.e., x and α satisfy $S_{\check{v}}(x) \leq \alpha \leq S_{\hat{v}}(x)$. There are two possible answers we want to have either $S_v(x) \geq \alpha$ or $S_v(x) \leq \alpha$. Hence we update the bounds of the set of capacities solution. In the first case the lower bound of the set of the capacities solution changes, while in the second one it is the upper bound.

– Suppose to have $S_v(x) \geq \alpha$. Let us denote the new lower bound by $\check{v}'$,

$$\check{v}'(A) = \check{v}(A) \vee \check{v}_{x,\alpha}(A) = \begin{cases} \max(\check{v}(A), \alpha) \text{ if } \{i|x_i \geq \alpha\} \subseteq A \\ \check{v}(A) \text{ otherwise} \end{cases}$$

Let us prove that $\check{v}' \leq \hat{v}$. We have $\check{v} \leq \hat{v}$ so we just need to prove that $\hat{v}(A) \geq \alpha$ if $\{i|x_i \geq \alpha\} \subseteq A$: We have

$$S_{\hat{v}}(x) = \max_{\beta \in L} \min(\hat{v}(\{i|x_i \geq \beta\}), \beta) \geq \alpha,$$

so there exists $\beta \geq \alpha$ such that $\hat{v}(\{i|x_i \geq \beta\}) \geq \beta$. We have $\{i|x_i \geq \beta\} \subseteq \{i|x_i \geq \alpha\}$ which entails $\hat{v}(\{i|x_i \geq \alpha\}) \geq \hat{v}(\{i|x_i \geq \beta\}) \geq \beta \geq \alpha$. We conclude using the monotonicity of $\hat{v}$.
– Suppose to have $S_v(x) \leq \alpha$. Let us denote the new upper bound by $\hat{v}'$,

$$\hat{v}'(A) = \hat{v}(A) \wedge \hat{v}_{x,\alpha}(A) = \begin{cases} \min(\hat{v}(A), \alpha) \text{ if } A \subseteq \{i|x_i > \alpha\} \\ \hat{v}(A) \text{ otherwise} \end{cases}$$

Let us prove that $\check{v} \leq \hat{v}'$. We have $\check{v} \leq \hat{v}$ so we just need to prove that $\check{v}(A) \leq \alpha$ if $A \subseteq \{i|x_i > \alpha\}$: We have

$$S_{\check{v}}(x) = \min_{\beta \in L} \max(\check{v}(\{i|x_i > \beta\}), \beta) \leq \alpha,$$

so there exists $\beta \leq \alpha$ such that $\check{v}(\{i|x_i > \beta\}) \leq \beta$. We have $\{i|x_i > \alpha\} \subseteq \{i|x_i > \beta\}$ which entails $\check{v}(\{i|x_i > \alpha\}) \leq \check{v}(\{i|x_i > \beta\}) \leq \beta \leq \alpha$. We conclude using the monotonicity of $\check{v}$.

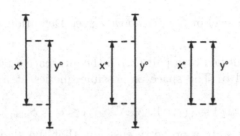

Fig. 1. The CSS1 strategy analyzes the different relative positions of the upper bounds and lower bounds of $x_{\mathcal{P}}^*$ and $y_{\mathcal{P}}^\circ$.

3.3   Strategies to Choose the Next Question

We now address the problem of choosing the next question. This is an important point since a good strategy for asking questions will reduce the length of the elic-itation process and as well mitigate the cognitive effort of the user. We consider different strategies to select the next question based on the current lattice of valid capacities. The effectiveness of these strategies are evaluated in simulation (see Sect. 4).

The Current solution strategy (CSS) uses the information about the current best recommendation $x_{\mathcal{P}}^*$, and the "adversary" $y_{\mathcal{P}}^\circ$ to derive a question to ask. We propose two versions of this idea:

- *CSS0* (simpler version): we simply choose to ask about $x_{\mathcal{P}}^*$ or $y_{\mathcal{P}}^\circ$ depending on which has the largest interval, and as level we pick the midpoint.
- *CSS1* (more elaborate): we evaluate candidate queries with respect to their capability of resolving the uncertainty about which between $x_{\mathcal{P}}^*$ and $y_{\mathcal{P}}^\circ$ has the highest Sugeno value. The discussion depends on how the intervals $[s_{\mathcal{P}}^\downarrow(x_{\mathcal{P}}^*), s_{\mathcal{P}}^\uparrow(x_{\mathcal{P}}^*)]$ and $[s_{\mathcal{P}}^\downarrow(y_{\mathcal{P}}^\circ), s_{\mathcal{P}}^\uparrow(y_{\mathcal{P}}^\circ)]$ relate to each other. It is worth notic-ing that we know that $s_{\mathcal{P}}^\downarrow(x_{\mathcal{P}}^*) \geq s_{\mathcal{P}}^\downarrow(y_{\mathcal{P}}^\circ)$ by definition of maximin.
  We inspect the order between $s_{\mathcal{P}}^\uparrow(x_{\mathcal{P}}^*)$ and $s_{\mathcal{P}}^\uparrow(y_{\mathcal{P}}^\circ)$ to decide which queries to consider. We then propose an heuristic in order to choose between these possible queries based on the length of the intervals $[s_{\mathcal{P}}^\downarrow(x_{\mathcal{P}}^*), s_{\mathcal{P}}^\uparrow(x_{\mathcal{P}}^*)]$ and $[s_{\mathcal{P}}^\downarrow(y_{\mathcal{P}}^\circ), s_{\mathcal{P}}^\uparrow(y_{\mathcal{P}}^\circ)]$ as depicted in Fig. 1. Note that, in the following discussion, we denote by $d(\alpha, \beta)$ the number of levels between α and β where α and β are elements on the scale L.

- *Case i):* $s_{\mathcal{P}}^{\uparrow}(y_{\mathcal{P}}^{\circ}) \leq s_{\mathcal{P}}^{\downarrow}(x_{\mathcal{P}}^{*})$, i.e., $\left[s_{\mathcal{P}}^{\downarrow}(y_{\mathcal{P}}^{\circ}) \; \left[s_{\mathcal{P}}^{\downarrow}(x_{\mathcal{P}}^{*}) \; s_{\mathcal{P}}^{\uparrow}(y_{\mathcal{P}}^{\circ}) \right] \; s_{\mathcal{P}}^{\uparrow}(x_{\mathcal{P}}^{*}) \right]$.

  The optimistic value of $y_{\mathcal{P}}^{\circ}$ is lower or equal than the pessimistic value of $x_{\mathcal{P}}^{*}$.

  In this case we could ask the user to compare alternative $x_{\mathcal{P}}^{*}$ and the level $s_{\mathcal{P}}^{\uparrow}(y_{\mathcal{P}}^{\circ})$. If the answer is that $S_v(x_{\mathcal{P}}^{*}) \geq s_{\mathcal{P}}^{\uparrow}(y_{\mathcal{P}}^{\circ})$, we know that $y_{\mathcal{P}}^{\circ}$ cannot be better than $x_{\mathcal{P}}^{*}$, and therefore we resolve the uncertainty between the two; this event happens if the true Sugeno value of $x_{\mathcal{P}}^{*}$ is between $s_{\mathcal{P}}^{\uparrow}(y_{\mathcal{P}}^{\circ})$ and $s_{\mathcal{P}}^{\uparrow}(x_{\mathcal{P}}^{*})$. We then quantify the "score" of this query as the proportion of the interval of $[s_{\mathcal{P}}^{\downarrow}(x_{\mathcal{P}}^{*}), s_{\mathcal{P}}^{\uparrow}(x_{\mathcal{P}}^{*})]$ that makes us certain that $x_{\mathcal{P}}^{*}$ is preferred to $y_{\mathcal{P}}^{\circ}$, i.e. the number of levels between $s_{\mathcal{P}}^{\uparrow}(x_{\mathcal{P}}^{*})$ and $s_{\mathcal{P}}^{\uparrow}(y_{\mathcal{P}}^{\circ})$ divided by the number of levels between $s_{\mathcal{P}}^{\uparrow}(x_{\mathcal{P}}^{*})$ and $s_{\mathcal{P}}^{\downarrow}(x_{\mathcal{P}}^{*})$. Hence the value of this query is $\frac{d(s_{\mathcal{P}}^{\uparrow}(x_{\mathcal{P}}^{*}), s_{\mathcal{P}}^{\uparrow}(y_{\mathcal{P}}^{\circ}))}{d(s_{\mathcal{P}}^{\uparrow}(x_{\mathcal{P}}^{*}), s_{\mathcal{P}}^{\downarrow}(x_{\mathcal{P}}^{*}))}$.

  Alternatively, we could also ask to compare alternative $y_{\mathcal{P}}^{\circ}$ and the level $s_{\mathcal{P}}^{\downarrow}(x_{\mathcal{P}}^{*})$. If the user states that the Sugeno value of $y_{\mathcal{P}}^{\circ}$ is lower than $s_{\mathcal{P}}^{\downarrow}(x_{\mathcal{P}}^{*})$, then we can also conclude that $y_{\mathcal{P}}^{\circ}$ cannot be better than $x_{\mathcal{P}}^{*}$. Reasoning as above, we score this query $\frac{d(s_{\mathcal{P}}^{\downarrow}(x_{\mathcal{P}}^{*}), s_{\mathcal{P}}^{\downarrow}(y_{\mathcal{P}}^{\circ}))}{d(s_{\mathcal{P}}^{\uparrow}(y_{\mathcal{P}}^{\circ}), s_{\mathcal{P}}^{\downarrow}(y_{\mathcal{P}}^{\circ}))}$.

  We ask the query (among the two) that has the highest "value".

- *Case ii):* $s_{\mathcal{P}}^{\downarrow}(x_{\mathcal{P}}^{*}) \leq s_{\mathcal{P}}^{\uparrow}(y_{\mathcal{P}}^{\circ})$, i.e., $\left[s_{\mathcal{P}}^{\downarrow}(y_{\mathcal{P}}^{\circ}) \; \left[s_{\mathcal{P}}^{\downarrow}(x_{\mathcal{P}}^{*}) \; s_{\mathcal{P}}^{\uparrow}(x_{\mathcal{P}}^{*}) \right] \; s_{\mathcal{P}}^{\uparrow}(y_{\mathcal{P}}^{\circ}) \right]$.

  (the optimistic value of $y_{\mathcal{P}}^{\circ}$ is at least the pessimistic value of $x_{\mathcal{P}}^{*}$).

  We can ask to compare alternative $y_{\mathcal{P}}^{\circ}$ and $s_{\mathcal{P}}^{\downarrow}(x_{\mathcal{P}}^{*})$, whose "score" is $\frac{d(s_{\mathcal{P}}^{\uparrow}(y_{\mathcal{P}}^{\circ}), s_{\mathcal{P}}^{\downarrow}(x_{\mathcal{P}}^{*}))}{d(s_{\mathcal{P}}^{\uparrow}(y_{\mathcal{P}}^{\circ}), s_{\mathcal{P}}^{\downarrow}(y_{\mathcal{P}}^{\circ}))}$, or, as in the previous case, to compare $y_{\mathcal{P}}^{\circ}$ and $s_{\mathcal{P}}^{\downarrow}(x_{\mathcal{P}}^{*})$, with value $\frac{d(s_{\mathcal{P}}^{\downarrow}(x_{\mathcal{P}}^{*}), s_{\mathcal{P}}^{\downarrow}(y_{\mathcal{P}}^{\circ}))}{d(s_{\mathcal{P}}^{\uparrow}(y_{\mathcal{P}}^{\circ}), s_{\mathcal{P}}^{\downarrow}(y_{\mathcal{P}}^{\circ}))}$. Since the denominator of both formulas is the same, the test reduces to checking $d(s_{\mathcal{P}}^{\uparrow}(y_{\mathcal{P}}^{\circ}), s_{\mathcal{P}}^{\downarrow}(x_{\mathcal{P}}^{*})) \geq d(s_{\mathcal{P}}^{\uparrow}(x_{\mathcal{P}}^{*}), s_{\mathcal{P}}^{\downarrow}(y_{\mathcal{P}}^{\circ}))$.

- *Case iii):* $s_{\mathcal{P}}^{\downarrow}(y_{\mathcal{P}}^{\circ}) = s_{\mathcal{P}}^{\downarrow}(x_{\mathcal{P}}^{*})$ and $s_{\mathcal{P}}^{\uparrow}(y_{\mathcal{P}}^{\circ}) = s_{\mathcal{P}}^{\uparrow}(x_{\mathcal{P}}^{*})$. In this case we ask about $x_{\mathcal{P}}^{*}$ and its midpoint level.

The Halve largest gap (HLG) asks the question about alternative x_H with the largest gap measured by the number of levels; $x_H = \arg\max_{x \in X} d(s_{\mathcal{P}}^{\downarrow}(x), s_{\mathcal{P}}^{\uparrow}(x))$ and the level α_H is the midpoint between $s_{\mathcal{P}}^{\uparrow}(x_H)$ and $s_{\mathcal{P}}^{\downarrow}(x_H)$.

The Random strategy chooses, as q query, an alternative x at random and a the midpoint between $s_{\mathcal{P}}^{\uparrow}(x)$ and $s_{\mathcal{P}}^{\downarrow}(x)$ (this strategy is considered as a baseline).

Example 1 (continued). We show how the different strategies will determine the next query to ask in our example. Remember that $x_{\mathcal{P}}^{*} = b$ and $y_{\mathcal{P}}^{\circ} = c$. The strategy CSS0 asks about alternative c and its midpoint level 0.6, since the interval $[s_{\mathcal{P}}^{\downarrow}(b), s_{\mathcal{P}}^{\uparrow}(b)] = [0.5, 0.7]$ is smaller (in terms of number of levels) that the interval $[s_{\mathcal{P}}^{\downarrow}(c), s_{\mathcal{P}}^{\uparrow}(c)] = [0.4, 0.8]$.

Since $s_\mathcal{P}^\uparrow(x_\mathcal{P}^*) = 0.7 < 0.8 = s_\mathcal{P}^\uparrow(y_\mathcal{P}^\circ)$ the analysis performed by CSS1 proceeds by considering the second case. CSS1 asks either to compare alternative c with level 0.5 or to compare alternative c with level 0.7 (their "score" is the same).

HLG asks about alternative d, that has the widest gap $[s_\mathcal{P}^\downarrow(d), s_\mathcal{P}^\uparrow(d)] = [0, 0.5]$, and its midpoint level 0.2. This query is not very informative since we know already that d cannot be strictly better than b.

Table 1. Simulation results (averaged over 30 runs) showing the number of queries that are needed in order to find a necessary winner.

Dataset	m	n	Capacity	Query strategies			
				CSS1	CSS0	HLG	Random
Tiny	7	4	WeightedMax	5.5	6.3	9.1	16.9
Tiny	7	4	WeightedMin	8.1	9.1	10.4	17.8
Tiny	7	4	Sampled	8.9	8.3	11.1	16.9
Synthetic	30	8	WeightedMax	14.7	10.2	45.2	56.2
Synthetic	30	8	WeightedMin	15.8	17.2	34.8	70.6
Synthetic	30	8	Sampled	27.3	22.3	54.7	100.6
Cars	80	6	WeightedMax	3.6	4.1	12.8	4.7
Cars	80	6	WeightedMin	12.6	12.2	26.8	42.7
Cars	80	6	Sampled	9.3	10.0	31.0	22.2

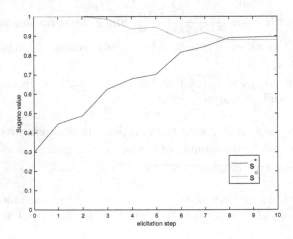

Fig. 2. The values s^* and s° as a function of the number of queries with CSS1 on the tiny dataset (averaged over 30 runs)

4   Experimental Results

We evaluate the proposed paradigm with numerical experiments where we simulate the elicitation process by assuming that the preferences of a decision-maker are consistent with a Sugeno integral with a capacity v. At each step of the simulation, a question of the type *"Is $S_v(x)$ lower or equal to α?"* is asked to decision maker. The simulated user answers such questions based on the "true" capacity v, and the answers are used by our algorithm to update the lattice of consistent capacities, determine the maximin recommendations and to select the question to ask next, as discussed before in Sect. 3.2.

In our tests we considered 3 different datasets: a very small dataset, dubbed "tiny" of 7 items and 4 criteria (with an evaluation scale of 20 levels), a randomly generated "synthetic" dataset (30 items, 8 criteria, 25 levels), and a dataset of "cars" (100 items, 6 criteria, 5 levels).

Simulated users answer queries according to capacities that are either WeightedMax, WeightedMin or generic capacities (note that the form of the capacity is not known to the elicitation algorithm). Capacities are randomly generated in the following way: for WeightedMax and WeightedMin the weight vector is uniformly sampled (one criteria forced to have weight 1_L). For generic capacities, we iteratively pick a random subset of criteria and assign it a random level (sampled uniformly) between 0_L and 1_L with subsets and supersets updated accordingly to monotonicity; the process is repeated a fixed number of times.

We compare the effectiveness of the heuristic strategies (presented in the Sect. 3.2) for choosing the next question to ask to the user. In Table 1 we show the average number of queries that are needed to find a necessary winner according to the different query strategies, in the different simulation settings (all experiments have been repeated 30 runs).

In Fig. 2 we provide, for one of the experiments, the detail about how the values $s_{\mathcal{P}}^*$ and $s_{\mathcal{P}}^\circ$ evolve over time: the former is monotonically non decreasing, while the latter decreases most of the time. Note that our protocol can be terminated early providing a "good" recommendation before a necessary winner is found.

The experimental results show the superiority of the CSS strategies with respect to the other heuristics, with both CSS0 and CSS1 performing quite well.

5   Discussion and Conclusions

The Sugeno integral is used as an aggregation method for multicriteria decision making. Despite its popularity, the elicitation of a Sugeno integral is still a problematic issue. In this paper we have provided a novel formalization for decision-making under capacity uncertainty using the maximin utility criterion. We have provided an incremental elicitation method for determining a necessary winner; at each step the user is asked to answer a query of the type *"Is the Sugeno value of item x at most α?"* where x and α are dynamically chosen to improve the knowledge about the capacity as much as possible. The algorithm maintains a representation of the lattice of consistent capacities that is updated after each answer. We provided an experimental validation of the approach with simulations comparing different heuristics for choosing the next question to ask.

Several directions for future research are possible: additional strategies for choosing the next question (for instance adapting the ideas of [14] in an ordinal setting), experimentation with real data, the extension to different type of questions (e.g. comparing two alternatives) and handling combinatorial domains. We are also interested in methods that support the interpretation of real data, for instance by using *if-then* rules based on Sugeno integrals. As the Sugeno integral represents a single threshold rule [6], an interesting direction is to adapt our procedure for Sugeno Utility Functionals (SUF) [4].

Another challenge is the prediction of preferences [9]; one idea is to compute a family of capacities on a training set of preferences and use it for prediction. The analogy-based method [3] seems as well to be an approach to consider.

References

1. Beliakov, G.: Construction of aggregation functions from data using linear programming. Fuzzy Sets Syst. **160**(1), 65–75 (2009). ISSN 0165-0114
2. Benabbou, N., Perny, P., Viappiani, P.: Incremental elicitation of Choquet capacities for multicriteria decision making. In: ECAI 2014, pp. 87–92 (2014)
3. Bounhas, M., Pirlot, M., Prade, H., Sobrie, O.: Comparison of analogy-based methods for predicting preferences. In: Ben Amor, N., Quost, B., Theobald, M. (eds.) SUM 2019. LNCS (LNAI), vol. 11940, pp. 339–354. Springer, Cham (2019). https://doi.org/10.1007/978-3-030-35514-2_25
4. Brabant, Q., Couceiro, M., Dubois, D., Prade, H., Rico, A.: Extracting decision rules from qualitative data via Sugeno utility functionals. In: Medina, J., et al. (eds.) IPMU 2018. CCIS, vol. 853, pp. 253–265. Springer, Cham (2018). https://doi.org/10.1007/978-3-319-91473-2_22
5. Davey, B., Priestley, H.: Introduction to Lattices and Order, 2nd edn. Cambridge University Press, Cambridge (2002)
6. Dubois, D., Durrieu, C., Prade, H., Rico, A., Ferro, Y.: Extracting decision rules from qualitative data using Sugeno integral: a case-study. In: Destercke, S., Denoeux, T. (eds.) ECSQARU 2015. LNCS (LNAI), vol. 9161, pp. 14–24. Springer, Cham (2015). https://doi.org/10.1007/978-3-319-20807-7_2
7. Dubois, D., Marichal, J.-L., Prade, H., Roubens, M., Sabbadin, R.: The use of the discrete Sugeno integral in decision-making: a survey. Int. J. Uncertainty Fuzziness Knowl. Based Syst. **9**(5), 539–561 (2001)
8. Grabisch, M., Nicolas, J.M.: Classification by fuzzy integral - performance and tests. Fuzzy Sets Syst. **65**, 255–271 (1994)
9. Hüllermeier, E., Fürnkranz, J.: Editorial: preference learning and ranking. Mach. Learn. **93**(2), 185–189 (2013). https://doi.org/10.1007/s10994-013-5414-z
10. Prade, H., Rico, A., Serrurier, M., Raufaste, E.: Elicitating Sugeno integrals: methodology and a case study. In: Sossai, C., Chemello, G. (eds.) ECSQARU 2009. LNCS (LNAI), vol. 5590, pp. 712–723. Springer, Heidelberg (2009). https://doi.org/10.1007/978-3-642-02906-6_61

11. Raufaste, E., Prade, H.: Sugeno integrals in subjective mental workload evaluation: application to flying personnel data. In: Proceedings of the 11th International Conference on Information Processing and Management of Uncertainty in Knowledge-Based Systems, IPMU 2006, pp. 564–570 (2006)
12. Rico, A., Grabisch, M., Labreuche, C., Chateauneuf, A.: Preference modelling on totally ordered sets by the Sugeno integral. Discrete Appl. Math. 147(1), 113–124 (2005)
13. Sugeno, M.: Theory of Fuzzy Integrals and its Applications. Ph.D. Thesis, Tokyo Institute of Technology (1974)
14. Viappiani, P., Boutilier, C.: Regret-based optimal recommendation sets in conversational recommender systems. In: Proceeding of the ACM Conference on Recommender Systems (RecSys), pp. 101–108 (2009)
15. Viappiani, P., Kroer, C.: Robust optimization of recommendation sets with the maximin utility criterion. In: Perny, P., Pirlot, M., Tsoukiàs, A. (eds.) ADT 2013. LNCS (LNAI), vol. 8176, pp. 411–424. Springer, Heidelberg (2013). https://doi.org/10.1007/978-3-642-41575-3_32

Computable Randomness Is About More Than Probabilities

Floris Persiau[✉], Jasper De Bock, and Gert de Cooman

FLip, ELIS, Ghent University, Technologiepark 125, 9052 Zwijnaarde, Belgium
floris.persiau@ugent.be

Abstract. We introduce a notion of computable randomness for infinite sequences that generalises the classical version in two important ways. First, our definition of computable randomness is associated with imprecise probability models, in the sense that we consider lower expectations—or sets of probabilities—instead of classical 'precise' probabilities. Secondly, instead of binary sequences, we consider sequences whose elements take values in some finite sample space. Interestingly, we find that every sequence is computably random with respect to at least one lower expectation, and that lower expectations that are more informative have fewer computably random sequences. This leads to the intriguing question whether every sequence is computably random with respect to a unique most informative lower expectation. We study this question in some detail and provide a partial answer.

Keywords: Computable randomness · Coherent lower expectations · Imprecise probabilities · Supermartingales · Computability

1 Introduction

When do we consider an infinite sequence $\omega = (x_1, \ldots, x_n, \ldots)$, whose individual elements x_n take values in some finite sample space $\mathcal{X}$, to be random? This is actually not a fair question, because randomness is never defined absolutely, but always relative to an uncertainty model. Consider for example an infinite sequence generated by repeatedly throwing a single fair die and writing down the number of eyes on each throw. In this case, we would be justified in calling this sequence random with respect to a precise probability model that assigns probability 1/6 to every possible outcome.

It is exactly such precise probability models that have received the most attention in the study of randomness [2,3,11]. Early work focused on binary sequences and the law of large numbers that such sequences, and computably selected subsequences, were required to satisfy: an infinite binary sequence of zeros and ones is called *Church random* if the relative frequencies in any computably selected subsequence converge to 1/2 [2]. Schnorr, inspired by the work of Ville, strengthened this definition by introducing a notion of computable randomness [11]. On his account, randomness is about betting. The starting point is that a precise probability model that assigns a (computable) probability p to 1 and $1 - p$ to 0 can be interpreted as stating that p is a fair price for bet $\mathbb{I}_1(X_i)$

© Springer Nature Switzerland AG 2020
J. Davis and K. Tabia (Eds.): SUM 2020, LNAI 12322, pp. 172–186, 2020.
https://doi.org/10.1007/978-3-030-58449-8_12

that yields 1 when $X_i = 1$ and 0 when $X_i = 0$, for every—a priori unknown—value X_i of a binary sequence $\omega = (x_1, \ldots, x_n, \ldots)$ of zeros and ones. Such a sequence is then considered to be *computably random* with respect to p if there is no computable betting strategy for getting rich without bounds along ω without borrowing, simply by betting according to this fair price. Notably, binary sequences that are computably random for $p = 1/2$ are also Church random. So here too, the relative frequency of any element $x \in X$ will converge to a limit frequency along ω—1/2 in the binary case for $p = 1/2$. In fact, this is typically true for any notion of randomness with respect to a precise probability model.

However, as has been argued extensively [7], there are various random phenomena where this stabilisation is not clearly present, or even clearly absent. Hence, only adopting precise probability models to define notions of random sequences is too much of an idealisation. Recently, this issue was addressed by De Cooman and De Bock for binary sequences by introducing a notion of computable randomness with respect to probability intervals instead of precise probability models, whose lower bounds represent supremum acceptable buying prices, and whose upper bounds represent infimum acceptable selling prices, again for the bet $\mathbb{I}_1(X_i)$ that, for every value x_i of a binary sequence $\omega = (x_1, \ldots, x_n, \ldots)$, yields 1 if $X_i = 1$ and 0 otherwise [5].

On this account, relative frequencies must not necessarily converge to a limit frequency along ω, but may fluctuate within the probability interval.

Here, we generalise the work done by De Cooman and De Bock [5] for binary sequences, and develop a similar concept for infinite sequences that take values in more general finite sample spaces. To this end, we consider an even more general framework for describing uncertainty: we use coherent lower expectations—or sets of probability mass functions—instead of probability intervals or probabilities. Loosely speaking, we say that an infinite sequence $\omega = (x_1, \ldots, x_n, \ldots)$ is *computably random* with respect to a (forecasting system of) lower expectation(s), when there is no computable betting strategy for getting rich without bounds along ω without borrowing and by only engaging in bets whose (upper) expected profit is non-positive or negative.[1]

This contribution is structured as follows. We start in Sect. 2 with a brief introduction to coherent lower expectations, and explain in particular their connection with probabilities and their interpretation in terms of gambles and betting. Next, in Sect. 3, we define a subject's uncertainty for an infinite sequence of variables $X_1, \ldots, X_n, \ldots$ by introducing forecasting systems that associate with every finite sequence $(x_1, \ldots, x_n)$ a coherent lower expectation for the variable X_{n+1}. This allows us to introduce corresponding betting strategies to bet on the infinite sequence of variables along a sequence $\omega = (x_1, \ldots, x_n, \ldots)$ in terms of non-negative (strict) supermartingales. After explaining in Sect. 4 when such a non-negative (strict) supermartingale is computable, we extend the existing notion of computable randomness from precise and interval probability models to coherent lower expectations in Sect. 5, and study its properties. The remainder of the paper focuses on

---

[1] A real number $x \in \mathbb{R}$ is called positive if $x > 0$, non-negative if $x \geq 0$, negative if $x < 0$ and non-positive if $x \leq 0$.

special cases. When we restrict our attention to stationary forecasting systems that forecast a single coherent lower expectation in Sect. 6, it turns out that every sequence ω is computably random with respect to at least one coherent lower expectation and that if ω is computably random for some coherent lower expectation, then it is also computably random for any coherent lower expectation that is less informative, i.e., provides fewer gambles. This makes us question whether there is a unique most informative coherent lower expectation for which ω is computably random. After inspecting some examples, it turns out that such a most informative coherent lower expectation sometimes exists, but sometimes does not. When it does not, our examples lead us to conjecture that it 'almost' exists. We conclude the discussion in Sect. 7 by introducing a derived notion of computable randomness with respect to a gamble f and an interval I by focusing on the behaviour of coherent lower expectations on a specific gamble f of their domain. It turns out that for every gamble f, a sequence ω is 'almost' computably random with respect to some smallest interval. To adhere to the page constraints, all proofs are omitted. They are available in an extended on-line version [9].

2   Coherent Lower Expectations

To get the discussion started, we consider a single uncertain variable X that takes values in some finite set $\mathcal{X}$, called the sample space. A subject's uncertainty about the unknown value of X can then be modelled in several ways. We will do so by means of a coherent lower expectation: a functional that associates a real number with every gamble, where a gamble $f : \mathcal{X} \to \mathbb{R}$ is a map from the sample space $\mathcal{X}$ to the real numbers. We denote the linear space of all gambles by $\mathcal{L}(\mathcal{X})$.

Definition 1. *A coherent lower expectation* $\underline{E} : \mathcal{L}(\mathcal{X}) \to \mathbb{R}$ *is a real-valued functional on* $\mathcal{L}(\mathcal{X})$ *that satisfies the following axioms. For all gambles* $f, g \in \mathcal{L}(\mathcal{X})$ *and all non-negative* $\alpha \in \mathbb{R}$:

C1. $\min f \leq \underline{E}(f)$ [boundedness]
C2. $\underline{E}(\alpha f) = \alpha \underline{E}(f)$ [non-negative homogeneity]
C3. $\underline{E}(f) + \underline{E}(g) \leq \underline{E}(f + g)$ [superadditivity]

We will use $\underline{\mathcal{E}}$ *to denote the set of all coherent lower expectations on* $\mathcal{L}(\mathcal{X})$.

As a limit case, for any probability mass function p on $\mathcal{X}$, it is easy to check that the linear expectation E_p, defined by $E_p(f) := \sum_{x \in \mathcal{X}} f(x)p(x)$ for all $f \in \mathcal{L}(\mathcal{X})$, is a coherent lower expectation, which corresponds to a maximally informative or least conservative model for a subject's uncertainty. More generally, a coherent lower expectation $\underline{E}$ can be interpreted as a lower envelope of such linear expectations. That is, there is always a (closed and convex) set $\mathcal{M}$ of probability mass functions such that $\underline{E}(f) = \min\{E_p(f) : p \in \mathcal{M}\}$ for all $f \in \mathcal{L}(\mathcal{X})$ [13]. In that sense, coherent lower expectations can be regarded as a generalisation of probabilities to (closed and convex) sets of probabilities. Alternatively, the lower expectation $\underline{E}(f)$ can be interpreted directly as a subject's supremum buying price for the uncertain reward f.

The particular interpretation that is adopted is not important for what we intend to do here. For our purposes, the only thing we will assume is that when a subject specifies a coherent lower expectation, every gamble $f \in \mathcal{L}(X)$ such that $\underline{E}(f) > 0$ is desirable to him and every gamble $f \in \mathcal{L}(X)$ such that $\underline{E}(f) \geq 0$ is acceptable to him. We think this makes sense under both of the aforementioned interpretations. Furthermore, as we will see in Sect. 5, the distinction between desirable and acceptable gambles does not matter for our definition of computable randomness. For now, however, we proceed with both notions.

Whenever a subject specifies a coherent lower expectation, we can consider an opponent that takes this subject up on a gamble f on the unknown outcome X in a betting game. Borrowing terminology from the field of game-theoretic probabilities [12], we will refer to our subject as Forecaster and to his opponent as Sceptic. Forecaster will only bet according to those gambles $f \in \mathcal{L}(X)$ that are acceptable to him ($\underline{E}(f) \geq 0$), or alternatively, those that are desirable to him ($\underline{E}(f) > 0$). This leads to an unknown reward $f(X)$ for Forecaster and an unknown reward $-f(X)$ for Sceptic. After Sceptic selects such a gamble, the outcome $x \in X$ is revealed, Forecaster receives the (possibly negative) reward $f(x)$, and Sceptic receives the reward $-f(x)$. Equivalently, when considering for any coherent lower expectation $\underline{E}$ the conjugate upper expectation $\overline{E}$, defined as $\overline{E}(f) := -\underline{E}(-f)$ for all $f \in \mathcal{L}(X)$, then Sceptic is allowed to bet according to any gamble $f \in \mathcal{L}(X)$ for which $\overline{E}(f) \leq 0$ (or $\overline{E}(f) < 0$), leading to an uncertain reward $f(X)$ for Sceptic and an uncertain reward $-f(X)$ for Forecaster. In what follows, we will typically take the perspective of Sceptic. The gambles that are available to her will thus be the gambles $f \in \mathcal{L}(X)$ with non-positive (or negative) upper expectation $\overline{E}(f) \leq 0$ ($\overline{E}(f) < 0$).

An important special case is the so-called vacuous coherent lower expectation $\underline{E}_v$, defined by $\underline{E}_v(f) := \min f$ for all $f \in \mathcal{L}(X)$. If Forecaster specifies $\underline{E}_v$, this corresponds to a very conservative attitude where he is only interested in gambles f that give him a guaranteed non-negative (or positive) gain, i.e., $\min f \geq 0$ ($\min f > 0$), implying that Sceptic has a guaranteed non-negative (or positive) loss, i.e., $\max f \leq 0$ ($\max f < 0$).

Example 2. Consider an experiment with three possible outcomes A, B and C, i.e., $X := \{A, B, C\}$, and three probability mass functions p_0, p_1 and p_2 defined by $(p_0(A), p_0(B), p_0(C)) := (0, 1/2, 1/2)$, $(p_1(A), p_1(B), p_1(C)) := (1/2, 0, 1/2)$ and $(p_2(A), p_2(B), p_2(C)) := (1/2, 1/2, 0)$. We can then define a coherent lower expectation $\underline{E}$ by $\underline{E}(f) := \min\{E_{p_0}(f), E_{p_1}(f), E_{p_2}(f)\}$ for every gamble $f \in \mathcal{L}(X)$. For the particular gamble f defined by $(f(A), f(B), f(C)) := (1, -2, 3)$, the value of this lower expectation then equals $\underline{E}(f) = \min\{1/2, 2, -1/2\} = -1/2$.   ◊

3   Forecasting Systems and Betting Strategies

We now consider a sequential version of the betting game in Sect. 2 between Forecaster and Sceptic, by considering a sequence of variables $X_1, \ldots, X_n, \ldots$, all of which take values in our finite sample space X.

On each round of the game, indexed by $n \in \mathbb{N}_0 := \mathbb{N} \cup \{0\}$, the a priori unknown finite sequence of outcomes $x_{1:n} = (x_1, \ldots, x_n)$ has been revealed and we assume that Forecaster's uncertainty about the next—as yet unknown— outcome $X_{n+1} \in \mathcal{X}$ is described by a coherent lower expectation. Hence, on each round of the game, Forecaster's uncertainty can depend on and be indexed by the past states.

All finite sequences $s = x_{1:n} = (x_1, \ldots, x_n)$—so-called situations—are collected in the set $\mathbb{S} := \mathcal{X}^* = \bigcup_{n \in \mathbb{N}_0} \mathcal{X}^n$. By convention, we call the empty sequence the initial situation and denote it by $\square$. The finite sequences $s \in \mathbb{S}$ form an event tree, and it is on this whole event tree that we will describe Forecaster's uncertainty, using a so-called forecasting system.

Definition 3. *A forecasting system $\underline{E}_{\bullet} \colon \mathbb{S} \to \underline{\mathcal{E}}$ is a map that associates with every situation $s \in \mathbb{S}$ a coherent lower expectation $\underline{E}_s \in \underline{\mathcal{E}}$. The collection of all forecasting systems is denoted by $\underline{\mathcal{E}}^{\mathbb{S}}$.*

Every forecasting system corresponds to a collection of bets that are available to Sceptic. That is, in every situation $s = x_{1:n}$, Sceptic is allowed to bet on the unknown outcome X_{n+1} according to any gamble $f \in \mathcal{L}(\mathcal{X})$ such that $\overline{E}_s(f) \leq 0$ (or $\overline{E}_s(f) < 0$). This leads to an uncertain reward $f(X_{n+1})$ for Sceptic and an uncertain reward $-f(X_{n+1})$ for Forecaster. Afterwards, when the outcome x_{n+1} is revealed, Sceptic gets the amount $f(x_{n+1})$, Forecaster gets the amount $-f(x_{n+1})$ and we move to the next round. To formalise this sequential betting game, we introduce the notion of a supermartingale, which is a special case of a so-called real process.

A real process $F \colon \mathbb{S} \to \mathbb{R}$ is a map that associates with every situation $s = x_{1:n} \in \mathbb{S}$ of the event tree, a real number $F(s)$. With every real process F there corresponds a process difference ΔF that associates with every situation $s \in \mathbb{S}$ a gamble $\Delta F(s)$ on $\mathcal{X}$, defined as $\Delta F(s)(x) := F(sx) - F(s)$ for every $s \in \mathbb{S}$ and $x \in \mathcal{X}$, where sx denotes the concatenation of s and x. We call a real process M a (strict) supermartingale if $\overline{E}_s(\Delta M(s)) \leq 0$ $(\overline{E}_s(\Delta M(s)) < 0)$ for every situation $s \in \mathbb{S}$. Note that a supermartingale is always defined relative to a forecasting system $\underline{E}_{\bullet}$. Similarly, a real process M is called a (strict) submartingale if $\underline{E}_s(\Delta M(s)) \geq 0$ $(\underline{E}_s(\Delta M(s)) > 0)$ for every $s \in \mathbb{S}$. Due to the conjugacy relation between upper and lower expectations, M is a (strict) supermartingale if and only if $-M$ is a (strict) submartingale. We collect the super- and submartingales in the sets $\overline{\mathbb{M}}(\underline{E}_{\bullet})$ and $\underline{\mathbb{M}}(\underline{E}_{\bullet})$, respectively. A supermartingale M is called non-negative (positive) if $M(s) \geq 0$ $(M(s) > 0)$ for all $s \in \mathbb{S}$.

From the previous discussion, it is clear that Sceptic's allowable betting behaviour corresponds to supermartingales or strict supermartingales, depending on whether we consider acceptable or desirable gambles, respectively. Indeed, in each situation $s = x_{1:n} \in \mathbb{S}$, she can only select a gamble $\Delta M(s)$ for which $\overline{E}_s(\Delta M(s)) \leq 0$ $(\overline{E}_s(\Delta M(s)) < 0)$ and her accumulated capital $M(x_{1:n}) = M(\square) + \sum_{k=0}^{n-1} \Delta M(x_{1:k})(x_{k+1})$, with $M(\square)$ being her initial capital, will therefore evolve as a (strict) supermartingale. As mentioned before, it will turn out not to matter whether we consider acceptable or desirable gambles, or equivalently, supermartingales or strict supermartingales. To be able to explain why that is, we will proceed with both. In particular, we will restrict Sceptic's allowed betting strategies to non-negative (strict) supermartingales, where the non-negativity is imposed to prevent her from borrowing money. Non-negative supermartingales M that start with unit capital $M(\square)$ are called test supermartingales.

Example 4. Consider a repetition of the experiment in Example 2, and a stationary forecasting system $\underline{E}_\bullet$ defined by $\underline{E}_s(f) = \underline{E}(f) = \min\{E_{p_0}(f), E_{p_1}(f), E_{p_2}(f)\}$ for every $s \in \mathbb{S}$ and $f \in \mathcal{L}(X)$, with p_0, p_1 and p_2 as in Example 2. An example of a non-negative (test) supermartingale M is then given by the recursion equation $\Delta M(s) = (\Delta M(s)(A), \Delta M(s)(B), \Delta M(s)(C)) := (-M(s)/2, M(s)/2, -M(s)/2)$ for every $s \in \mathbb{S}$, with $M(\square) := 1$. E.g., for $s = A$, it follows that $M(A) = M(\square) + \Delta M(\square)(A) = M(\square) - M(\square)/2 = M(\square)/2 = 1/2$. It is easy to see that M is non-negative by construction and, for every $s \in \mathbb{S}$, it holds that $\overline{E}_s(\Delta M(s)) = \max\{0, -M(s)/2, 0\} = 0$.   ◊

In what follows, we will use Sceptic's allowed betting strategies—so non-negative (strict) supermartingales—to introduce a notion of computable randomness with respect to a forecasting system. We denote the set of all infinite sequences of states—or so-called paths—by $\Omega := X^\mathbb{N}$ and, for every such path $\omega = (x_1, \ldots, x_n, \ldots) \in \Omega$, we let $\omega^n := (x_1, \ldots, x_n)$ for all $n \in \mathbb{N}_0$.

However, not all betting strategies within the uncountable infinite set of all allowed betting strategies are implementable. We will therefore restrict our attention to those betting strategies that are computable, as an idealisation of the ones that can be practically implemented.

4   A Brief Introduction to Computability

Computability deals with the ability to compute mathematical objects in an effective manner, which means that they can be approximated to arbitrary precision in a finite number of steps. In order to formalise this notion, computability theory uses so-called recursive functions as its basic building blocks [8,10].

A function $\phi: \mathbb{N}_0 \to \mathbb{N}_0$ is recursive if it can be computed by a Turing machine, which is a mathematical model of computation that defines an abstract machine. By the Church–Turing thesis, this is equivalent to the existence of an algorithm that, upon the input of a natural number $n \in \mathbb{N}_0$, outputs the natural number $\phi(n)$. The domain $\mathbb{N}_0$ can also be replaced by any other countable set whose elements can be expressed by adopting a finite alphabet, which for example allows us to consider recursive functions from $\mathbb{S}$ to $\mathbb{N}_0$ or from $\mathbb{S} \times \mathbb{N}_0$ to $\mathbb{N}_0$. Any set of recursive functions is countable, because the set of all algorithms, which are finite sequences of computer-implementable instructions, is countable.

We can also consider recursive sequences of rationals, recursive rational processes and recursive nets of rationals. A sequence $\{r_n\}_{n \in \mathbb{N}_0}$ of rational numbers is called recursive if there are three recursive maps a, b, σ from $\mathbb{N}_0$ to $\mathbb{N}_0$ such that $b(n) \neq 0$ for all $n \in \mathbb{N}_0$ and $r_n = (-1)^{\sigma(n)} \frac{a(n)}{b(n)}$ for all $n \in \mathbb{N}_0$. By replacing the domain $\mathbb{N}_0$ with $\mathbb{S}$, we obtain a recursive rational process. That is, a rational process $F: \mathbb{S} \to \mathbb{Q}$ is called recursive if there are three recursive maps a, b, σ from $\mathbb{S}$ to $\mathbb{N}_0$ such that $b(s) \neq 0$ for all $s \in \mathbb{S}$ and $F(s) = (-1)^{\sigma(s)} \frac{a(s)}{b(s)}$ for all $s \in \mathbb{S}$.

In a similar fashion, a net of rationals $\{r_{s,n}\}_{s\in\mathbb{S},n\in\mathbb{N}_0}$ is called recursive if there are three recursive maps a, b, σ from $\mathbb{S}\times\mathbb{N}_0$ to $\mathbb{N}_0$ such that $b(s,n) \neq 0$ for every $s\in\mathbb{S}$ and $n\in\mathbb{N}_0$, and $r_{s,n} = (-1)^{\sigma(s,n)}\frac{a(s,n)}{b(s,n)}$ for all $s\in\mathbb{S}$ and $n\in\mathbb{N}_0$.

Using these recursive objects, we now move on to define the following mathematical objects that can be computed in an effective manner: computable reals, computable real gambles, computable probability mass functions and, finally, computable real processes such as non-negative supermartingales.

We say that a sequence $\{r_n\}_{n\in\mathbb{N}_0}$ of rational numbers converges effectively to a real number $x\in\mathbb{R}$ if $|r_n - x| \leq 2^{-N}$ for all $n, N\in\mathbb{N}_0$ such that $n\geq N$. A real number x is then called computable if there is a recursive sequence $\{r_n\}_{n\in\mathbb{N}_0}$ of rationals that converges effectively to x. Of course, every rational number is a computable real. A gamble $f\colon X \to \mathbb{R}$ and a probability mass function $p\colon X \to [0,1]$ are computable if $f(x)$ or $p(x)$ is computable for every $x\in X$, respectively. After all, finitely many algorithms can be combined into one.

However, a real process $F\colon \mathbb{S} \to \mathbb{R}$ may not be computable even if each of its individual elements $F(s)$ is, with $s\in\mathbb{S}$, because there may be no way to combine the corresponding infinite number of algorithms into one finite algorithm. For that reason, we will look at recursive nets of rationals instead of recursive sequences of rationals. We say that a net $\{r_{s,n}\}_{s\in\mathbb{S},n\in\mathbb{N}_0}$ of rational numbers converges effectively to a real process $F\colon \mathbb{S} \to \mathbb{R}$ if $|r_{s,n} - F(s)| \leq 2^{-N}$ for all $s\in\mathbb{S}$ and $n, N\in\mathbb{N}_0$ such that $n\geq N$. A real process F is then called computable if there is a recursive net $\{r_{s,n}\}_{s\in\mathbb{S},n\in\mathbb{N}_0}$ of rationals that converges effectively to F. Of course, every recursive rational process is also a computable real process. Observe also that, clearly, for any computable real process F and any $s\in\mathbb{S}$, $F(s)$ is a computable real number. Furthermore, a constant real process is computable if and only if its constant value is.

To end this section, we would like to draw attention to the fact that the set of all real processes is uncountable, while the set of all computable real (or recursive rational) processes is countable, simply because the set of all algorithms is countable. In the remainder, we will denote by $\overline{\mathbb{M}}_C(\underline{E}_{\bullet})$ the set of all computable non-negative supermartingales for the forecasting system $\underline{E}_{\bullet}$.

5   Computable Randomness for Forecasting Systems

At this point, it should be clear how Forecaster's uncertainty about a sequence of variables $X_1, \ldots, X_n, \ldots$ can be represented by a forecasting system $\underline{E}_{\bullet}$, and that such a forecasting system gives rise to a set of betting strategies whose corresponding capital processes are non-negative (strict) supermartingales. We will however not allow Sceptic to select any such betting strategy, but will require that her betting strategies should be effectively implementable by requiring that the corresponding non-negative (strict) supermartingales are computable. In this way, we restrict Sceptic's betting strategies to a countably infinite set. We will now use these strategies to define a notion of computable randomness with respect to a forecasting system $\underline{E}_{\bullet}$. The definition uses supermartingales rather than strict supermartingales, but as we will see shortly, this makes no difference.

Loosely speaking, we call a path ω computably random for $\underline{E}_\bullet$ if there is no corresponding computable betting strategy M that allows Sceptic to become rich without bounds along ω, i.e., $\sup_{n \in \mathbb{N}_0} M(\omega^n) = +\infty$, without borrowing.

Definition 5. *A path ω is computably random for a forecasting system $\underline{E}_\bullet$ if there is no computable non-negative real supermartingale $M \in \overline{\mathbb{M}}_C(\underline{E}_\bullet)$ that is unbounded along ω. We denote the collection of all forecasting systems for which ω is computably random by $\underline{\mathcal{E}}^S_C(\omega)$.*

It turns out that our definition is reasonably robust with respect to the particular types of supermartingales that are considered.

Proposition 6. *A path ω is computably random for a forecasting system $\underline{E}_\bullet$ if and only if there is no recursive positive rational strict test supermartingale $M \in \overline{\mathbb{M}}_C(\underline{E}_\bullet)$ such that $\lim_{n \to \infty} M(\omega^n) = +\infty$.*

As a consequence, whenever we restrict Sceptic's allowed betting strategies to a set that is smaller than the one in Definition 5, but larger than the one in Proposition 6, we obtain a definition for computably random sequences that is equivalent to Definition 5. Consequently, it indeed does not matter whether we restrict Sceptic's allowed betting strategies to supermartingales or strict supermartingales.

If we consider binary sequences and restrict Sceptic's betting behaviour to non-negative computable test supermartingales, our definition of computable randomness coincides with the one that was recently introduced by De Cooman and De Bock for binary sequences [5]. The equivalence is not immediate though because the forecasting systems in Ref. [5] specify probability intervals rather than coherent lower expectations. Nevertheless, it does hold because in the binary case, for every coherent lower expectation, the corresponding closed convex set of probability mass functions on $\mathcal{X} = \{0, 1\}$—see Sect. 2—is completely characterised by the associated probability interval for the outcome 1. Furthermore, in the case of binary sequences and stationary, precise, computable forecasting systems, it can also be shown that our definition of computable randomness coincides with the classical notion of computable randomness w.r.t. computable probability mass functions [11].

Next, we inspect some properties of computably random sequences ω and the set of forecasting systems $\underline{\mathcal{E}}^S_C(\omega)$ for which ω is computably random. We start by establishing that for every forecasting system $\underline{E}_\bullet$, there is at least one path $\omega \in \Omega$ that is computably random for $\underline{E}_\bullet$.

Proposition 7. *For every forecasting system $\underline{E}_\bullet$, there is at least one path ω such that $\underline{E}_\bullet \in \underline{\mathcal{E}}^S_C(\omega)$.*

Consider now the vacuous forecasting system $\underline{E}_{\bullet,v}$ defined by $\underline{E}_{s,v} := \underline{E}_v$ for every $s \in \mathbb{S}$. Our next result shows that the set of forecasting systems $\underline{\mathcal{E}}^S_C(\omega)$ for which ω is computably random is always non-empty, as it is guaranteed to contain this vacuous forecasting system.

Proposition 8. *All paths are computably random for the vacuous forecasting system:* $\underline{E}_{\bullet,v} \in \underline{\mathcal{E}}_C^S(\omega)$ *for all* $\omega \in \Omega$.

Furthermore, if a path ω is computably random for a forecasting system $\underline{E}_\bullet$, then it is also computably random for every forecasting system that is more conservative.

Proposition 9. *If* ω *is computably random for a forecasting system* $\underline{E}_\bullet$, *i.e., if* $\underline{E}_\bullet \in \underline{\mathcal{E}}_C^S(\omega)$, *then* ω *is also computably random for any forecasting system* $\underline{E}'_\bullet$ *for which* $\underline{E}'_\bullet \leq \underline{E}_\bullet$, *meaning that* $\underline{E}'_s(f) \leq \underline{E}_s(f)$ *for all situations* $s \in \mathbb{S}$ *and gambles* $f \in \mathcal{L}(\mathcal{X})$.

The following result establishes an abstract generalisation of frequency stabilisation, on which early notions of randomness—like Church randomness—were focused [2]. It states that if we systematically buy a gamble f for its coherent lower expectation $\underline{E}(f)$, then in the long run we will not lose any money. The connection with frequency stabilisation will become apparent further on in Sect. 6, where we present an intuitive corollary that deals with running averages of a gamble f along the infinite sequence ω and its computable infinite subsequences.

Theorem 10. *Consider a computable gamble* f, *a forecasting system* $\underline{E}_\bullet$ *for which* $\underline{E}_\bullet(f)$ *is a computable real process, a path* $\omega = (x_1, \ldots, x_n, \ldots) \in \Omega$ *that is computably random for* $\underline{E}_\bullet$, *and a recursive selection process* $S : \mathbb{S} \to \{0, 1\}$ *for which* $\lim_{n\to+\infty} \sum_{k=0}^n S(x_{1:k}) = +\infty$. *Then*

$$\liminf_{n\to+\infty} \frac{\sum_{k=0}^{n-1} S(x_{1:k})[f(x_{k+1}) - \underline{E}_{x_{1:k}}(f)]}{\sum_{k=0}^{n-1} S(x_{1:k})} \geq 0.$$

6   Computable Randomness for Lower Expectations

We now introduce a simplified notion of imprecise computable randomness with respect to a single coherent lower expectation; a direct generalisation of computable randomness with respect to a probability mass function. We achieve this by simply constraining our attention to stationary forecasting systems: forecasting systems $\underline{E}_\bullet$ that assign the same lower expectation $\underline{E}$ to each situation $s \in \mathbb{S}$. In what follows, we will call ω computably random for a coherent lower expectation $\underline{E}$ if it is computably random with respect to the corresponding stationary forecasting system. We denote the set of all coherent lower expectations for which ω is computably random by $\underline{\mathcal{E}}_C(\omega)$.

Since computable randomness for coherent lower expectations is a special case of computable randomness for forecasting systems, the results we obtained before carry over to this simplified setting. First, every coherent lower expectation has at least one computably random path.

Corollary 11. *For every coherent lower expectation* $\underline{E}$, *there is at least one path* ω *such that* $\underline{E} \in \underline{\mathcal{E}}_C(\omega)$.

Secondly, $\underline{\mathcal{E}}_C(\omega)$ is non-empty as every path ω is computably random for the vacuous coherent lower expectation $\underline{E}_v$.

Corollary 12. *All paths are computably random for the vacuous coherent lower expectation:* $\underline{E}_v \in \underline{\mathcal{E}}_C(\omega)$ *for all* $\omega \in \Omega$.

Thirdly, if a path ω is computably random for a coherent lower expectation $\underline{E} \in \underline{\mathcal{E}}_C(\omega)$, then it is also computably random for any coherent lower expectation $\underline{E}'$ that is more conservative.

Corollary 13. *If* ω *is computably random for a coherent lower expectation* $\underline{E}$, *then it is also computably random for any coherent lower expectation* $\underline{E}'$ *for which* $\underline{E}' \leq \underline{E}$, *meaning that* $\underline{E}'(f) \leq \underline{E}(f)$ *for every gamble* $f \in \mathcal{L}(\mathcal{X})$.

And finally, for coherent lower expectations, Theorem 10 turns into a property about running averages. In particular, it provides bounds on the limit inferior and superior of the running average of a gamble f along the infinite sequence ω and its computable infinite subsequences. Please note that unlike in Theorem 10, we need not impose computability on the gamble f nor on the real number $\underline{E}(f)$.

Corollary 14. *Consider a path* $\omega = (x_1, \dots, x_n, \dots) \in \Omega$, *a coherent lower expectation* $\underline{E} \in \underline{\mathcal{E}}_C(\omega)$, *a gamble* f *and a recursive selection process* S *for which* $\lim_{n \to +\infty} \sum_{k=0}^{n} S(x_{1:k}) = +\infty$. *Then*

$$\underline{E}(f) \leq \liminf_{n \to +\infty} \frac{\sum_{k=0}^{n-1} S(x_{1:k}) f(x_{k+1})}{\sum_{k=0}^{n-1} S(x_{1:k})} \leq \limsup_{n \to +\infty} \frac{\sum_{k=0}^{n-1} S(x_{1:k}) f(x_{k+1})}{\sum_{k=0}^{n-1} S(x_{1:k})} \leq \overline{E}(f).$$

When comparing our notion of imprecise computable randomness with the classical precise one, there is a striking difference. In the precise case, for a given path ω, there may be no probability mass function p for which ω is computably random (for example, when the running frequencies do not converge). But, if there is such a p, then it must be unique (because a running frequency cannot converge to two different numbers). In the imprecise case, however, according to Corollary 12 and 13, every path ω is computably random for the vacuous coherent lower expectation, and if it is computably random for a coherent lower expectation $\underline{E}$, it is also computably random for any coherent lower expectation $\underline{E}'$ that is more conservative—or less informative—than $\underline{E}$. This leads us to wonder whether for every path ω, there is a least conservative—or most informative—coherent lower expectation $\underline{E}_\omega$ such that ω is computably random for every coherent lower expectation $\underline{E}$ that is more conservative than or equal to $\underline{E}_\omega$, but not for any other. Clearly, if such a least conservative lower expectation exists, it must be given by

$$\underline{E}_\omega(f) := \sup\{\underline{E}(f) : \underline{E} \in \underline{\mathcal{E}}_C(\omega)\} \quad \text{for all } f \in \mathcal{L}(\mathcal{X}),$$

which is the supremum value of $\underline{E}(f)$ over all coherent lower expectations $\underline{E}$ for which ω is computably random. The crucial question is whether this $\underline{E}_\omega$ is coherent (C1 and C2 are immediate, but C3 is not) and whether ω is computably random with respect to $\underline{E}_\omega$. If the answer to both questions is yes, then $\underline{E}_\omega$ is the least conservative coherent lower expectation for which ω is computably random.

The following example illustrates that there are paths ω for which this is indeed the case. It also serves as a nice illustration of some of the results we have obtained so far.

Example 15. Consider any set $\{p_0, \ldots, p_{M-1}\}$ of M pairwise different, computable probability mass functions, and any path ω that is computably random for the non-stationary precise forecasting system $\underline{E}_\bullet$, defined by $\underline{E}_s := E_{p_{n \bmod M}}$ for all $n \in \mathbb{N}_0$ and $s = x_{1:n} \in \mathbb{S}$; it follows from Proposition 7 that there is at least one such path. Then as we are about to show, ω is computably random for a coherent lower expectation $\underline{E}'$ if and only if $\underline{E}' \le \underline{E}$, with $\underline{E}(f) := \min_{k=0}^{M-1} E_{p_k}(f)$ for all $f \in \mathcal{L}(X)$.

The 'if'-part follows by recalling Proposition 9 and noticing that for all $s = x_{1:n} \in \mathbb{S}$ and all $f \in \mathcal{L}(X)$:

$$\underline{E}'(f) \le \underline{E}(f) = \min\{E_{p_0}(f), \ldots, E_{p_{M-1}}(f)\} \le E_{p_{n \bmod M}}(f) = \underline{E}_s(f).$$

For the 'only if'-part, consider for every $i \in \{0, \ldots, M-1\}$ the selection process $S_i \colon \mathbb{S} \to \{0, 1\}$ that assumes the value $S_i(x_{1:n}) = 1$ whenever $n \bmod m = i$ and 0 elsewhere. Clearly, these selection processes are recursive and $\lim_{n \to \infty} \sum_{k=0}^n S_i(x_1, \ldots, x_n) = +\infty$ along the path $\omega = (x_1, \ldots, x_n, \ldots)$—and any other path, in fact. Furthermore, due to the computability of the probability mass functions p_i, it follows that $\underline{E}_\bullet(f)$ is a computable real process for any computable gamble $f \in \mathcal{L}(X)$. For any computable gamble $f \in \mathcal{L}(X)$, it therefore follows that

$$\underline{E}'(f) \le \liminf_{n \to \infty} \sum_{k=0}^{n-1} \frac{f(x_{i+kM})}{n} \le \limsup_{n \to \infty} \sum_{k=0}^{n-1} \frac{f(x_{i+kM})}{n} \le E_{p_i}(f),$$

where the first and third inequality follow from Corollary 14 and Theorem 10, respectively, and the second inequality is a standard property of limits inferior and superior. Since (coherent lower) expectations are continuous with respect to uniform convergence [13], and since every gamble on a finite set X can be uniformly approximated by computable gambles on X, the same result holds for non-computable gambles as well. Hence, for any gamble $f \in \mathcal{L}(X)$ we find that $\underline{E}'(f) \le E_{p_i}(f)$. As this is true for every $i \in \{0, \ldots, M-1\}$, it follows that $\underline{E}'(f) \le \underline{E}(f)$ for all $f \in \mathcal{L}(X)$.

Hence, ω is indeed computably random for $\underline{E}'$ if and only if $\underline{E}' \le \underline{E}$. Since $\underline{E}$ is clearly coherent itself, this also implies that ω is computably random with respect to $\underline{E}$ and—therefore—that $\underline{E}_\omega = \underline{E}$. So for this particular path ω, $\underline{E}_\omega = \underline{E}$ is the least conservative coherent lower expectation for which ω is computably random. ◊

However, unfortunately, there are also paths for which this is not the case. Indeed, as illustrated in Ref. [5], there is a binary path ω—so with $X = \{0, 1\}$—that is not computably random for $\underline{E}_\omega$ with $\underline{E}_\omega(f) := \frac{1}{2} \sum_{x \in \{0,1\}} f(x)$ for every gamble $f \in \mathcal{L}(X)$.

Interestingly, however, in the binary case, it has also been shown that while ω may not be computably random with respect to $\underline{E}_\omega$, there are always coherent lower expectations $\underline{E}$ that are infinitely close to $\underline{E}_\omega$ and that do make ω computably random [5].[2] So one could say that ω is 'almost' computably random with respect to $\underline{E}_\omega$. Whether a similar result continuous to hold in our more general—not necessarily binary—context is an open problem. We conjecture that the answer is yes.

Proving this conjecture is beyond the scope of the present contribution though. Instead, we will establish a similar result for expectation intervals.

7   Computable Randomness for Expectation Intervals

As a final specialisation of our notion of computable randomness, we now focus on a single gamble f on X and on expectation intervals $I = [\underline{E}(f), \overline{E}(f)]$ that correspond to lower expectations for which ω is computably random. We will denote the set of all closed intervals $I \subseteq [\min f, \max f]$ by $\mathcal{I}_f$.

Definition 16. *A path ω is computably random for a gamble $f \in \mathcal{L}(X)$ and a closed interval I if there is a coherent lower expectation $\underline{E} \in \mathcal{E}_C(\omega)$ for which $\underline{E}(f) = \min I$ and $\overline{E}(f) = \max I$. For every gamble $f \in \mathcal{L}(X)$, we denote the set of all closed intervals for which ω is computably random by $\mathcal{I}_f(\omega)$.*

Note that if ω is computably random for a gamble f and a closed interval I, it must be that $I \in \mathcal{I}_f$; so $\mathcal{I}_f(\omega) \subseteq \mathcal{I}_f$. This follows directly from C1 and conjugacy. We can also prove various properties similar to the ones in Sect. 5 and 6. The following result is basically a specialisation of Corollaries 11–13.

Proposition 17. *Consider any gamble $f \in \mathcal{L}(X)$. Then*

(i) *for every $I \in \mathcal{I}_f$, there is at least one $\omega \in \Omega$ for which $I \in \mathcal{I}_f(\omega)$;*
(ii) *for every $\omega \in \Omega$, $\mathcal{I}_f(\omega)$ is non-empty because $[\min f, \max f] \in \mathcal{I}_f(\omega)$;*
(iii) *for every $\omega \in \Omega$, if $I \in \mathcal{I}_f(\omega)$ and $I \subseteq I' \in \mathcal{I}_f$, then also $I' \in \mathcal{I}_f(\omega)$.*

Moreover, as an immediate consequence of Corollary 14, if ω is computably random for a gamble f and a closed interval $I \in \mathcal{I}_f$, then the limit inferior and limit superior of the running averages of the gamble f along the path ω and its computable infinite subsequences, lie within the interval I.

The properties in Proposition 17 lead to a similar question as the one we raised in Sect. 6, but now for intervals instead of lower expectations. Is there, for every path ω and every gamble $f \in \mathcal{L}(X)$, a smallest interval such that ω is computably random or 'almost' computably random for this gamble f and all intervals that contain this smallest interval, but for no other. The following result is the key technical step that will allow us to answer this question positively. It establishes that when ω is computably random for a gamble f and two intervals I_1 and I_2, then it is also computably random for their intersection.

---

[2] This result was established in terms of probability intervals; we paraphrase it in terms of coherent lower expectations, using our terminology and notation.

Proposition 18. *For any $\omega \in \Omega$ and $f \in \mathcal{L}(X)$ and for any two closed intervals I and I' in $\mathcal{I}_f$: if $I \in \mathcal{I}_f(\omega)$ and $I' \in \mathcal{I}_f(\omega)$, then $I \cap I' \neq \emptyset$ and $I \cap I' \in \mathcal{I}_f(\omega)$.*

Together with Proposition 17 and the fact that $\mathcal{I}_f(\omega)$ is always non-empty, this result implies that $\mathcal{I}_f(\omega)$ is a filter of closed intervals. Since the intersection of a filter of closed intervals in a compact space—such as $[\min f, \max f]$—is always closed and non-empty [1], it follows that the intersection $\bigcap \mathcal{I}_f(\omega)$ of all closed intervals I for which ω is computably random with respect to f and I, is non-empty and closed, and is therefore a closed interval itself. Recalling the discussion in Sect. 6, it furthermore follows that $\bigcap \mathcal{I}_f(\omega) = [\underline{E}_\omega(f), \overline{E}_\omega(f)]$. Similar to what we saw in Sect. 6, it may or may not be the case that ω is computably random for the gamble f and the interval $[\underline{E}_\omega(f), \overline{E}_\omega(f)]$; that is, the—possibly infinite—intersection $\bigcap \mathcal{I}_f(\omega)$ may not be an element of $\mathcal{I}_f(\omega)$. However, in this interval case, there is a way to completely characterise the models—in this case intervals—for which ω is computably random. To that end, we introduce the following two subsets of $[\min f, \max f]$:

$$L_f(\omega) := \{\min I : I \in \mathcal{I}_f(\omega)\} \text{ and } U_f(\omega) := \{\max I : I \in \mathcal{I}_f(\omega)\}.$$

Due to Proposition 17(iii), these sets are intervals: on the one hand $L_f(\omega) = [\min f, \underline{E}_\omega(f)]$ or $L_f(\omega) = [\min f, \underline{E}_\omega(f))$ and on the other hand $U_f(\omega) = [\overline{E}_\omega(f), \max f]$ or $U_f(\omega) = (\overline{E}_\omega(f), \max f]$. As our final result shows, these two intervals allow for a simple characterisation of whether a path ω is computably random for a gamble f and a closed interval I.

Proposition 19. *Consider a path ω, a gamble $f \in \mathcal{L}(X)$ and a closed interval I. Then $I \in \mathcal{I}_f(\omega)$ if and only if $\min I \in L_f(\omega)$ and $\max I \in U_f(\omega)$.*

So we see that while ω may not be computably random for f and the interval $[\underline{E}_\omega(f), \overline{E}_\omega(f)]$, it will definitely be 'almost' computably random, in the sense that it is surely random for f and any interval $I \in \mathcal{I}_f$ such that $\min I < \underline{E}_\omega(f)$ and $\max I > \overline{E}_\omega(f)$. In order to get some further intuition about this result, we consider an example where $L_f(\omega)$ and $U_f(\omega)$ are closed, and where ω is therefore computably random for f and $[\underline{E}_\omega(f), \overline{E}_\omega(f)]$.

Example 20. Consider two probability mass functions p_0 and p_1, and let the coherent lower expectation $\underline{E}$ be defined by $\underline{E}(f) := \min\{E_{p_0}(f), E_{p_1}(f)\}$ for all $f \in \mathcal{L}(X)$. Then, as we have seen in Example 15, there is a path ω for which $\underline{E}$ is the least conservative coherent lower expectation that makes ω random. Clearly, for any fixed $f \in \mathcal{L}(X)$, if we let $I := [\underline{E}(f), \overline{E}(f)]$, it follows that $\bigcap \mathcal{I}_f(\omega) = I \in \mathcal{I}_f(\omega)$, and therefore also that $L_f(\omega) = [\min f, \min I]$ and $U_f(\omega) = [\max I, \max f]$. Note that in this example, by suitably choosing p_0 and p_1, I can be any interval in $\mathcal{I}_f$, including the extreme cases where $I = [\min f, \max f]$ or I is a singleton.   ◊

8   Conclusions and Future Work

We have introduced a notion of computable randomness for infinite sequences that take values in a finite sample space X, both with respect to forecasting systems and with respect to two related simpler imprecise uncertainty models: coherent lower expectations and expectation intervals. In doing so, we have generalised the imprecise notion of computable randomness of De Cooman and De Bock [5], from binary sample spaces to finite ones.

An important observation is that many of their ideas, results and conclusions carry over to our non-binary case. On our account as well as theirs, and in contrast with the classical notion of (precise) computable randomness, every path ω is for example computably random with respect to at least one uncertainty model, and whenever a path ω is computably random for a certain uncertainty model, it is also computably random for any uncertainty model that is more conservative—or less informative.

For many of our results, the move from the binary to the non-binary case was fairly straightforward, and our proofs then mimic those in Ref. [5]. For some results, however, additional technical obstacles had to be overcome, all related to the fact that coherent lower expectations are more involved than probability intervals. Proposition 18, for example, while similar to an analogous result for probability intervals in Ref. [5], eluded us for quite a while. The key step that made the proof possible is our result that replacing computable (real) betting strategies with recursive (rational) ones leads to an equivalent notion of computable randomness; see Proposition 6.

In our future work, we would like to extend our results in Sect. 7—that for every path ω and every gamble f, ω is 'almost' computably random for a unique smallest expectation interval—from expectation intervals to coherent lower expectations. That is, we would like to prove that every path ω is 'almost' computably random for a unique maximally informative coherent lower expectation. We are convinced that, here too, Proposition 6 will prove essential.

Furthermore, we would like to develop imprecise generalisations of other classical notions of randomness, such as Martin-Löf and Schnorr randomness [2], and explore whether these satisfy similar properties. Moreover, we want to explore whether there exist different equivalent imprecise notions of computable randomness in terms of generalised randomness tests, bounded machines etc. [6] instead of supermartingales. We also wonder if it would be possible to define notions of computable randomness with respect to uncertainty models that are even more general than coherent lower expectations, such as choice functions [4].

Finally, we believe that our research can function as a point of departure for developing completely new types of imprecise learning methods. That is, we would like to create and implement novel algorithms that, given a finite sequence of data out of some infinite sequence, estimate the most informative expectation intervals or coherent lower expectation for which the infinite sequence is computably random. In this way, we obtain statistical methods that are reliable in the sense that they do not insist anymore on associating a single precise probability mass function, which is for example, as was already mentioned in the introduction, not defensible in situations where relative frequencies do not converge.

Acknowledgements. Floris Persiau's research was supported by BOF grant BOF19/DOC/196. Jasper De Bock and Gert de Cooman's research was supported by H2020-MSCA-ITN-2016 UTOPIAE, grant agreement 722734. We would also like to thank the reviewers for carefully reading and commenting on our manuscript.

References

1. Aliprantis, C.D., Border, K.C.: Infinite Dimensional Analysis: A Hitchhiker's Guide. Springer, London (2006). https://doi.org/10.1007/3-540-29587-9
2. Ambos-Spies, K., Kucera, A.: Randomness in computability theory. Contemp. Math. **257**, 1–14 (2000)
3. Bienvenu, L., Shafer, G., Shen, A.: On the history of martingales in the study of randomness. Electron. J. Hist. Probab. Stat. **5**, 1–40 (2009)
4. De Bock, J., de Cooman, G.: Interpreting, axiomatising and representing coherent choice functions in terms of desirability. PMLR **103**, 125–134 (2019)
5. De Cooman, G., De Bock, J.: Computable randomness is inherently imprecise. PMLR **62**, 133–144 (2017)
6. Downey, R.G., Hirschfeldt, D.R.: Algorithmic Randomness and Complexity. Springer, New York (2010). https://doi.org/10.1007/978-0-387-68441-3
7. Gorban, I.I.: The Statistical Stability Phenomenon. ME. Springer, Cham (2017). https://doi.org/10.1007/978-3-319-43585-5
8. Li, M., Vitányi, P.: An Introduction to Kolmogorov Complexity and Its Applications. TCS. Springer, New York (2008). https://doi.org/10.1007/978-0-387-49820-1
9. Persiau, F., De Bock, J., De Cooman, G.: Computable randomness is about more than probabilities (2020). http://arxiv.org/abs/2005.00471
10. Pour-El, M.B., Richards, J.I.: Computability in Analysis and Physics. Cambridge University Press, Cambridge (2016)
11. Rute, J.: Computable randomness and betting for computable probability spaces. Math. Logic Q. **62**, 335–366 (2012)
12. Shafer, G., Vovk, V.: Game-Theoretic Foundations for Probability and Finance. Wiley, Hoboken (2019)
13. Walley, P.: Statistical Reasoning with Imprecise Probabilities. Chapman and Hall, London (1991)

Equity in Learning Problems: An OWA Approach

Juliette Ortholand, Sébastien Destercke[⊠], and Khaled Belahcene

Université de Technologie de Compiègne, Heudiasyc, Compiègne, France
{juliette.ortholand,sebastien.destercke,khaled.belahcene}@hds.utc.fr

Abstract. It is well-known in computational social choice that the weighted average does not guarantee any equity or fairness in the share of goods. In a supervised learning problem, this translates into the fact that the empirical risk will lead to models that are good in average, but may have terrible performances for under-represented populations. Such a behaviour is quite damaging in some problems, such as the ones involving imbalanced data sets, in the inputs or the outputs (default prediction, ethical issues, ...). On the other hand, the OWA operator is known in computational social choice to be able to correct this unfairness. This paper proposes a means to transpose this feature to the supervised learning setting.

1 Introduction

The typical way to learn a predictive model from data is to search for the model that minimizes the average loss of the predictions made by this model on a set of training data. However, minimizing the average loss may well lead to poor results on some under-represented populations.

This is a well known fact, that happens in several settings that have proposed different solutions to the issue: in class imbalanced data sets, concerning for instance rare diseases or default (of payment, of production), the classical solution is to modify the sample sizes, for instance by over-sampling instances of the under-represented class [9]; in fairness issues [10], where the goal can be to protect sensitive populations or minorities, often by modifying not the sample but the loss function adequately; in extreme statistics [11], where one must guarantee that rare instances will be well predicted, for instance by learning a model specifically dedicated to them.

In this paper, we look at another aspects of misrepresentation of some data in the learning problem. Namely, we want to ensure that the loss incurred for data poorly represented in the feature space (whatever their class is) is not high. This is yet a different kind of under-representation of some population, whose closest related problem is the previously mentioned one of extreme statistics [11]. Our goal here is to propose a method ensuring that under-represented data will not suffer from a too high loss, while preserving a good average accuracy. To perform such a task, we will modify the classical expected loss by using the notion of ordered weighted averaging, an often used notion in fairness problems within computational social choice [13].

© Springer Nature Switzerland AG 2020
J. Davis and K. Tabia (Eds.): SUM 2020, LNAI 12322, pp. 187–199, 2020.
https://doi.org/10.1007/978-3-030-58449-8_13

More precisely, we will propose to give more weight to unknown zones. The paper is organised as follows: the formal mathematical framework can be found in Sect. 2, where we provide reminders, notations and preliminaries, and in Sect. 3, where we describe our proposal. This is followed by the experiments in Sect. 4 then by some related works in the Sect. 5. The paper ends with some conclusion and discussion on our work in Sect. 6.

2   Preliminaries

We consider a standard supervised problem where we have observations $(x_i, y_i) \in \mathcal{X} \times \mathcal{Y}$, $i = 1, \ldots, n$ where x_i are some inputs, and y_i the observed outputs.

2.1   Supervised Classification via Empirical Risk Minimization

The general goal of supervised machine learning is to estimate a predictive model $h^* : \mathcal{X} \to \mathcal{Y}$, issued from a space $\mathcal{H}$ of hypothesis, such that the model delivers good prediction in average. This principle is most often translated by choosing the model minimizing the empirical risk, i.e.,

$$h^* = \arg \min_{h \in \mathcal{H}} R_{emp}(h)$$

where

$$R_{emp}(h) = \sum_{i=1}^{n} \ell(h(x_i), y_i) \tag{1}$$

with $\ell(h(x_i), y_i)$ the loss of predicting $h(x)$ when y is the observed value. This empirical loss serves as an estimate of the true loss, i.e., $R(h) = \int_{\mathcal{X} \times \mathcal{Y}} \ell(h(x), y) dp(x, y)$, that is inaccessible as we do not know $p(x, y)$. Also, in many cases, $\mathcal{H}$ is a parametric family with parameters $\theta \in \Theta$, and in this case we will denote by h_θ the predictive function having θ for parameter.

2.2   Some Shortcomings Due to the Averaging of the Risk

Guaranteeing a low average loss does not prevent from having large losses for poorly represented groups of values, and even in some cases promote such large discrepancies [10].

Example 1. Figure 1 displays two different classes (i.e., $y \in \{0, 1\}$) represented in red and blue, that suffer from the problem we consider in this paper. Indeed, the two classes balanced (there are about as much red as blue), but some region of the input space are less represented than others. More precisely, the data corresponding to each class have been generated by the following distributions:

$$X|y = 0 \sim \pi_0^1 \, \mathcal{N}(\begin{bmatrix} 1 \\ 1 \end{bmatrix}, \begin{bmatrix} 3 & 0 \\ 0 & 0.5 \end{bmatrix}) + \pi_0^2 \, \mathcal{N}(\begin{bmatrix} 16 \\ 6 \end{bmatrix}, \begin{bmatrix} 1 & 0 \\ 0 & 0.5 \end{bmatrix})$$

$$X|y = 1 \sim \pi_0^1 \, \mathcal{N}(\begin{bmatrix} 0 \\ 2 \end{bmatrix}, \begin{bmatrix} 3 & 0 \\ 0 & 0.5 \end{bmatrix}) + \pi_0^2 \, \mathcal{N}(\begin{bmatrix} 15 \\ 7 \end{bmatrix}, \begin{bmatrix} 1 & 0 \\ 0 & 0.5 \end{bmatrix})$$

with $\pi_0^1 = \pi_1^1 = 0.95$ and $\pi_0^2 = \pi_1^2 = 0.05$, meaning that the upper-right region is much less represented than the lower-left in Fig. 1. The frontier in this region corresponds to the one obtained by a logistic regression trained according to Eq. 1. It is easy to see that the model does a very bad job at predicting the data in the upper-right corner, as could be expected.

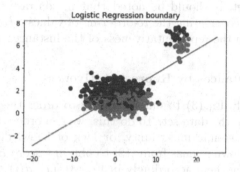

Fig. 1. Logistic Regression boundary (Color figure online)

2.3 A Short Introduction to Ordered Weighted Averages (OWA)

The OWA operators, initially introduced by Yager [14], apply a weighted aggregation functions on ordered values. Usually, the OWA is applied to a vector $\mathbf{a} = (a_1, \ldots, a_n)$ of real-valued quantities $a_i \in \mathbb{R}$, and is defined by a set $(w_1, \ldots, w_n)$ of positive weights ($w_i \geq 0$) summing up to one ($\sum_i w_i = 1$). Formally speaking, the OWA consists in first permuting the values a_i in ascending order, i.e., such that for $i < j$, we have $a_{\sigma(i)} \leq a_{\sigma(j)}$, with σ denoting the permutation. The classical OWA operator is then

$$OWA(a_1, \ldots, a_n) = \sum_{i=1}^{n} w_i a_{\sigma(i)}. \qquad (2)$$

The OWA operator therefore allows to put different strength on lower or higher values, irrespectively of where they appear in the vector $\mathbf{a}$. We retrieve the arithmetic mean for $w_i = 1/n$, and kth percentiles when we have $w_{k/n} = 1$ for some i. In particular, the minimum and maximum values are respectively retrieved for $w_1 = 1$ and $w_n = 1$. They characterize what Yager called extreme behaviour of "andness" and "orness", respectively.

3 Our Proposal

In this paper, we consider the use of OWA operators [8], that propose to make weighted averages not on the initial observations (x_i, y_i) and their associated losses, but on a re-ordering of them, with the idea that a higher weight should be given to poorly represented instances, to be ranked first in our proposal. We will denote by $(x_{\sigma(i)}, y_{\sigma(i)})$ the corresponding permutation on observations. More precisely, we propose not to optimise $R_{emp}(h)$, but rather $R_{OWA}(h)$, where

$$R_{OWA}(h) = \sum_{i=1}^{n} w_i \ell(h(x_{\sigma(i)}), y_{\sigma(i)}) \tag{3}$$

with the idea that when $i \leq j$, the instances $x_{\sigma(i)}$ is not as well represented as $x_{\sigma(j)}$ in the data set. It should be noted that in contrast to usual OWA and Eq. (3), we will not considering the re-ordering of values $\ell(h(x_{\sigma(i)}), y_{\sigma(i)})$, but a re-ordering based on the representativeness of the instances x_i.

3.1   Ranking Instances by Representativeness

A first task to apply Eq. (3) to our setting is to order the instances by their representativeness in the data set. To do this, we can order them by measuring, for instance, the epistemic uncertainty, or lack of knowledge concerning each instance x_i, e.g., following ideas from [12] to obtain a score E_i for each instance x_i, and then ordering them accordingly in Eq. (3), i.e., $\sigma(i) \leq \sigma(j)$ if $E_i \geq E_j$.

One simple idea that we will apply in the following is to measure the density of points around a neighbourhood of fixed size around x_i to compute its associated epistemic uncertainty. For this, an easy technique one can use is to simply perform a kernel density estimation through the formula

$$f(x) = \frac{1}{n} \sum_{i=1}^{n} K_\epsilon(x - x_i)$$

with K_ϵ a kernel function. Common choices of kernel functions include:

- the Parzen window, defined as $K_\epsilon(x - x_i) = \frac{1}{2\epsilon} 1_{|x-x_i|<\epsilon}$, that simply comes down to count the number of points that are at a distance below a certain ϵ;
- the triangle kernel, defined as $K_\epsilon(x - x_i) = (\frac{1}{\epsilon} - |x - x_i|) 1_{|x-x_i|<\epsilon}$, for which weights decrease linearly from one to zero depending on the distance;
- the normal window, defined as $K_\epsilon(x - x_i) = \frac{1}{\epsilon\sqrt{2\pi}} e^{\frac{-(x-x_i)^2}{2\epsilon^2}}$, for which the weights for the points depend on a normal distribution around the chosen point with a mean of zero and a standard deviation of ϵ.

The three kernels are pictured in Fig. 2a.

Once a kernel is chosen, we can then simply use $f(x_i) = E_i$ as a score quantifying epistemic uncertainty. Note that in our case, it is not important to have a reliable estimate of the density (a very difficult problem), but to just have a reliable ordering between the different points, as $f(x_i)$ will only be used to order values in OWA operators.

Figure 2b represents the distribution of the epistemic uncertainty of points given in Fig. 1, computed for a triangular kernel with ϵ being the mean distance between points. One can readily see that the most uncertain points (hence the first in the re-ordering) are those in the upper right corner, that is precisely those for which we would like to increase accuracy, followed by the ones on the border of the big cluster.

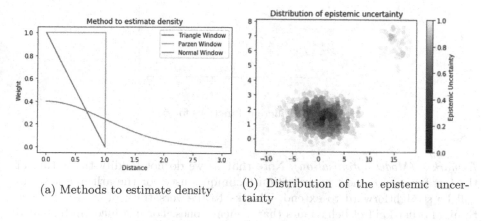

(a) Methods to estimate density

(b) Distribution of the epistemic uncertainty

Fig. 2. Epistemic uncertainty

3.2 OWA Weights to Induce Equity

The next step is how we can choose the weights in Eq. (3) so as to balance the accuracy in the model between well-represented and poorly represented instances. Clearly, if we pick $w_i = 1/n$, this re-ordering is a useless step and leads us to the usual solution. However, we can easily pick weights that will enforce giving more importance to poorly represented instances. More precisely, we can pick a function $\phi : [0,1] \to [0,1]$ and take as weights

$$w_i = \phi(i/n) - \phi(i-1/n)$$

if $\phi(x) = x$, then we retrieve the weighted average. If ϕ is concave, then we start giving more weights to first ordered instances, and less weight to last ordered instances. In terms of OWA, we increase the "andness" of the function, that we can then parameterize to be more or less fair. Ideally, this number of parameters should not be too high, and example of such functions include:

- The L_p norm on with $p \in [0,1]$, which function is $\phi(x) = x^p$. The lower p, the more we increase the "andness"
- piece-wise linear functions made of two linear parts, that can be define with two parameters p and $prop$ as follows:

$$\phi(x) = \begin{cases} px & if \quad x \le prop \\ \frac{1-p\times prop}{1-prop}x + \frac{prop(p-1)}{1-prop} & if \; prop \le x \le 1 \end{cases} \tag{4}$$

where p defines the slope before the abscissa value $prop$.

Both are represented on Fig. 3 with the $L_{1/2}$ norm and the linear by part function with $prop = 0.1$ and $p = 4$.

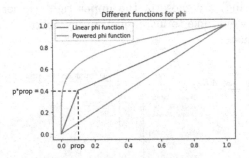

Fig. 3. Different functions for ϕ

Remark 1 (Model optimisation). Note that as we do not modify the nature of the loss function ℓ, most optimisation techniques used for the arithmetic mean will be straightforward to extend. Thanks to the versatility of ϕ, we can also think of other kind of behaviours than concave ones. For instance, an S-shaped function would amount to try to balance between being quite good on poorly represented as well as quite good of very-well represented groups, thus protecting minorities and majorities.

4   Experiments

This section presents some experiments using our approach to try to augment the accuracy on the poorly represented data, that we will call *minority*, while preserving a good average accuracy. After quickly describing the chosen model, we will provide results first on synthetic data sets, second on real-world data sets. In this latter case, since there are no benchmark data sets focusing on the problem we try to solve, we will try to adapt common UCI data sets [6] to our setting.

4.1   Implementing the Proposal

We will apply our approach to standard regularized logistic regression in binary classification problems, with the output class $Y \in \{0, 1\}$. Let us simply recall that in this case, we learn a probabilistic model of the shape

$$h_\theta(x) = \frac{1}{1 + \exp -\theta x} \tag{5}$$

with $h_\theta(x) = p(y = 1|x)$. The associated loss is

$$\ell(y, h_\theta(x)) = -(y \log(h_\theta(x)) + (1 - y)\log(1 - h_\theta(x)) + \theta^2 \tag{6}$$

that corresponds to a logarithmic loss with a L_2 regularisation term. In experiments, we use python sklearn package to fit the different models, with which it is straightforward to add weights to samples.

In the experiments, we used the triangle kernel applied to data with an Euclidean distance computed between them. The reason for this choice is that it gives no ties between values $f(x_i)$ in practice (while Parzen windows delivers the same value when having the same number of data within it), and that it has a finite support, therefore being more coherent than the normal kernel with the fact that epistemic uncertainty is mostly a local property. However, our tests with other kernel functions show no significant differences.

Regarding the parametric shape of the OWA, we picked the shape given by Eq. (4), as in our experiments the use of the L_p norm tended to give too quickly too much importance to poorly represented data, introducing sometimes important discontinuities in our results for small changes of the parameter. This can already be seen in Fig. 3. Thus, the ϕ function depends on two parameters p, the slope of the first linear part of the function and $prop$, the abscissa of the slope breaking point.

As our aim is to improve accuracy on minorities while keeping a good average accuracy, this means that our performances will be measures according to two values: the average accuracy on the minority samples only, $acc \in [0, 1]$, and the classical average accuracy, $ACC \in [0, 1]$. For this reasons, we will present our experimental results a Pareto front on the space $[0, 1] \times [0, 1]$, as for a given couple $(p, prop)_k$ of the proposal, we will obtain a pair (acc_k, ACC_k). This means that we will present the results for all non-dominated values, that is all (acc_k, ACC_k) such that there will be no other pairs $(p, prop)_{k'}$ with $acc_{k'} \geq acc_k$ and $ACC_{k'} \geq ACC_k$.

4.2 Synthetic Data Set

In the first set of experiments, we simply consider the same distributions as the ones described in Example 1 for a binary problem.

As in Example 1, minorities of each class represent about 5% of the total quantity of samples from that class. For each set of experiments, we generate 1000 points for the training set, and as much for the testing set.

In the experiments, we proceeded with a simple grid search to fix p and $prop$. We let p vary between 1 and 5 with a 0.5 increment, and $prop$ between $[0.1, 0.3]$ with a step of 0.05. Every test is made on 10 different sets of data and the mean is taken to obtain reliable estimates. The total accuracy and the accuracy of the minority are studied.

Figure 4 illustrates the results obtained for a particular run and give the model obtained for the parameters $p = 5.0$ and $prop = 0.2$. One can easily see that the obtained model is much more relevant on the minorities, as it starts to discriminate the two classes in this region.

One question though is to know whether this potential benefit on the minority region does not alter too much the overall accuracy. The answer is provided by Fig. 5 that displays the obtained Pareto front as well as the results of the basic

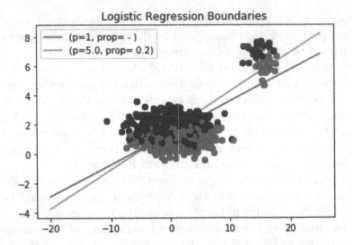

Fig. 4. Logistic Regression Boundaries for extreme values

model $(1, -)$. One can observe that the accuracy on the minority region can increase by more than 10%, going from 0.53 to 0.66, while the loss on the average accuracy is below 3%. Note that the Pareto allows a possible decision maker to finely choose the trade-off between minority protection and overall performances.

4.3   UCI Data Sets

As we said, there are to our knowledge no benchmark data set that explicitly deals with the problem of within-class imbalance, the situation described by our synthetic example. For this reason, we tried to apply it to UCI data sets susceptible to display similar behaviours.

To test whether this is the case, a simple procedure is adopted: we split the data set into training and test sets, and order the elements of the test set according to their epistemic uncertainty, computed by using the samples of the training set. A logistic regression is then fitted to the data, and we check the difference between the global average accuracy (ACC) and the average accuracy of the first $\alpha\%$ of the ordered test samples (acc). If the difference $ACC - acc$ is big enough, we retain the data set.

In our experiment, we fixed the value α to 10%, and similarly to the previous case, proceeded to apply our method by letting p vary between 1 and 5 with a 0.5 increment, and *prop* between $[0.1, 0.3]$ with a step of 0.05. Each training/testing experiment is made by taking 50% of the data set as training, and the process is repeated a hundred times for each configuration, the mean being kept as a representative point.

Perhaps surprisingly, it proved quite hard in this manner to find suitable data sets. Finally, we retained three binary classification data sets that are summarised in Table 1.

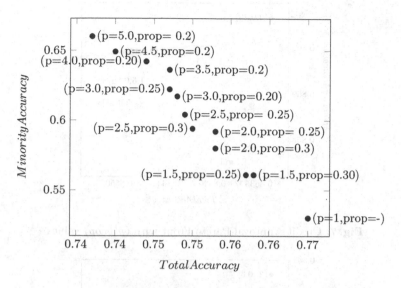

Fig. 5. Pareto front on synthetic data set

Table 1. Data set descriptions

Data set	Samples	Percentage of positive class
Istanbul Stock Exchange [4]	536	50%
Credit Approval [5]	653	45%
Vertebral Column [5]	310	68%

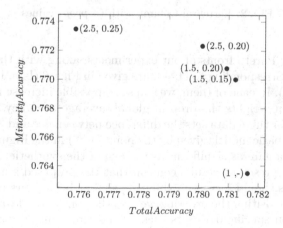

Fig. 6. Istanbul Stock Exchange Pareto Front with $(p, prop)$ values

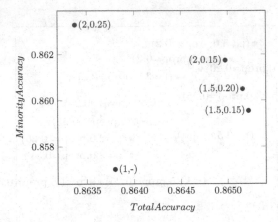

Fig. 7. Credit Approval Pareto Front with $(p, prop)$ values

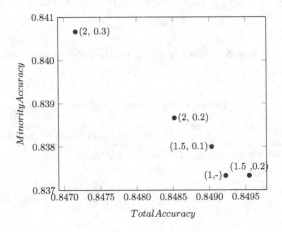

Fig. 8. Vertebral Column with $(p, prop)$ values

The resulting Pareto fronts of our experiments, along with the accuracies of the base model corresponding to $(1, -)$, are given in Figs. 6 (Istanbul), 7 (Credit) and 8 (Vertebral). In each of them, we can see a possible increase in the minority accuracy that out-weights the drop in global, average accuracy. However, it is clear that since for these data sets the difference between acc and ACC is already quite low for the basic model (given by the point $(1, -)$ in the Figures), we cannot hope to achieve a gain as significant as the one of the synthetic data sets.

So, while the presented results confirm that the proposed approach is working, future works should focus on exhibiting similar behaviours in existing data sets, maybe by revisiting the ordering we use, the considered learning algorithm or by focusing on specific data sets such as class imbalanced data sets, hoping that the imbalance in the classes is transferred to the input space.

5   Links with Other Learning and Estimation Approaches

To our knowledge, the learning approach presented here is quite original, in the sense that applying OWA to learning problems in order to solve inequities has, to our knowledge, not been done before.

5.1   From OWA to Probability Sets

A now well-established trend in the learning literature is the so-called distributionally robust approach [1,7] that consider the problem of finding the minimax model over a possible set $\mathcal{P}$ of probability distributions, mostly defined as a neighbourhood of the empirical distribution of the observations (x_i, y_i). Such approaches have been applied, for instance, to fairness issues in machine learning [10] or to transfer learning problems [2].

Since it is well-known that the OWA operator correspond to apply a Choquet integral with a specific Choquet measure [8], and that such Choquet integral can correspond to lower/upper expectations computed for specific probability sets, it would be interesting to study under which conditions and to which extent our current approach could be interpreted as the solution of a minimax problem for some specific set $\mathcal{P}$.

5.2   From OWA Loss to Weighted Likelihood

Another classical way to learn a model, and particularly probabilistic models, is through the maximisation of a likelihood function. In such a case, each parameter value θ determine a probability distribution of a random variable X. First recall that the likelihood of a parameter value θ, given a set of observations $x_i \in \mathcal{X}$ is $\mathcal{L}(\theta|x_i) = \prod_i p_\theta(x_i)$.

Provided we consider the logarithmic loss in Eq. (3), one can easily express our weighting scheme in terms of likelihood. For this, it is sufficient to consider the log-likelihood

$$\mathcal{C}(\theta|x_i) = -ln(\mathcal{L}(\theta|x_i)) = -\sum_i ln(p_\theta(x_i)) \tag{7}$$

where the loss is $ln(p_\theta(x_i))$. We can then apply the OWA loss instead of the current loss to get $\mathcal{C}_{OWA}(\theta|x_i) = -\sum_i w_i \ln(p_\theta(x_{\sigma(i)}))$ with w_i the OWA weights. It is then possible to go back to the formula of the likelihood, obtaining

$$\mathcal{L}(X) = \prod_i p_\theta(x_{\sigma(i)})^{w_i}$$

as our new, weighted likelihood. Thus the OWA transformation on the loss which corresponds to multiply the terms by weights, is equivalent to a exponent operation with weight for the likelihood. While such an exponent weight may appear odd at first, it should be noticed that it has been proposed and used before like in [3], where it has been used to down weight anomalous point in Bayesian prediction.

6 Conclusion

In this paper, we have presented an original approach, based on OWA operators, to obtain more balanced and equitable classifiers in those problems where data can be scarce in some regions of the input space. Such an approach aims at ensuring that even poorly represented instances will be treated fairly, in the sense that we will not allow them to suffer huge losses, while keeping an average loss comparable to the one obtained without including such equity requirement.

Our illustrative experiments on synthetic data sets indeed show that the method is appropriate, and allows one to obtain a more balanced model. We have also shown that the same observation can be made on UCI data sets, albeit the improvement is here much more general, due to the fact that there is no benchmark data sets explicitly suffering from the problem we have considered in this paper.

We nevertheless believe that the idea of using aggregation operators issued from the social choice literature to solve inequities and unfairness in supervised learning procedure is a promising idea, that needs to be developed. This study is simply a first proposal going in this direction, and many aspects remain to be studied, such as the nature of the ordering between instances or whether there are algorithms where our approach can make a bigger difference, notably in the case of multi-class problems, as we only used logistic regression on binary problems here.

References

1. Abadeh, S.S., Esfahani, P.M.M., Kuhn, D.: Distributionally robust logistic regression. In: Advances in Neural Information Processing Systems, pp. 1576–1584 (2015)
2. Abadeh, S.S., Nguyen, V.A., Kuhn, D., Esfahani, P.M.M.: Wasserstein distributionally robust Kalman filtering. In: Advances in Neural Information Processing Systems, pp. 8474–8483 (2018)
3. Agostinelli, C., Greco, L.: Weighted likelihood in Bayesian inference. In: Proceedings of the 46th Scientific Meeting of the Italian Statistical Society, pp. 746–757 (2012)
4. Akbilgic, O., Bozdogan, H., Balaban, M.E.: A novel hybrid RBF neural networks model as a forecaster. Stat. Comput. **24**(3), 365–375 (2014)
5. Dua, D., Graff, C.: UCI Machine Learning Repository (2017)
6. Frank, A., Asuncion, A.: UCI Machine Learning Repository (2010)
7. Gabrel, V., Murat, C., Thiele, A.: Recent advances in robust optimization: an overview. Eur. J. Oper. Res. **235**(3), 471–483 (2014)
8. Grabisch, M.: OWA operators and nonadditive integrals. In: Yager, R.R., Kacprzyk, J., Beliakov, G. (eds.) Recent Developments in the Ordered Weighted Averaging Operators: Theory and Practice. Studies in Fuzziness and Soft Computing, vol. 265. Springer, Heidelberg (2011). https://doi.org/10.1007/978-3-642-17910-5_1
9. Han, H., Wang, W.-Y., Mao, B.-H.: Borderline-SMOTE: a new over-sampling method in imbalanced data sets learning. In: Huang, D.-S., Zhang, X.-P., Huang, G.-B. (eds.) ICIC 2005. LNCS, vol. 3644, pp. 878–887. Springer, Heidelberg (2005). https://doi.org/10.1007/11538059_91

10. Hashimoto, T., Srivastava, M., Namkoong, H., Liang, P.: Fairness without demographics in repeated loss minimization. In: International Conference on Machine Learning, pp. 1929–1938 (2018)
11. Jalalzai, H., Clémençon, S., Sabourin, A.: On binary classification in extreme regions. In: Advances in Neural Information Processing Systems, pp. 3092–3100 (2018)
12. Senge, R., et al.: Reliable classification: learning classifiers that distinguish aleatoric and epistemic uncertainty. Inf. Sci. **255**, 16–29 (2014)
13. Shams, P., Beynier, A., Bouveret, S., Maudet, N.: Minimizing and balancing envy among agents using ordered weighted average (2019)
14. Yager, R.R.: On ordered weighted averaging aggregation operators in multicriteria decisionmaking. IEEE Trans. Syst. Man Cybern. **18**(1), 183–190 (1988)

Conversational Recommender System by Bayesian Methods

Francesca Mangili, Denis Broggini, and Alessandro Antonucci[✉]

IDSIA, Lugano, Switzerland
{francesca,denisb,alessandro}@idsia.ch

Abstract. We present a Bayesian approach to conversational recommender systems. After any interaction with the user, a probability mass function over the items is updated by the system. The conversational feature corresponds to a sequential discovery of the user preferences based on questions. Information-theoretic criteria are used to optimally shape the interactions and decide when the conversation ends. Most probable items are consequently recommended. Dedicated elicitation techniques for the prior probabilities of the parameters modelling the interactions are derived from basic structural judgements based on logical compatibility and symmetry assumptions. Such prior knowledge is combined with data for better item discrimination. Our Bayesian approach is validated against matrix factorization techniques for cold-start recommendations based on metadata using the popular benchmark data set *MovieLens*. Results show that the proposed approach allows to considerably reduce the number of interactions while maintaining good ranking performance.

Keywords: Conversational recommender systems · Bayesian methods

1 Introduction

The task of selecting from a collection of items the one which is in some sense optimal for a specific user is a classic AI problem. Several algorithms have been explored to perform tasks of this kind and automated recommendations are currently provided by all major e-commerce platforms [9,18]. In standard setups, no explicit interaction with the users is considered, and the recommendation system bases its decision on some background information (also called *metadata*) about the user, historical records of her choices and those of other similar users, and, more recently, contextual information [1,13].

Modern technologies such as chatbots or personal assistants need instead systems capable of supporting and modelling dynamic and sequential interactions, thus requiring a substantial re-design of traditional approaches. Here we focus on such a newer class of recommendation systems, called here *conversational*, as we term conversation a sequence of dynamically customized interactions between the user and the system, before the latter returns an item recommendation [4]. This class of systems is typically based on knowledge-based recommendation techniques requiring a strong interaction with the user.

© Springer Nature Switzerland AG 2020
J. Davis and K. Tabia (Eds.): SUM 2020, LNAI 12322, pp. 200–214, 2020.
https://doi.org/10.1007/978-3-030-58449-8_14

Recommendations based on conversations are particularly relevant when the goal is to support the user in purchasing high-involvement products. In such situations, indeed, the user wants to be involved in the decision process, and thus is not bothered by the need to interact with the system [10].

To achieve that, we take inspiration from existing approaches in the field of computer testing [3,14], whose goal is to determine the skills of a student on the basis of the answers to the questions of a test. In particular we focus on *adaptive* approaches, where the system selects the next question to ask to the student from a given set on the basis of the previous answers by information-theoretic scores, that are also used to decide when the test should be ended. This can be easily obtained by generative probabilistic models such as Bayesian networks [16], to be sequentially updated any time a new answer is collected.

The adaptive concept is reshaped here as a conversational approach which lends itself to the (future) development of a dynamic generation of questions and richer interaction models. The question selection is consequently reduced to an inference in the corresponding generative model. The major issue consists therefore in a reliable and scalable elicitation strategy of the model parameters. This is achieved by combining prior structural judgements and historical data.

The paper is organized as follows. We first review the basics of probabilistic approaches to recommender systems in Sect. 2. Elicitation strategies based on structural judgements such as logical compatibility and symmetry are in Sect. 3. A procedure to cope also with non-exclusive answers is in Sect. 4. Our conversational approach driven is indeed detailed in Sect. 5. In Sect. 6, we show how our elicitation techniques can be merged with historical data and how to prevent a fragmentation of these data over the items. Finally, we discuss our experiments in Sect. 7 and the conclusions in Sect. 8.

2 Bayesian Recommendations

Consider a recommendation system based on a *catalogue* $\mathcal{I}$ of n items, say $\mathcal{I} := \{i_1, \ldots, i_n\}$. The system is intended to support the user in selecting a single item from the catalogue on the basis of a conversational process. The system outcome can however be used to rank the items. We assume that, after the conversation, the user always picks one item from $\mathcal{I}$. The uncertain quantity I denotes the element of $\mathcal{I}$ to be picked by the user. We consider a Bayesian setup and model the subjective probabilities of picking the different items before the beginning of the conversational process as a *prior* probability mass function $P(I)$.

We call *questions*, the interactions between the system and the user. Let us start with a static approach to elicitation based on a list of m questions $\mathbf{Q} = \{Q_1, \ldots, Q_m\}$, called here a *questionnaire*. A generic question is denoted as an uncertain quantity Q taking its possible values (*answers*) from $\mathcal{Q}$. We assume a finite set of possible answers for Q, say $\mathcal{Q} := \{q^1, \ldots, q^r\}$. The selection of the optimal item to be suggested at the end of the conversation might be based on the conditional probabilities $P(i|\mathbf{q})$ for each item $i \in \mathcal{I}$ given the collected answers, denoted as $\mathbf{q}$.

To improve the quality of the user experience, it seems desirable to minimize the number of questions required to safely identify the optimal suggestion. To this goal, the list of questions in **Q** should be built dynamically. Such a customized list of questions is called *conversation*. Before discussing this dynamic approach, we will present the set up that allows for the computation of $P(i|\mathbf{q})$, be $\mathbf{q}$ the output of a conversation or a whole questionnaire.

Let us start from the case of a single question Q. As a model of the relation between Q and I, we might be able to elicit a *conditional probability table* $P(Q|I)$, whose columns are indexed by the answers, and whose rows are associated to the items. After an answer $Q = q \in Q$ is received from the user, the probability mass function over the items can be updated by Bayes rule, i.e.,

$$P(i|q) = \frac{P(i)\,P(q|i)}{\sum_{i \in \mathcal{I}} P(i)\,P(q|i)}.$$ (1)

This shows that the impact of an answer $Q = q$ on the choice I only depends on the relative proportions of the values of $P(q|i)$ for the different values of $i \in \mathcal{I}$. In particular, setting an element of the conditional probability table equal to zero implies a logical constraint for which the answer associated with column $Q = q$ makes the choice of the item associated with row $I = i$ impossible. To model a whole conversation made of m questions, we formulate a naive-like assumption about the conditional independence of the questions given the item:

$$P(q_1, \ldots, q_m|i) = \prod_{j=1}^{m} P(q_j|i).$$ (2)

Under the assumption in Eq. (2), if a conditional probability table $P(Q|I)$ is available for each $Q \in \mathbf{Q}$, the probability $P(i|\mathbf{q})$ can be obtained by recursively applying Eq. (1) to update the probability $P(i|q_1, \ldots, q_k)$ after the first k answers by conditioning also on the next answer q_{k+1}. The following example will be used through the paper to illustrate the features of our method.

Example 1. Users of a platform for online booking of entertainers are invited to answer a number of questions about the kind of entertainment they are looking for. The two questions Q_1 and Q_2 asked to the users and a catalogue of three entertainers-items are in Table 1.

Table 1. A catalogue and two questions of a questionnaire

$\mathcal{I}$	Description	Q_1 Which artist?		Q_2 Which event?	
i_1	DJ for all events	q_1^1	DJ	q_2^1	Wedding
i_2	Band for weddings and corporate events	q_1^2	Band	q_2^2	Corporate event
i_3	Magician for birthdays and parties	q_1^3	Musician	q_2^3	Birthday
		q_1^4	Entertainer	q_2^4	Party for kids

3 Structural Judgements for Prior Elicitation

In this section we describe a general elicitation procedure for the conditional probability table $P(Q|I)$. This is based on logical arguments, discussed in Sect. 3.1, to be integrated by symmetry considerations, discussed in Sect. 3.2.

3.1 Logical Compatibility

Consider the natural language item descriptions in Table 1. It is straightforward to deduce from this information a number of *incompatibility* relations between the entertainers and the answers to the two questions in Table 1. E.g., the magician is not available for weddings, this meaning that answer q_2^1 is not compatible with item i_3. Assuming such a compatibility analysis available for all the items and answers, we might define an indicator function $\delta(i, q)$, for each $i \in \mathcal{I}$ and $q \in \mathcal{Q}$, and for each $Q \in \mathbf{Q}$, that returns one if i is compatible with q and zero otherwise. An item i is therefore characterized by its set of compatible answers or *support*. We will denote as $\mathcal{Q}_j(i)$ the support of i associated with Q_j, i.e.,

$$\mathcal{Q}_j(i) := \{q \in \mathcal{Q}_j : \delta(i, q) = 1\}. \tag{3}$$

For instance, in Example 1, $\mathcal{Q}_1(i_2) = \{q_1^2\}$ and $\mathcal{Q}_2(i_2) = \{q_2^1, q_2^2\}$. Notation $\mathcal{Q}(i)$ is used instead for the union of all the supports of i associated to the different questions in $\mathbf{Q}$. E.g., in Example 1, $\mathcal{Q}(i_2) = \{q_1^2, q_2^1, q_2^2\}$.

At the probabilistic level, we translate logical incompatibilities into zero-probability statements as follows:

$$q \notin \mathcal{Q}(i) \Rightarrow P(q|i) = 0, \tag{4}$$

i.e., if the right item to recommend is not compatible with an answer, that answer is impossible. If $q \notin \mathcal{Q}(i)$, also the joint probability $P(i, q)$ is zero as $P(i, q) = P(q|i) \, P(i)$. On the other hand, if $q \in \mathcal{Q}(i)$, the elicitation of $P(q|i)$ might need additional assumptions. This is not the case of the following example.

Example 2. In the setup of Example 1, for question Q_1, $\mathcal{Q}_1(i_1) = \{q_1^1\}$, $\mathcal{Q}_1(i_2) = \{q_1^2\}$ and $\mathcal{Q}_1(i_3) = \{q_1^4\}$. In other words, each item admits only a single consistent answer. Thus, because of Eq. (4), for each i, $P(q|i)$ is zero for any q apart from the one in $\mathcal{Q}_1(i)$. $P(Q_1|I)$ takes therefore the values in Table 2 (left).

Table 2. Elicitation of $P(Q_1|I)$ and $P(Q_2|I)$ (two alternatives)

$P(q_1|i)$	q_1^1	q_1^2	q_1^3	q_1^4	$P(q_2|i)$	q_2^1	q_2^2	q_2^3	q_2^4	$P(q_2|i)$	q_2^1	q_2^2	q_2^3	q_2^4	$\tilde{q}_2$
i_1	1	0	0	0	i_1	$\frac{1}{4}$	$\frac{1}{4}$	$\frac{1}{4}$	$\frac{1}{4}$	i_1	$\frac{1}{4}$	$\frac{1}{4}$	$\frac{1}{4}$	$\frac{1}{4}$	0
i_2	0	1	0	0	i_2	$\frac{1}{2}$	$\frac{1}{2}$	0	0	i_2	$\frac{1}{4}$	$\frac{1}{4}$	0	0	$\frac{1}{2}$
i_3	0	0	0	1	i_3	0	0	$\frac{1}{2}$	$\frac{1}{2}$	i_3	0	0	$\frac{1}{4}$	$\frac{1}{4}$	$\frac{1}{2}$

The above notion of logical compatibility is sufficient to elicit the conditional probability table $P(Q|I)$ in the special case where each item is compatible only with a single answer, i.e., $|\mathcal{Q}(i)| = 1$ for each $i \in \mathcal{I}$, where symbol $|\cdot|$ denotes cardinality. If this is the case, $P(q|i) = 1$ if $q \in \mathcal{Q}(i)$ and zero otherwise.

For questions of this kind, after an answer $Q = q \in \mathcal{Q}$, the updating as in Eq. (1) corresponds to a uniform *scaling* of the prior probability for the items compatible with q, whereas all the incompatible ones receive zero mass:

$$P(i|q) = \begin{cases} P(i)[\sum_{i':q\in\mathcal{Q}(i')} P(i')]^{-1} & \text{if } q \in \mathcal{Q}(i) \\ 0 & \text{otherwise} \end{cases} \tag{5}$$

Note that in this setting the Bayesian approach acts as a logical filter, making the items incompatible with the answers impossible.

3.2   Symmetry Statements

Consider a question $Q \in \mathbf{Q}$ such that the assumptions in the previous section are not satisfied, i.e., there are items compatible with more than a single answer. In this case, the elicitation of $P(Q|I)$ requires further assumptions. We adopt a strategy forcing answers to have the same impact on all compatible items. Such a prior assumption will be relaxed when coping with historical data in Sect. 6). Other strategies are in [15]. Let us introduce the discussion by an example.

Example 3. In the setup of Example 1, for question Q_2, we have $\mathcal{Q}_2(i_1) = \{q_2^1, q_2^2, q_2^3, q_2^4\}$, $\mathcal{Q}_2(i_2) = \{q_2^1, q_2^2\}$ and $\mathcal{Q}_2(i_3) = \{q_2^3, q_2^4\}$. Assuming the same probability for the answers compatible with a given item completely determines the conditional probability table $P(Q_2|I)$, that is depicted in Table 2 (middle).

The procedure considered in Example 3 allows to elicit a conditional probability table in the general case of items compatible with multiple answers. This basically corresponds to set a uniform probability $P(q|i) = |\mathcal{Q}(i)|^{-1}$ for all $q \in \mathcal{Q}(i)$ and zero otherwise. After answer $Q = q$, the updating in Eq. (1), differently from Eq. (5), is a non-uniform rescaling:

$$P(i|q) = P(i)\frac{|\mathcal{Q}(i)|^{-1}}{\sum_{i':q\in\mathcal{Q}(i')} P(i')|\mathcal{Q}(i')|^{-1}}, \tag{6}$$

if $q \in \mathcal{Q}(i)$ and zero otherwise. Yet, when supports have different cardinalities, this approach might lead to questionable results as in the following example.

Example 4. Consider the same setup of Example 2 with a uniform prior over the items, i.e., $P(i_k) = \frac{1}{3}$ for $k = 1, 2, 3$. After the answer $Q_2 = q_2^1$, following Eq. (6), we obtain $P(i_1|q_2^1) = \frac{1}{3}$ and $P(i_2|q_2^1) = \frac{2}{3}$. In other words, both the band and the DJ are available for the wedding, but despite the uniform prior, the band has twice the probability of being selected, the reason being the lower cardinality of the support of the band compared to the DJ.

According to Example 4, the strategy of Example 2 leading to Table 2 (middle) has a bias towards items with smaller supports (i.e., lower compatibility). To prevent such bias, we propose an alternative elicitation as in this example.

Example 5. In the setup of Example 2 add to question Q_2 a fifth answer $\tilde{q}_2$ compatible with all the items. Consider the elicitation of $P(I|Q_2)$ in Table 2 (right). With a uniform prior over I as in Example 4, $P(i_1|q_2^1) = P(i_2|q_2^1) = \frac{1}{2}$.

The procedure in Example 5 can be generalized by setting the parameters of the conditional probability table $P(Q|I)$ as follows:

(i) $P(q|i) = \kappa^{-1}$ for each $q \in \mathcal{Q}$ and $i \in \mathcal{I}$,
(ii) $P(\tilde{q}|i) = 1 - \kappa^{-1}|\mathcal{Q}(i)|$.

The value of $\kappa > 0$ is arbitrary, provided that the table is a proper stochastic matrix, i.e., its elements have non-negative values and normalized rows. Setting $\kappa := \max_{i \in \mathcal{I}} |\mathcal{Q}(i)|$ guarantees non-negative probability of the *dummy* state $\tilde{q}$. Non-negativity of other elements follows from (i) and normalization from (ii).

Considering this approach to the elicitation of $P(Q|I)$, after a (non-dummy) answer $Q = q$, the updated probability is given by Eq. (5) as all valid $P(i|q)$ have the same value. This proves that posterior inferences are not affected by the choice of κ. Moreover, if i' and i'' are two items consistent with q:

$$\frac{P(i'|q)}{P(i''|q)} = \frac{P(i')}{P(i'')}, \tag{7}$$

i.e., even in general, this approach does not suffer from the bias in Example 4.

4   Coping with Non-exclusive Answers

In some cases it might be reasonable to allow users to select multiple answers to a question, thus violating the original assumption about answers being described as mutually exclusive events. For instance, in the setup of Example 1, if we assume the answers to Q_1 non-exclusive, an user might return a set-valued answer such as $\mathcal{Q}_1^* = \{q_1^1, q_1^2\}$, meaning *(DJ, band)*. Different semantics can be considered to model set-valued observations. Following the random set literature [5], we distinguish between *ontic* (or conjunctive) and *epistemic* (or disjunctive) semantics. From an epistemic perspective, the set reflects an incomplete knowledge about the actual value of the answer (which has a single value). In our example, this assumption implies that the user is eventually looking for a DJ or for a band and certainly not for a musician or an entertainer. Yet, for some reason (e.g., limited capability of introspection or limited amount of time to answer), she is not able to isolate a single option (which should however exists). The ontic view, instead, regards the set of answers as a precise component of a more complex entity, e.g., the combination *(DJ, band)* is exactly the option the user wants to select. The two semantics requires different modelling strategies.

In the ontic case, we replace the set of possible answers $\mathcal{Q}$ with its partition set in order to include all the $2^{|\mathcal{Q}|-1}$ possible set-valued options (remember that the user should always provide at least an answer). Afterwards, the elicitation and updating procedures remain as in the previous section.

In the epistemic case, in terms of elicitation nothing changes as the question Q is assumed to take single elements of $\mathcal{Q}$ as values. Consequently the quantification of $P(Q|I)$ can be achieved as in Sect. 3.2. The difference arises at the updating level as the procedure in Eq. (5) cannot directly cope with set-valued answers. To update item probabilities we should instead intend a set-valued answer $\mathcal{Q}^* \subseteq \mathcal{Q}$ as a *virtual evidence* [17], that is not able to reveal the actual state of Q in a reliable way. A virtual evidence describes an observation in terms of its likelihoods for the different (actual) states of Q, these values being defined up to a non-negative constant. Here, we set to zero the likelihoods of the answers not selected by the user, while giving the same value to the selected ones, i.e., those in $\mathcal{Q}^*$. This leads to the following updating rule:

$$P(i|\mathcal{Q}^*) := \frac{\sum_{q \in \mathcal{Q}^*} P(q|i)\, P(i)}{\sum_{q \in \mathcal{Q}^*} P(q)}, \tag{8}$$

where $P(q) = \sum_{i \in \mathcal{I}} P(q|i)\, P(i)$. Let us see how this can be achieved in practice.

Example 6. Consider the same setup of Example 5 but with a set-value answer $\mathcal{Q}_1^* = \{q_1^1, q_1^2\}$, i.e., the user replied DJ *and* band. Following Eq. (8):

$$P(i|\mathcal{Q}_2^*) \propto [P(q_2^1|i) + P(q_2^2|i)]P(i) \tag{9}$$

for each $i \in \mathcal{I}$, where the proportionality constant is obtained by normalization.

5 Adaptive Conversations

We detail here the high-level procedure used to create a conversation by selecting sequentially the question and driven by decision theoretic criteria. This is inspired by the similar approaches considered for computer adaptive testing [14].

Question Selection. In classical recommendation systems the assessment of the user preferences with respect to the different items is based on a static block of background user information. Such information can be already available in the system or directly obtained from the user after some kind of reduced interaction, e.g., a predefined questionnaire. However, as discussed in the introduction, in modern setups the process of collecting information about the user preferences with respect to the catalogue should be based on a sequence of dynamic interactions. In this view, the questionnaire approach leaves the place to a conversational process taking the form of a personalized sequence of questions dynamically picked from a larger set of questions. The prior probability mass function $P(I)$ is thus sequentially updated each time a new answer is collected, and the updated probability $P(I|\mathbf{Q})$ is used to select the most informative next question. The choice between a possibly huge set of candidate questions/interactions can be driven by information-theoretic criteria making any sequence potentially different from the other. In particular, taking inspiration from the literature in the field of adaptive testing, we pick the question that minimizes the conditional entropy (and hence maximizes the expected information gain), i.e., the adaptive conversational process selects the question Q_j^* such that:

$$j^* = \arg \min_{j=1,\dots,m} H(I|Q_j), \tag{10}$$

where $H(I|Q) := \sum_{q \in \mathcal{Q}} H(I|q)P(q)$ and $H(I|q)$ is the entropy of the posterior mass function over I after the answer $q \in \mathcal{Q}$ to the question Q, and $\{Q_j\}_{j=1}^m$ is the set of questions the system can choose from.

Stopping Rule. This procedure is iterated after every answer and the conversation ends if the posterior entropy $H(I|q)$ decreases under a fixed threshold. As the entropy of a mass function $P(I)$ is defined as $H(I) := -\sum_{i \in \mathcal{I}} P(i) \log_{|\mathcal{I}|} P(i)$, a natural threshold H_τ^* is the entropy of a mass function over I which is uniform on τ items, while the other ones have zero probability, i.e., $H_\tau^* := -\log_{|\mathcal{I}|} \tau^{-1}$. Setting this value in the stopping rule forces the system to halt when most of the posterior probability mass is concentrated on the τ most probable items.

6   Coping with Data

The elicitation procedure in Sect. 3 is based on structural judgements about logical compatibility for items and answers merged with symmetry (i.e., indifference) statements. Yet, this might be insufficient to capture actual preferences as outlined by the following example.

Example 6. Consider the setup in Example 5. In Table 2 (right), $P(q_2^1|i_2) = P(q_2^2|i_2)$, i.e., the band is equally likely to perform at weddings and corporate events. Yet, in past activities, the band was more likely to attend weddings.

Data about the users, in the form of their answers to the questions and the items eventually picked, might help to achieve a better quantification of the model parameters. In this section we describe how to combine data with the elicitation procedure described in Sect. 3. Let us first show how this can be directly achieved in the conditional probability table $P(Q|I)$.

6.1   Bayesian Learning

Bayesian approaches allow to combine the elicited parameters with data about items and questions. The probability tables can be used to parametrize a multinomial Dirichlet prior, to be merged with the likelihood of the observed data. The posterior expected value might increase the discriminative power of the system and the consequent quality of the recommendations.

Let $\alpha_j^k := P(q^k|i_j)$ for each $q^k \in \mathcal{Q}$ and $i_j \in \mathcal{I}$, where $P(Q|I)$ has been elicited as in Sect. 3.2. Assume that n_j i.i.d. observations about users who picked item $I = i_j$ are available, of which n_j^k reported answer q^k, for $k = 1, \dots, |\mathcal{Q}|$. We update the values of $P(Q|i)$ by integrating the product of a Dirichlet prior with weights $s\,\alpha_j^k$ and the likelihood based on the counts [2]. This gives the posterior:

$$P'(q^k|i_j) = \frac{s\alpha_j^k + n_j^k}{s + n_j},\tag{11}$$

for each $j = 1,\ldots,|\mathcal{Q}|$, where $s > 0$ is the equivalent sample size of the Dirichlet prior. Note that the parameters of the Dirichlet distribution are required to be strictly positive, while the elements of $P(Q|I)$ can be zero. If the latter is the case, we take the limit $\alpha_j^k \to 0$ and consider Eq. (11) still valid.

Example 7. In the setup of Example 5 assume data about the answers provided to question Q_2 for the band i_2 available. The counts are $n_2^1 = 5$, $n_2^2 = 3$, $n_2^3 = 1$, $n_2^4 = 0$. For $s = 1$, with the prior parameters as in Table 2 (right), Eq. (11) gives $P(q_2^1|i_2) = 0.525$ and $P(q_2^2|i_2) = 0.325$, which means that higher number of weddings compared to corporate events breaks the prior equal probability of the two options. Note also that, despite its (prior) incompatibility with the band the birthday option (q_2^3) has a non-zero count. In such a prior-data conflict [7], the option takes a non-zero probability $P(q_2^3|i_2) = 0.1$, while the incompatible and unobserved answer q_2^4 gets sharp zero probability.

A separate discussion should be provided for the updated probabilities of the dummy state $\tilde{q}$. This answer has zero counts by definition, but its prior might be non-zero, thus leading to a non-zero posterior (e.g., in the previous example, $P(\tilde{q}_2|i_2) = 0.05$). As discussed in Sect. 3.2, these non-zero values are not affecting those of the updated mass functions $P(I|q)$, and hence those of the corresponding entropies $H(I|q)$. This is not the case for conditional entropy $H(I|Q)$ because of weighted average based on $P(Q)$. A temporary *revision* of the marginal probability mass function $P(Q)$, that sets $P(\tilde{q}) = 0$ and rescales the other values, only for the choice of the next question, might fix the issue.

6.2   Preventing Data Fragmentation

The above procedure separately processes data corresponding to different items. For large catalogues this induces a data *fragmentation*, thus making some estimator unreliable. To prevent this we introduce the notion of *latent question*. Given question Q, its latent counterpart $\hat{Q}$ is a second variable taking its values (called *latent answers*) from the same set of possible states $\mathcal{Q}$ and sharing the same supports (see Sect. 3.1). The latent question $\hat{Q}$ is a (stochastic) function of $I = i$ and $Q = q$. Unlike Q, $\hat{Q}$ is always consistent with the support of i, being a random element of it or its unique value, or q in the case also Q is consistent. Therefore, the support of an item i can be defined in terms of the *latent* compatible answers and indicated as $\hat{\mathcal{Q}}(i)$. We assume conditional independence between Q and I given $\hat{Q}$, this meaning that if the answer to the latent question is known, the answer to the non-latent question does not depend on the selected item, i.e., $P(q|\hat{q},i) = P(q|\hat{q})$, for each $q, \hat{q} \in \mathcal{Q}$ and $i \in \mathcal{I}$. By this statement and total probability, the elements of table $P(Q|I)$ can be expressed as:

$$P(q|i) = \sum_{\hat{q} \in \mathcal{Q}} P(q|\hat{q}) \, P(\hat{q}|i) \,. \tag{12}$$

Equation (12) offers a strategy to probability tables elicitation less prone to fragmentation issues. To achieve that, for $P(\hat{Q}|I)$, we adopt the strategy in Sect. 3.2. The data are used instead for the quantification of $P(Q|\hat{Q})$, for which we adopt a diagonal Dirichlet prior to be combined with the data likelihood. Note that the values of $\hat{Q}$ can be generated from those about Q and I. To reduce variance, we replace sampled outcomes of $\hat{Q}$ with a collection of uniform fractional counts for the different elements of $\hat{\mathcal{Q}}(i)$. Following a Bayesian scheme analogous to that described in Sect. 6.1, we indeed obtain the posterior estimator for $P(Q|\hat{Q})$. Finally, marginalizing the latent question as in Eq. (12) we obtain $P(Q|I)$. Note also that, without data, $P(Q|\hat{Q})$ acts as the identity matrix, and the approach proposed here corresponds to that in Sect. 3.2.

Example 9. In the setup of Example 1, consider the data about Q_1 and I in Table 3 (left). The DJ and the band exhibit some inconsistency with respect to their support, e.g., once the DJ performed even if the request was for a musician. This is modelled by the latent question $\hat{Q}_1$. The definition of $\hat{Q}_1$ allows to complete the data about the latent question as in Table 3 (middle) and hence $P(\hat{Q}|Q)$. Following Eq. (12), we combine this probability table with $P(\hat{Q}_1|I)$ elicited as in Table 2 and obtain the quantification in Table 3 (right).

Table 3. Counts for Q_1, $\hat{Q}_1$ and I (left) and corresponding $P(Q_1|I)$ (right).

$n(Q_1, I)$	q_1^1	q_1^2	q_1^3	q_1^4
i_1	4	0	1	0
i_2	2	6	3	0
i_3	0	0	0	5

$n(\hat{Q}_1, Q_1)$	q_1^1	q_1^2	q_1^3	q_1^4
$\hat{q}_1^1$	4	0	1	0
$\hat{q}_1^2$	2	6	3	0
$\hat{q}_1^3$	0	0	0	0
$\hat{q}_1^4$	0	0	0	5

$P(Q_1|I)$	q_1^1	q_1^2	q_1^3	q_1^4
i_1	$\frac{5}{6}$	0	$\frac{1}{6}$	0
i_2	$\frac{1}{6}$	$\frac{7}{12}$	$\frac{1}{4}$	0
i_3	0	0	0	1

In the above procedure, the latent question and its operational definition can be intended as a rule for the clustering of data associated to different items, the latent answers indexing the clusters, i.e., all items associated to the data in cluster $\hat{q}$ include the latent answer $\hat{q}$ in their support. As we act in a probabilistic setting, *soft clustering* can be obtained [12], e.g., when the item associated to an observation has $|\mathcal{Q}(i)| > 1$ and the answer $q \notin \mathcal{Q}(i)$. Nothing prevents us from considering other data-based clustering algorithms relaxing the one-to-one correspondence between questions and latent questions.

In this section we have focused on models with a single question. To cope with multiple questions, the naive-like assumption formulated in Eq. (2) for (non-latent) questions should be reformulated in terms of the latent ones and we assume conditional independence of the latent questions given the item.

Finally, if a more general notion of latent question is adopted, so that (observable) questions and latent questions may take their values from different sets, we could also consider the case of multiple (*observable*) questions associated to the same latent one. This can be easily achieved by maintaining the assumption of conditional independence between the item and all the questions given their latent question, that corresponds in this case to the conditional independence between the questions of the same latent question given the latent question itself. In practice, if Q_a and Q_b are the two questions and $\hat{Q}$ their common latent question, we have Q_a and Q_b independent given $\hat{Q}$, we separately learn $P(Q_a|\hat{Q})$ and $P(Q_b|\hat{Q})$, and finally obtain:

$$P(q_a, q_b|i) = \sum_{\hat{q} \in \hat{\mathcal{Q}}} P(q_a|\hat{q})P(q_b|\hat{q})P(\hat{q}|i) . \qquad (13)$$

Moreover, with the aim of preserving the naive independence structure defined in Eq. (2), which is not necessary but simplifies the implementation of the conversational approach discussed in Sect. 5, the above model may be approximated by one assuming conditional independence of Q_a and Q_b and having for all $q_a \in \mathcal{Q}_a$

$$P(q_a|i) = \sum_{q_b \in \mathcal{Q}_b} P(q_a, q_b|i) . \qquad (14)$$

7   Experiments

For validation we consider the classical *MovieLens* database [8]. The dataset contains information about 13'816 movies. Each record is characterized by a list of genres taken from nineteen pre-defined categories, a period (before 1980, from 1980 to 1999, from 2000 to 2016, after 2016, in our discretization) and a *tag* relevance score for each of the tags included in *MovieLens* database Tag Genome. The tag relevance score is a number between zero and one, describing how much the tag reflects the movie characteristics. E.g., tag *violence* has relevance 0.991 for the movie *Reservoir Dogs* and 0.234 for *Toy Story*. We only consider the 200 tags most used by the users, out of the 1'128 available.

Tags together with genre and period are regarded as questions and used to simulate a conversation based on a static questionnaire whose goal is to detect the right item/movie. No set-valued answers are allowed for these questions. We instead allow sets for answers about the genre in the learning phase, for which the ontic semantics discussed in Sect. 4 is adopted. Each element of the set is treated as a fractional observation with weight inversely proportional to the cardinality of set of genres provided. The (unobserved) value assumed by the latent genres is assigned consistently with the support of the corresponding item using the procedure described in Sect. 6.2. On the other end, to simplify the procedure, and at the same time increasing the flexibility of the adaptive approach by augmenting the number of questions in the pool, each genre is separately asked by nineteen independent questions along the conversation.

For each movie in the database, a complete list of questionnaire answers is generated as follows. For the question *period* we might corrupt the actual period by changing it into the previous or the next with probability p_y. For the genre, we take the list of genres assigned to a movie and corrupt each of them (independently) with probability p_g. Corruption consists either in removing the genre from the list, or permuting the genre by replacing it by a similar one. Finally, answers to tag-related questions are simulated as follows. A positive answer to a question about a tag is assigned with probability equal to the tag relevance score of the movie. Subsequently, positive tag answers can be corrupted with probability p_t by setting them to zero. This corruption step, simulates a user who has, in general, a number of desiderata smaller than the total number of tags that could properly describe a movie. In our experiments we use $p_y = p_g = p_t = p$ and consider two noise levels, namely *low* ($p = 10\%$) and *high* ($p = 50\%$).

We select 2'000 movies, used as the catalogue, from which we have generated the test set for the experiments. The remaining movies are used to generate the training dataset from which the conditional probability tables $P(Q|\hat{Q})$ for the genre, period and tag questions given the actual genre, period and tags of the selected movie (latent question) are estimated as described in Sect. 6. The tables $P(\hat{Q}|I)$ for the genre and period are defined by compatibility and similarity constraints as in Sect. 3. Concerning tags, instead, the probability $P(\hat{Q} = 1|i)$ of a tag being compatible with item i is set equal to the tag relevance score for item i. Finally, a uniform prior over the items is assumed. The goal of the system is to detect the right movie on the basis of the answers to the questions. In the conversational case, questions are selected following the adaptive strategy detailed in Sect. 5. The algorithm in charge of the elicitation and question selection has been implemented in Python.[1]

The performances of our approach are compared against that of traditional algorithms based on measures of similarity between the item and user features, such as the *LightFM* algorithm [11]. Here the item features, i.e., the (eventually probabilistic) supports $Q_j(i)$ for all questions Q_j, $j = 1, \ldots, m$ and the user features, i.e., their (simulated) noisy answers, take values from the same space. Thus, there is no need of learning linear transformations to a common latent space, as one can simply compute the items-user similarity in the original space. In the experiments below, the *cosine similarity* is used to measure the affinity between items and users requests. Note that, as we are considering a situation where each user has only a single conversation with the system, we cannot learn users and items similarities from historical interactions as done in traditional collaborative filtering. The fact that the similarity-based approach does not take advantage of the training dataset to improve its item selection strategy limits its accuracy, especially when the noise is large. However, as no standard solutions could be found in the literature (the field of conversational recommendations is relatively new) and since our focus was the ability of the Bayesian model to drive an adaptive conversation (rather than a possible improvement in the accuracy of the final rank), we have been satisfied with validating its accuracy against

---

[1] https://github.com/IDSIA/bayesrecsys.

such simple baseline, yet quite popular in the context of neighbourhood-based filtering [6].

As a first experiment, we evaluate the benefits of the adaptive approach in the question selection process with respect to a random strategy. Figure 1 shows the average probability of the movie selected by the user as a function of the number of questions for the two strategies with the two different noise levels. In both cases increasing the number of questions makes the right movie more probable, but the growth is quicker if the questions are selected on the basis of their expected information gain. with the gap between the two approaches increasing unless half of the questions in the questionnaire has been asked.

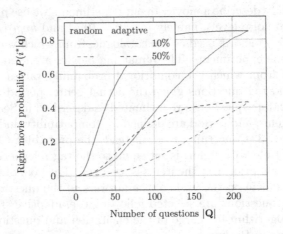

Fig. 1. Adaptive vs. random selection for two noise levels.

For the validation against the similarity-based approach, which does not provide probabilities, we consider the average rank of the right movie as a function of the number of questions. As in the previous experiment we consider our Bayesian approach with adaptive selection of the questions, and the same approach with random selection of the questions. The two methods are compared against the similarity-based approach. As we have no standard method available to select the order of the questions in this case, we test the similarity-based approach using both the order selected by the Bayesian conversation and the random order. Note that, as a consequence, the improvement in the performance of the similarity-based approach in the adaptive setting compared to the random one, are only due to the Bayesian question selection strategy. Results are shown in Fig. 2 (top left), with a zoom of the tails in Fig. 2 (top right). As in the previous analysis the advantages of the adaptive approach compared to the random one are clear. Moreover, results show that the Bayesian model, by correctly modelling the users behaviours from the historical data, outperforms the similarity approach the more the largest is the noise corrupting the data. In Fig. 2 (top left) we also report as a black dotted line the ranks of the Bayesian adaptive approach based on structural judgments without learning from the data. The higher ranks compared to the approach based on data advocates the learning procedure discussed in Sect. 6.

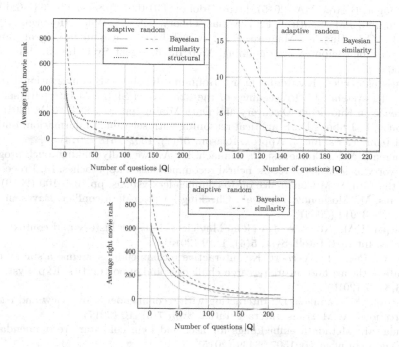

Fig. 2. Average recommendation rank as a function of the number of questions for low (10%, top left and right) and high (50%, bottom) noise levels.

8   Conclusions

A new approach to automatic recommendations which assumes a dynamic interaction between the system and the user to provide customized and self-adaptive recommendations has been developed on the basis of a pure Bayesian approach. The framework introduced in this paper sets the ground to several future developments, among which the dynamic generations of questions in order to improve the conversational nature of the system. This could be based on a natural language generation system interacting with the structured probabilistic description of item properties and elicitation of user needs. Also, this work has focused on a setting where each user interacts with the system a single time. Therefore, collaborative filtering approaches that learn users' taste and items similarities from historical data were not applicable. Such setting looks appropriate for highly involved decisions, such as the selection of an entertainer for a special event.

References

1. Adomavicius, G., Tuzhilin, A.: Context-aware recommender systems. In: Ricci, F., Rokach, L., Shapira, B. (eds.) Recommender Systems Handbook, pp. 191–226. Springer, Boston, MA (2015). https://doi.org/10.1007/978-1-4899-7637-6_6
2. Bernardo, J.M., Smith, A.F.: Bayesian Theory, vol. 405. Wiley, Hoboken (2009)
3. Butz, C.J., Hua, S., Maguire, R.B.: A web-based Bayesian intelligent tutoring system for computer programming. Web Intell. Agent Syst. Int. J. 4(1), 77–97 (2006)
4. Christakopoulou, K., Radlinski, F., Hofmann, K.: Towards conversational recommender systems. In: Proceedings of the 22nd ACM SIGKDD International Conference on Knowledge Discovery and Data Mining, pp. 815–824. ACM (2016)
5. Couso, I., Dubois, D.: Statistical reasoning with set-valued information: ontic vs. epistemic views. Int. J. Approx. Reason. 55(7), 1502–1518 (2014)
6. Dacrema, M.F., Cremonesi, P., Jannach, D.: Are we really making much progress? A worrying analysis of recent neural recommendation approaches. In: Proceedings of the 13th ACM Conference on Recommender Systems, pp. 101–109 (2019)
7. Evans, M., Moshonov, H., et al.: Checking for prior-data conflict. Bayesian Anal. 1(4), 893–914 (2006)
8. Harper, F.M., Konstan, J.A.: The Movielens datasets: history and context. ACM Trans. Interact. Intell. Syst. 5(4), 1–19 (2015)
9. He, C., Parra, D., Verbert, K.: Interactive recommender systems: a survey of the state of the art and future research challenges and opportunities. Exp. Syst. Appl. 56, 9–27 (2016)
10. Jugovac, M., Jannach, D.: Interacting with recommenders - overview and research directions. ACM Trans. Interact. Intell. Syst. 7(3), 10 (2017)
11. Kula, M.: Metadata embeddings for user and item cold-start recommendations. arXiv preprint arXiv:1507.08439 (2015)
12. Lewicki, M.S.: Bayesian modeling and classification of neural signals. In: Heidbreder, G.R. (ed.) Maximum Entropy and Bayesian Methods. Fundamental Theories of Physics (An International Book Series on The Fundamental Theories of Physics: Their Clarification, Development and Application), vol. 62. Springer, Dordrecht (1996)
13. Lu, J., Wu, D., Mao, M., Wang, W., Zhang, G.: Recommender system application developments: a survey. Decis. Support Syst. 74, 12–32 (2015)
14. Mangili, F., Bonesana, C., Antonucci, A.: Reliable knowledge-based adaptive tests by credal networks. In: Antonucci, A., Cholvy, L., Papini, O. (eds.) ECSQARU 2017. LNCS (LNAI), vol. 10369, pp. 282–291. Springer, Cham (2017). https://doi.org/10.1007/978-3-319-61581-3_26
15. Mangili, F., Broggini, D., Antonucci, A., Alberti, M., Cimasoni, L.: A Bayesian approach to conversational recommendation systems. In: AAAI 2020 Workshop on Interactive and Conversational Recommendation Systems (WICRS) (2020)
16. Millán, E., Trella, M., Pérez-de-la-Cruz, J., Conejo, R.: Using Bayesian networks in computerized adaptive tests. In: Ortega, M., Bravo, J. (eds.) Computers and Education in the 21st Century, pp. 217–228. Springer, Dordrecht (2000). https://doi.org/10.1007/0-306-47532-4_20
17. Pearl, J.: Probabilistic Reasoning in Intelligent Systems: Networks of Plausible Inference. Elsevier (1988)
18. Sarwar, B., Karypis, G., Konstan, J., Riedl, J., et al.: Analysis of recommendation algorithms for e-commerce. In: EC 2000, pp. 158–167 (2000)

Short Papers

Dealing with Atypical Instances in Evidential Decision-Making

Benjamin Quost[1,2(✉)], Marie-Hélène Masson[1,3], and Sébastien Destercke[1,4]

[1] UMR UTC-CNRS 7253 Heudiasyc, Compiègne cedex, France
benjamin.quost@utc.fr
[2] Sorbonne Universités, Université de Technologie de Compiègne, Compiègne cedex, France
[3] Université de Picardie Jules Verne, Amiens, France
[4] Centre National de la Recherche Scientifique, Paris, France

Abstract. When classifying an example on the basis of an observed population of (training) samples, at least three kinds of situations can arise where picking a single class may be difficult: high aleatory uncertainty due to the natural mixing of classes, high epistemic uncertainty due to the scarcity of training data, and non-conformity or atypicality of the example with respect to observations made so far. While the two first kinds of situations have been explored extensively, the last one still calls for a principled analysis. This paper is a first proposal to address this issue within the theory of belief function.

Keywords: Belief functions · Supervised classification · Epistemic and aleatoric uncertainty · Atypicality management · Novelty detection

1 Introduction

In a classification problem, it is classical to make a (precise, crisp) decision by assigning a test instance to a single class based on a set of training instances. However, various kinds of uncertainty may prove to be a hindrance to this process: the classes can be mixed and none seems to prevail (aleatoric uncertainty or ambiguity); the training data can be scarce (epistemic uncertainty); or the example to be classified can differ from the training observations (non-conformity or atypicality). Although related to epistemic uncertainty, atypicality cannot necessarily be reduced by obtaining additional labeled training data. It is central in novelty, anomaly or outlier detection [2].

The theory of belief functions, introduced in [3,6], and then further developed by Smets [9], provides a suitable framework for representing uncertainties. In this paper, we study how atypical instances can be accounted for in the framework of belief functions, in addition to situations of ambiguous or scarce data. Our ultimate purpose is to propose a safe, cautious decision-making process in presence of uncertainty, by taking into account the quantity and quality of data based on which the decision is to be made.

Note that atypicality has been already accounted for in different settings, such as distance rejection [5] or conformal predictions [7], and was also addressed using belief functions (e.g., [1]) for specific kinds of atypicality. Yet, to our knowledge, no principled, generic way to deal with atypicality has been proposed in the belief function

ⓒ Springer Nature Switzerland AG 2020
J. Davis and K. Tabia (Eds.): SUM 2020, LNAI 12322, pp. 217–225, 2020.
https://doi.org/10.1007/978-3-030-58449-8_15

framework. This paper can be seen as a preliminary contribution to this issue. We establish our basic setting in Sect. 2. Section 3 then discusses some desirable properties when accounting for atypicality, for which Sect. 4 proposes several strategies.

2 Basic Setting

We recall here basic material on the theory of belief functions, a rich and flexible framework for managing uncertainty, which will be required in the rest of the paper.

2.1 Preliminaries on Belief Functions

Let us consider a variable ω taking values in a finite unordered set $\Omega = \{\omega_1, \ldots, \omega_M\}$ called the frame of discernment. Partial knowledge regarding the actual value taken by ω is represented by a mass function [6] $m : 2^\Omega \to [0; 1]$ such that

$$\sum_{A \subseteq \Omega} m(A) = 1. \tag{1}$$

The sets $A \subseteq \Omega$ such that $m(A) > 0$ are called *focal sets* of m. If $m(A) = 1$ for some $A \subseteq \Omega$, m is said to be *categorical* and is denoted by m_A (if $A = \Omega$, m_Ω represents complete ignorance). It is often required that $m(\emptyset) = 0$; otherwise, $m(\emptyset)$ may have various interpretations, such as the degree of conflict after inconsistent pieces of information were aggregated, or the degree of belief that $\omega \notin \Omega$ (*open world assumption*).

Any mass function can be equivalently represented by a belief function *bel*, and a plausibility function *pl* defined, respectively, for all $A \subseteq \Omega$ by:

$$bel(A) = \sum_{B \subseteq A} m(B), \quad pl(A) = \sum_{B \cap A \neq \emptyset} m(B). \tag{2}$$

Various strategies have been proposed for making decisions based on a belief function—see, e.g., [4]. Hereafter, we will denote by δ any decision operator to be applied to a mass function defined over the set of classes. For instance, the *interval dominance* operator may result in a cautious decision, as it may provide a set of classes.

Definition 1 (Interval dominance). *Given a mass m, ω_i is said to dominate ω_j, noted $\omega_i \succ \omega_j$, if $bel(\{\omega_i\}) > pl(\{\omega_j\})$. The interval dominance rule consists in computing the set of non-dominated classes:*

$$\delta_{ID}(m) = \{\omega_i : pl(\{\omega_i\}) \geq bel(\{\omega_j\}) \text{ for all } j \neq i\}. \tag{3}$$

2.2 Uncertain Class Membership Model

We assume that a source provides us with information regarding the actual class of a test instance $\mathbf{x}$ to be classified, in the form of a mass function m. This mass function is usually derived from a sample of N training instances $\mathbf{x}_i$ ($i = 1, \ldots, \mathbf{x}_N$) observed in the same region than $\mathbf{x}$, and to which $\mathbf{x}$ is assumed to be similar. For example, when using decision trees, $\mathbf{x}$ is classified using the training data falling into the same leaf node; in the K-NN algorithm, the decision is made based on the K closest training instances.

Aleatoric uncertainty (due to mixed classes) can straightforwardly be modeled and transferred to the test instance via a multinomial model. Let $(n_1, n_2, \ldots, n_M)$ denote the class counts in the observed training sample of size N (with $\sum_k n_k = N$); then,

$$m(\{\omega_i\}) = \frac{n_i}{N} \quad \forall i = 1, \ldots, M.$$

In cautious classification, it is desirable to take the quantity of information carried out by the training sample into account in the decision process: then, should the available information be scarce, cautious strategies may be deployed, such as retaining a set of plausible classes instead of a single one. The imprecise Dirichlet model (IDM) makes it possible to integrate this notion of quantity of available information in the model of class frequencies. Rather than the mass function provided above, it would result in

$$m^{IDir}(\Omega) = \frac{s}{N+s}, \quad m^{IDir}(\{\omega_i\}) = \frac{n_i}{N+s} \quad \forall i = 1, \ldots, M, \tag{4}$$

where the user-defined parameter s can be interpreted as a number of additional unknown observations interfering with estimating the probabilities of the classes. This mass function produces in turn the bounds

$$\left[bel^{IDir}(\{\omega_i\}) = \frac{n_i}{N+s} ; pl^{IDir}(\{\omega_i\}) = \frac{n_i+s}{N+s} \right] \quad \forall i = 1, \ldots, M, \tag{5}$$

which account for both aleatoric uncertainty (which occurs if $n_1, \ldots, n_M$ take similar values) and epistemic uncertainty (in which case the width of the intervals will increase when $s/(N+s)$ increases).

2.3 Conformity

Although the IDM integrates both the aleatoric and epistemic uncertainty due to the training sample, it does not take into account the typicality of a test instance of interest, that is, the extent to which it is similar to one of the training instances from which m^{IDir} is obtained. We assume here that this information is provided by a separate source, in the form of a conformity score $C \in [0; 1]$: we have $C = 0$ for a completely unusual instance, and $C = 1$ for a normal one. How this level of typicality may be assessed is beyond the scope of this preliminary work, and thus left aside for now.

Figure 1 displays two situations where an instance $\mathbf{x}$ is to be classified into one of three classes $\{\omega_1, \omega_2, \omega_3\} = \Omega$, based on a set of four training instances (with known classes). The same IDM would be built (the class counts being the same), but the level of typicality of $\mathbf{x}$ with respect to the four instances is very different. Note that here, C can be derived from the distance of $\mathbf{x}$ to its first neighbour in the training sample.

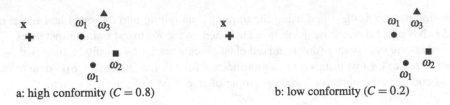

a: high conformity ($C = 0.8$)   b: low conformity ($C = 0.2$)

Fig. 1. Two situations with identical class counts but different levels of typicality

The purpose of this paper is to determine how a mass function m related to the class of the instance **x** can be revised according to its level of typicality C. To this end, we introduce the notion of *conformity operator* Cf, which updates m into a new mass function Cf $[m, C]$. This operator may be required to have various properties (see Sect. 3), according to which different operators may be proposed (see Sect. 4).

3   Desirable Properties of Conformity Operators

Hereafter, by abuse of notation, Cf $[pl, \cdot]$ (respectively, Cf $[bel, \cdot]$) will stand for the plausibility function (resp., belief function) obtained from a revised mass function Cf $[m, \cdot]$.

Property 1 (Class preference preservation). A conformity operator Cf preserves the preferential information over the classes if, for any $C \in [0; 1]$,

$$pl(\{\omega_i\}) \leq bel(\{\omega_j\}) \Rightarrow \text{Cf} \, [pl, C] \, (\{\omega_i\}) \leq \text{Cf} \, [bel, C] \, (\{\omega_j\}). \qquad (6)$$

Plainly put, it means that taking into account conformity does not alter interval dominance between classes (see Definition (1)).

The example in Fig. 2 displays the decision boundaries of a decision tree applied to the Iris dataset (two features were kept for illustrative purpose). Instances 1, 2 and 3 are atypical. For instance 1, class Setosa clearly dominates both others, an information which may reasonably be kept by the revision process. However, it is more questionable for instance 2, which seems closer to class Versicolor than class Setosa once class dispersion is taken into account: then, its seems legitimate to discard the information brought by the training subset associated with the leaf of the tree. Overall, keeping the preference information inferred from the reference population seems reasonable if the model is unlikely to confuse atypicality with another source of uncertainty. In Fig. 2, instance 2 is equally far from the Versicolor and Setosa classes, which by nature the decision tree cannot detect.

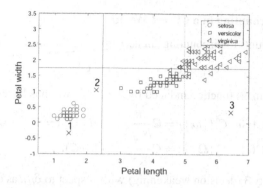

Fig. 2. Iris dataset example and some non-conformal examples

Property 2 (Decision strengthening). A conformity operator Cf strengthens the decisions made with a strategy δ if

$$C \leq C' \Rightarrow \delta(\text{Cf}\,[m,C]) \subseteq \delta\left(\text{Cf}\,[m,C']\right). \tag{7}$$

In other terms, the set of plausible classes for an instance should grow with its level of conformity: as it becomes atypical, classes previously deemed likely may be dropped off. This is similar to assuming an open-world, since known classes are discarded, possibly ending up with an empty set at the limit, similarly to conformal predictions.

Property 3 (Decision weakening). A conformity operator Cf weakens the decisions made with a strategy δ if

$$C \leq C' \Rightarrow \delta\left(\text{Cf}\,[m,C']\right) \subseteq \delta(\text{Cf}\,[m,C]). \tag{8}$$

Conversely to Property 2, Property 3 is more in line with a closed world assumption, where Ω is assumed to necessarily contain all classes, but where the information related to an atypical example may seem too weak to provide a reliable prediction.

In the example above corresponding to Fig. 2, assume that the decision for instance 3 is $\delta(m_3) = \{\text{Versicolor, Virginica}\}$. Then, requesting Property 2 would amount to discarding Versicolor, this class being too far; whereas Property 3 would rather leave us with complete ignorance, Setosa being then added to the set of plausible classes.

Other properties can also be proposed, for instance to specify the desired behaviour of the decision rule for extremely non-conformal examples (when $C \to 0$). In Sect. 4, we mention a few operators which satisfy some of the properties described above.

4   Some Conformity Operators and Their Decision Rule

This section investigates various belief-theoretic conformity operators in the light of the aforementioned properties. In a nutshell, they consist in computing a linear transformation of the initial mass according to the level of non-conformity.

4.1   Classical Discounting in a Closed World

A first strategy amounts to discounting[1] m according to the level $1 - C$ of atypicity:

$$Cf_1[m,C] = Cm + (1-C)m_\Omega. \tag{9}$$

In the case of a mass function induced by the IDM, we thus have

$$\begin{cases} Cf_1[m,C](\{\omega_i\}) = C\dfrac{n_i}{N+s}, & \text{for all } i = 1,\ldots,M; \\ Cf_1[m,C](\Omega) = C\dfrac{s}{N+s} + (1-C). \end{cases} \tag{10}$$

Cf_1 satisfies Property 3 (decision weakening) with respect to δ_{ID}, as discounting makes the belief-plausibility intervals wider, leading to the extreme case $\delta_{ID}(Cf_1[m,0]) = \Omega$. On the contrary, and for the same reason, Cf_1 does not satisfy Property 1 (class preference preservation). Such a rule therefore appears to be more consistent with a closed world assumption, where atypical instances are treated as being hard to characterize: complete atypicity should therefore be associated with complete ignorance.

4.2   Open World with an "Unknown" Class ω_u

Our second operator Cf_2 considers the open world assumption via an "unknown" class ω_u: that is, $Cf_2[m,C]$ is now a mass function defined on a frame $\Theta = \Omega \cup \omega_u$:

$$Cf_2[m,C] = Cm^{\uparrow\Theta} + (1-C)m_{\omega_u}, \tag{11}$$

where the *vacuous extension* $m^{\uparrow\Theta}$ of m onto Θ [8] is such that $m^{\uparrow\Theta}(A) = m(A)$ for any $A \subseteq \Omega$ and $m^{\uparrow\Theta}(A) = 0$ for $A \not\subseteq \Omega$; and where $m_{\omega_u}(\{\omega_u\}) = 1$.

When applied to a mass function m^{IDir} generated by the IDM, this operator gives

$$\begin{cases} Cf_2[m^{IDir},C](\{\omega_i\}) = C\dfrac{n_i}{N+s} & \text{for all } \omega_i \in \Omega, \\ Cf_2[m^{IDir},C](\{\omega_u\}) = 1 - C, \\ Cf_2[m^{IDir},C](\Omega) = C\dfrac{s}{N+s}; \end{cases} \tag{12}$$

then, for any $\omega_i \in \Omega$, we have the following belief and plausibility values:

$$Cf_2[bel^{IDir},C](\{\omega_i\}) = C\frac{n_i}{N+s}, \quad Cf_2[pl^{IDir},C](\{\omega_i\}) = C\frac{n_i+s}{N+s}, \tag{13}$$

and

$$Cf_2[bel^{IDir},C](\{\omega_u\}) = Cf_2[pl^{IDir},C](\{\omega_u\}) = 1 - C. \tag{14}$$

Applying δ_{ID} to an updated mass function $Cf_2[m,C]$ (defined on Θ) satisfies Properties 1 and 2, with the extreme case $\delta_{ID}(Cf_2[m,0]) = \{\omega_u\}$. Also note that

$$\delta_{ID}(Cf_2[m,C]) \ni \omega_u \quad \Leftrightarrow \quad \max_{\omega_j \in \Omega} Cf_2[bel,C](\{\omega_j\}) \le 1 - C, \tag{15}$$

$$\delta_{ID}(Cf_2[m,C]) = \{\omega_u\} \quad \Leftrightarrow \quad \max_{\omega_j \in \Omega} Cf_2[pl,C](\{\omega_j\}) < 1 - C. \tag{16}$$

---

[1] The discounting $^\varepsilon m$ of m by a factor ε is defined by $^\varepsilon m(A) = (1-\varepsilon)m(A)$, for all $A \neq \Omega$; and $^\varepsilon m(\Omega) = (1-\varepsilon)m(\Omega) + \varepsilon$.

As a consequence, the set of decisions will include $\{\omega_u\}$ only if the degree of support to each class is low. This inspires an alternative strategy, where ω_u is left aside when computing non-dominated classes, and added post-hoc when it is non-dominated.

Definition 2 (interval dominance with atypicity trigger). *Given a mass m defined on $\Theta = \Omega \cup \{\omega_u\}$, the* interval dominance with atypicity trigger *rule is defined by*

$$\delta_{ID:AT}(m) = \begin{cases} \delta_{ID}(m[\Omega]) & \text{if } \min\limits_{\omega_j \in \delta_{ID}(m[\Omega])} bel(\{\omega_j\}) > 1 - C, \\ \delta_{ID}(m[\Omega]) \cup \{\omega_u\} & \text{otherwise}, \end{cases} \tag{17}$$

where the conditioning $m[\Omega]$ *of m on Ω [8] is such that $m[\Omega](A) = \sum_{B \subseteq \Theta:B\cap\Omega=A} m(B)$, for any $A \subseteq \Omega$.*

In a nutshell, the set of non-dominated classes is determined from well-identified classes (i.e., associated with an identified subpopulation), and a warning trigger is sent if the instance is deemed atypical. This strategy satisfies Property 1 and includes ω_u when $C = 0$: in particular, $\delta_{ID:AT}(\text{Cf}_1[m,0]) = \Theta$, and $\delta_{ID:AT}(\text{Cf}_2[m,0]) = \omega_u$.

4.3 Classical Discounting in an Open World

Finally, we propose a third operator where the mass m is first vacuously extended onto Θ and then discounted according to the level of atypicity C:

$$\text{Cf}_3[m,C] = C m^{\uparrow\Theta} + (1-C) m_\Theta. \tag{18}$$

In the case of masses m^{IDir} obtained via the IDM, we thus obtain:

$$\begin{cases} \text{Cf}_3[m^{IDir},C](\{\omega_i\}) = C\dfrac{n_i}{N+s} & \text{for all } \omega_i \in \Omega, \\ \text{Cf}_3[m^{IDir},C](\Omega) = C\dfrac{s}{N+s}, \\ \text{Cf}_3[m^{IDir},C](\Theta) = 1-C; \end{cases} \tag{19}$$

therefore, for any $\omega_i \in \Omega$, we have the following belief and plausibility values:

$$\text{Cf}_3[bel^{IDir},C](\{\omega_i\}) = C\frac{n_i}{N+s}, \quad \text{Cf}_3[pl^{IDir},C](\{\omega_i\}) = C\frac{n_i+s}{N+s}+1-C, \tag{20}$$

and

$$\text{Cf}_3[bel^{IDir},C](\{\omega_u\}) = 0, \quad \text{Cf}_3[pl^{IDir},C](\{\omega_u\}) = 1-C. \tag{21}$$

Note that applying δ_{ID} to $\text{Cf}_3[m,\cdot]$ satisfies Property 3, since $\delta_{ID}(\text{Cf}_3[m,\cdot]) = \Theta$.

Remark 1 (Open world assumption). The "unknown" class ω_u introduced above plays in spirit a role very similar to $\emptyset$ in the "canonical" open-world assumption (where $m(\emptyset)$ quantifies the belief that the instance is from a class outside Ω). However, introducing ω_u makes it possible to 1) distinguish between this degree of belief and the degree of conflict arising from combining belief masses, and 2) properly handle this degree of belief when it comes to decision making.

Table 1. Summary of the conformity operators and their associated decision strategies

Conformity operator	Mass function combined to m	Associated decision rule	Set of decisions for $C = 0$	Properties satisfied			Type of frame	
				Prop. 1	Prop. 2	Prop. 3	Open	Closed
Cf_1	m_Ω	δ_{ID}	Ω			×	×	
Cf_2	m_{ω_u}	δ_{ID}	ω_u	×	×			×
Cf_2	m_{ω_u}	$\delta_{ID:AT}$	$\delta_{ID}(m[\Omega]) \cup \{\omega_u\}$	×		×	×	×
Cf_3	$m_{\Omega \cup \omega_u}$	δ_{ID}	$\Omega \cup \omega_u$			×		×

5   Conclusion and Perspectives

Table 1 summarizes the properties of the conformity operators with their associated decision strategies presented in this paper. We recall the mass function used in each conformity operator, the set of decisions retrieved by the strategy when $C = 0$, the properties satisfied (class preference preservation, decision strengthening, decision weakening), and the frame assumptions associated with the operator. Note that a conformity operator cannot satisfy all properties at the same time, some of them being antgonist.

When defining a conformity operator, whether a property should be satisfied— or, equivalently, whether the associated frame assumption should be accounted for— depends on the application considered. For instance, in novelty detection, the open world assumption is clearly at work; this is not so clear in outlier or anomaly detection problems, where the set of possible classes (or states) is assumed to be known, a deviance to which the user is interested in detecting.

Future work will be conducted into two directions. First, we will study whether further properties should be required or desirable, for which we may propose accordingly additional conformity operators. Besides, we will compare our strategies based on conformity operators to other approaches dealing with atypical examples, such as conformal predictions. For this purpose, we will define a thorough experimental evaluation process, so as to assess the interest of the various properties and operators proposed.

Acknowledgements. This work was carried out in the framework of Labex MS2T (ANR-11-IDEX-0004-02), which were funded by the French Government, through the program "Investments for the future" managed by the National Agency for Research.

References

1. Aregui, A., Denoeux, T.: Novelty detection in the belief functions framework. In: Proceedings of IPMU, vol. 6, pp. 412–419 (2006)
2. Carreño, A., Inza, I., Lozano, J.A.: Analyzing rare event, anomaly, novelty and outlier detection terms under the supervised classification framework. Artif. Intell. Rev. **53**(5), 3575–3594 (2019). https://doi.org/10.1007/s10462-019-09771-y
3. Dempster, A.: Upper and lower probabilities induced by a multivalued mapping. Ann. Math. Stat. **38**, 325–339 (1967)
4. Denoeux, T.: Decision-making with belief functions: a review. Int. J. Approx. Reason. **109**, 87–110 (2019)

5. Dubuisson, B., Masson, M.: A statistical decision rule with incomplete knowledge about classes. Patt. Recogn. **26**(1), 155–165 (1993)
6. Shafer, G.: A Mathematical Theory of Evidence. Princeton University Press, New Jersey (1976)
7. Shafer, G., Vovk, V.: A tutorial on conformal prediction. J. Mach. Learn. Res. **9**, 371–421 (2008)
8. Smets, P.: Belief functions: the disjunctive rule of combination and the generalized Bayesian theorem. Int. J. Approx. Reason. **9**, 1–35 (1993)
9. Smets, P., Kennes, R.: The transferable belief model. Artif. Intell. **66**, 191–234 (1994)

Evidence Theory Based Combination of Frequent Chronicles for Failure Prediction

Achraf Ben Chrayet[1], Ahmed Samet[2(✉)], and Mohamed Anis Bach Tobji[1]

[1] Univ. Manouba, ESEN, Manouba, Tunisia
{achraf.chrayet,mohamed.tobji}@esen.tn
[2] ICUBE/SDC Team (UMR CNRS 7357)-INSA Strasbourg, Strasbourg, France
ahmed.samet@insa-strasbourg.fr

Abstract. A chronicle is a kind of temporal pattern mined from a set of sequences made-up of time-stamped events. It has been shown recently that such knowledge is effective in sketching machines' behaviours in industry. However, chronicles that describe a same new sequence of events could be multiple and conflictual. To predict nature and time interval of future events, we need to consider all the chronicles that match a new sequence. In this paper, we introduce a new approach, called *FCP*, that uses the evidence theory and chronicle mining to classify sequences. The approach has been evaluated on both synthetic and real-world data sets and compared to baseline state-of-the-art approaches.

Keywords: Chronicle mining · Prediction maintenance · Evidence theory

1 Introduction

In industry 4.0, predictive maintenance relies on analysing sequential data containing time-stamped events. Therefore, data mining and particularly pattern mining techniques [1] turned to be very effective to understand failure sequences [9] by finding recurrent abnormal behaviours before any prediction task.

One type of pattern stands out thanks to its information richness and it is called *chronicle*. A chronicle is a pattern that represents a sequence of events that happened enough frequently to be extracted. Introduced in [6], this new kind of sequences is enriched with the time interval that separates each pair of events, making it possible to predict that an event B will probably happen at a time interval $[t1, t2]$ if event A occurs. If the event B requires an intervention, such as a machine failure, then maintenance may be performed on time avoiding cascading troubles.

Chronicles are complex but highly expressive patterns that enable to take into account the quantitative temporal dimension of the data contrary to classical sequential patterns. Dousson et al. [5] introduced what is called later chronicle

© Springer Nature Switzerland AG 2020
J. Davis and K. Tabia (Eds.): SUM 2020, LNAI 12322, pp. 226–233, 2020.
https://doi.org/10.1007/978-3-030-58449-8_16

mining. They proposed an incomplete algorithm (which does not generate all the patterns) called FACE (Frequency Analyzer for Chronicle Extraction). Then, Cram et al. [3] introduced another complete algorithm to mine the complete set of chronicles. Sellami et al. have introduced a new approach called FADE [9] that mines failure chronicles (chronicles that end with failure event).

In this paper, we tackle the problem of sequences' classification and failure time prediction in the context of predictive maintenance. We aim to understand and predict failures for a target maintenance. Once the failure is predicted, the maintenance is scheduled using failure criticality assessment [2]. Chronicle mining algorithm is used to extract knowledge from the data set: normal and abnormal behaviour patterns. Assuming a set of chronicles representing different machine behaviours with certain level of reliability, a major task is how to classify new incoming sequence.

Therefore, we propose to combine the use of evidence theory and chronicle mining to classify sequence in the context of predictive maintenance. The evidence theory, is a strong mathematical framework that allows to model uncertain knowledge and combine information for decision making. To summarize, this paper introduces two contributions: *(i)* using both normal and failure chronicles for sequence classification and time to failure prediction and finally *(ii)*, a new algorithm called FCP that uses the mined chronicles and evidence theory framework to combine information and predict if a new sequence will lead to a machine failure, and if yes, in which time interval the crash will occur.

2 Background

2.1 Evidence Theory

The evidence theory also called the belief function theory was introduced by Dempster [4]. In this section, we present the main concepts of this theory. The frame of discernment is the set of N possible answers for a treated problem and generally denoted θ. It is composed of exhaustive and exclusive hypotheses: $\theta = (H_1, H_2, \ldots, H_N)$.

These elements are assumed to be mutually exclusive and exhaustive. From the frame of discernment θ, we deduce the set 2^θ containing all the 2^N subsets A of θ: $2^\theta = \{A, A \subseteq \theta\} = \{H_1, H_2, \ldots, H_N, H_1 \cup H_2, \ldots, \theta\}$. A Basic Belief Assignment (BBA) m is the mapping from elements of the power set 2^θ onto $[0, 1]$, having as constraints:

$$\begin{cases} \sum_{A \subseteq \theta} m(A) = 1 \\ m(\emptyset) = 0. \end{cases} \tag{1}$$

The belief function offers many advantages. One of its proposed asset is the information fusion allowing extracting the more veracious proposition from a multi-source context. This benefit is granted by the Dempster rule of combination [4] defined as follows:

$$m_{\oplus}(A) = m_1 \oplus m_2(A) = \frac{1}{1 - \sum_{B \cap C = \emptyset} m_1(B) * m_2(C)} \sum_{B \cap C = A} m_1(B) * m_2(C); \forall A \subseteq \theta, A \neq \emptyset \tag{2}$$

The pignistic transformation allows the decision from a BBA by distributing equiprobably the mass of a proposition A on its sub-hypotheses, formally:

$$BetP(H_n) = \sum_{A \subseteq \theta} \frac{|H_n \cap A|}{|A|} * m(A); \forall H_n \in \theta \tag{3}$$

2.2 Chronicle Mining

To give formal definition of chronicles, this section starts by introducing the concept of event [3].

Definition 1 (Event). *Let $\mathbb{E}$ be a set of event types, and $\mathbb{T}$ a time domain such that $\mathbb{T} \subseteq \mathbb{R}$. $\mathbb{E}$ is assumed totally ordered and is denoted $\leq_{\mathbb{E}}$. According to [3], an event is a couple (e, t) where $e \in \mathbb{E}$ is the type of the event and $t \in \mathbb{T}$ is its time.*

Definition 2 (Sequence). *Let $\mathbb{E}$ be a set of event types, and $\mathbb{T}$ a time domain such that $\mathbb{T} \subseteq \mathbb{R}$. $\mathbb{E}$ is assumed totally ordered and is denoted $\leq_{\mathbb{E}}$. According to the definition in [3], a sequence is a couple $\langle SID, \langle (e_1, t_1), (e_2, t_2), ..., (e_n, t_n) \rangle \rangle$ such that $\langle (e_1, t_1), (e_2, t_2), ..., (e_n, t_n) \rangle$ is a sequence of events, and SID its identifier. For all $i, j \in [1, n], i < j \Rightarrow t_i \leq t_j$. If $t_i = t_j$ then $e_i <_{\mathbb{E}} e_j$ where $<_{\mathbb{E}}$ is the lexical order.*

When the events are time-stamped, how to describe the quantitative time intervals among different events is very important for the prediction of possible future events. To achieve this goal, the notion *temporal constraints* is introduced.

Definition 3 (Temporal constraint). *A temporal constraint is a quadruplet (e_1, e_2, t^-, t^+), denoted $e_1[t^-, t^+]e_2$, where $e_1, e_2 \in \mathbb{E}$, $e_1 \leq_{\mathbb{E}} e_2$ and $t^-, t^+ \in \mathbb{T}$.*

t^- and t^+ are two integers which are called lower and upper bounds of the time interval, such that $t^- \leq t^+$. A couple of events (e_1, t_1) and (e_2, t_2) are said to satisfy the temporal constraint $e_1[t^-, t^+]e_2$ iff $t_2 - t_1 \in [t^-, t^+]$. It is defined that $e_1[a, b]e_2 \subseteq e_1'[a', b']e_2'$ iff $[a, b] \subseteq [a', b']$, $e_1 = e_1'$, and $e_2 = e_2'$. The concept of chronicles [3] is defined as follows.

Definition 4 (Chronicle). *A chronicle is a pair $C = (\mathcal{E}, \mathcal{T})$ such that:*

1. *$\mathcal{E} = \{e_1 ... e_n\}$, where $\forall i, e_i \in \mathcal{E}$ and $e_i \leq_{\mathbb{E}} e_{i+1}$,*
2. *$\mathcal{T} = \{t_{ij}\}_{1 \leq i < j \leq |\mathcal{E}|}$ is a set of temporal constraints on $\mathcal{E}$ such that for all pairs (i, j) satisfying $i < j$, t_{ij} is denoted by $e_i[t_{ij}^-, t_{ij}^+]e_j$.*

Definition 5 (Chronicle support). *An occurrence of a chronicle C in a sequence S is a set $(e_1, t_1) \ldots (e_n, t_n)$ of events of the sequence S that satisfies all temporal constraints defined in C. The support of a chronicle C, denoted Supp(.) in the sequence S is the number of its occurrences in a data set of sequences [9]. In this paper, we assume that a sequence could contain at most only one occurrence of any chronicle.*

3   Chronicle Mining and Evidence Theory for Failure Prediction

In this section, we define the notions we use in our approach to combine chronicles for prediction.

Definition 6 (Chronicle cover). *Assuming a sequence $S = \langle (e_1, t_1), (e_2, t_2), \ldots, (e_n, t_n) \rangle$ and a frequent chronicle C. We say that C covers the sequence S, denoted by $C <\cdot S$, if and only if the events represented by the chronicle belong to the sequence as well as the time intervals between these events in the sequence belong to the temporal constraints extracted by the chronicle, i.e.,*

$$C <\cdot S \Leftrightarrow \forall e_i [t^-, t^+] e_j \in C, \exists ((e, t), (e', t')) \in S \wedge e = e_i, e' = e_j \wedge t' \quad t \in [t^-, t^1]. \quad (4)$$

Let $\mathscr{C}$ be a set of frequent chronicles, $C_T \subset \mathscr{C}$, such that $T \in \{F, N\}$ and where $\bar{F} = N$. C_F denotes the set of chronicles that point to the failure event, where C_N is the set of chronicles that do not, and so match normal sequences.

Definition 7 (BBA modeling). *Assuming a chronicle $C_i \in C_T$ that covers a sequence S, we model the BBA m_i of C_i in $\theta = \{T, \bar{T}\}$ as follows:*

$$\begin{cases} m_i(T) = Supp(C_i) \\ m_i(\bar{T}) = 0 \\ m_i(\theta) = 1 - Supp(C_i) \end{cases} \quad (5)$$

Definition 8 (Chronicles combination). *Assuming N chronicles C_i that cover a sequence S, with m_i, $i \in [1, N]$, the mass function relative to the ith chronicle. The joint mass function that combines all the m_i mass functions of the chronicles C_i that cover S using the Dempster Rule of combination is defined as follows:*

$$m_\oplus(A) = m_1 \oplus \ldots \oplus m_N(A); \forall A \subseteq \theta \quad (6)$$

To make the decision, we compute the pignistic probability $BetP$ for failure (F) and normal (N). The final decision is obtained by retaining the hypothesis that maximized the pignistic probability as follows:

$$x = argmax BetP_{x_i \in \theta}(x_i). \quad (7)$$

For the prediction task, we developed the FCP method (Fusion of Chronicles for Prediction). It consists in comparing the input sequence (to predict) with every chronicle in terms of events and time constraints. To each matching chronicle, we model a BBA that measures to which degree the chronicle expresses the failure (F) and normal (N) behaviour classes. The level of uncertainty is retained using the support of the chronicle. Once all matching chronicles are modelled, we use the Dempster rule of combination to combine all the BBAs. The joint BBA shows the membership of the input sequence to both classes. The final class is computed using the argmax function. If the final class is failure, we display the failure time by aggregating the time constraints of all matching failure chronicles. Algorithm 1 performs the combination of the covering chronicles to predict the status of a sequence using all aforementioned notions.

Algorithm 1 Fusion of Chronicle for Prediction

Require: S: sequence, $\mathcal{C}$: chronicles set
Ensure: R: result, $min\_time\_failure$: minimum time to failure, $max\_time\_failure$: maximum time to failure
1: $C_M \leftarrow \{\}$
2: **for all** $C \in \mathcal{C}$ **do**
3:    **if** (coverage(S,C)) **then**
4:       $C_M \leftarrow C_M \cup C$
5: **for all** $C \in C_M$ **do**
6:    **if** $(C.Type == F)$ **then**
7:       $m \begin{cases} m(F) = Supp(C) \\ m(N) = 0 \\ m(\theta) = 1 - m(F) \end{cases}$
8:    **else**
9:       $m \begin{cases} m(F) = 0 \\ m(N) = Supp(C) \\ m(\theta) = 1 - m(N) \end{cases}$

10:    $m_\oplus \leftarrow m_\oplus \oplus m$
11: $R \leftarrow argmax_{x_i \in \theta} BetP(x_i)$
12: **if** $(R == F)$ **then**
13:    Init(C_M,$min\_time\_failure$,$max\_time\_failure$)
14:    ▷ Initialize $min\_time\_failure$ and $max\_time\_failure$
15:    **for all** $C \in C_M$ **do**
16:       **if** ($min\_time\_failure$ > $C.min\_time$) **then**
17:          $min\_time\_failure$ $\leftarrow$ $C.min\_time$
18:          **if** ($max\_time\_failure$ < $C.max\_time$) **then**
19:             $max\_time\_failure$ $\leftarrow$ $C.max\_time$
20:    **return** $R, min\_time\_failure, max\_time\_failure$
21: **else**
22:    **return** R

---

4 Experiments and Results

Two kinds of data sets are used to validate our approach. The first one is generated synthetically according to several parameters, such as the number of sequences, the mean size (i.e. width) of a sequence and the number of items (events).[1] In addition, data are generated following a failure model sequence that represents 5% of the entire produced data set. Even such kind of data sets do not include natural patterns of failure/normal events, they are interesting in the way they allow the evaluation of our approach when we vary the data features, which is infeasible with real data sets whose parameters are fixed.

The second experiment is made on an industrial real data set, denoted SECOM (semi-conductor manufacturing process), introduced in [8]. It's a data set that records 1567 measurements of 590 sensors installed in manufacturing

---

[1] Reader may refer to https://gitlab.inria.fr/tguyet/pychronicles for further details about data sets generation.

machines. Each record has a timestamp (the instant at which the 590 measurements are taken), and also a general state; 1 for a normal state, and -1 for a failure.

4.1   The Performance Evaluation

The performance of our approach is evaluated on different synthetic data sets to assess the effect of several parameters mainly on the run-time and the memory usage. Figure 1 shows the execution time of FCP according to the number of sequences and the vocabulary size. The execution time increases when the number of sequences and their sizes increase. Indeed, when number and size of sequences are large, the number of extracted frequent chronicles increases accordingly. The Dempster rule of combination is the most consuming part of our approach. The more we find matching chronicles, the more we model and combine BBAs.

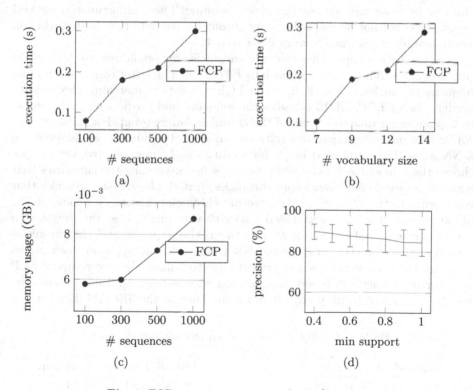

Fig. 1. FCP experiments on synthetic data sets

As part of the performance evaluation, we also assessed the memory consumption of both algorithms. Figure 1 shows the amount of memory used according to the sequence number variation. For FCP, the use of memory increases when the number of covering chronicles increases, especially because of the operation of combination that uses matrix structures within the evidence theory for mass functions.

4.2   The Prediction Quality Experiments

To evaluate the prediction quality of our approach, we used the 10-fold-cross-validation method [10] to compute the precision, the recall and the F-measure. A failure sequence is considered correctly classified, if we predict the failure state and also the time interval into which the breakdown will occur. A normal sequence is correctly classified if we predict the normal state.

First, we evaluate the prediction quality of our approach *FCP* on different synthetic data sets. We carried out experiments to assess the precision of *FCP* by varying the minimum support, denoted *minsup*, of the mining algorithm [9]. The Fig. 1 (d) pictures the results. It shows that the precision decreases as long as *minsup* increases. In fact, precision and *minsup* are both linked. Indeed, the more we increase *minsup*, less chronicles we mine. Then, unfortunately, several sequences could not be covered by any chronicle. We note that best prediction results are observed when *minsup* is set to 0.4.

The prediction approaches that use chronicles for prediction are limited. In this paper, we compare our approach to *FADE* [9]. The latter consists in mining frequent chronicles. Then, it uses the highest support matching chronicle to predict. As for *FCP*, *FADE* classifies the sequence and predicts when it is going to happen using time constraint of the chronicle failure event. For these reasons *FADE* is a natural comparative reference to *FCP*. In addition, we adapted the k-NN algorithm introduced in [7] for evaluation. In our adapted version, we choose the k most similar chronicles to our sequence among the chronicles that cover it. So we do not consider all chronicles, just the k nearest chronicles that correspond to the top 30% of the covering chronicles. Second, we combine the obtained classes using the weighted majority vote method, so the weight of a class is proportional to the distance between the sequence and the chronicles that represent the class in question. Knowing that, our approach uses mined patterns, classifies sequences and predict time to failure. We also compare FCP to other neural network based classification approaches [11]. Table 1 shows the results in terms of recall, precision and F-measure on the SECOM data set.

Table 1. Quality of prediction on the SECOM data set

Approach	Parameters	Recall	Precision	F-measure
FCP	minsup $= 0.4$	**0.78**	**0.81**	**0.79**
FADE [9]	minsup $= 0.4$	0.72	0.70	0.71
k-NN [7]	k equivalent to 30% of chronicles	0.69	0.71	0.69
LSTM [11]	1 shared layer; 2 prediction layers	0.73	0.74	0.73

5   Conclusion

In this paper, we are interested in prediction of failures as well as their time of occurrence, in the context of predictive maintenance of industrial machines. To resolve this problem, we rely on frequent chronicle mining, which allows not only the extraction of patterns, but also the time constraint between events for each sequence in the data set. We used evidence theory to combine chronicles. Experiments show that our FCP approach is more effective than existing methods. As future work, we intend to work on improving prediction of the occurrence time. As current works predict a large time interval, we intend to be more precise by predicting the most probable instant of occurrence.

Acknowledgements. This work has received funding from INTERREG Upper Rhine (European Regional Development Fund) and the Ministries for Research of Baden-Wrttemberg, Rheinland-Pfalz (Germany) and from the Grand Est French Region in the framework of the Science Offensive Upper Rhine HALFBACK project.

References

1. Borgelt, C.: Frequent itemset mining. In: Wiley Interdisciplinary Reviews: Data Mining and Knowledge Discovery, pp. 35–50, November 2012
2. Cao, Q., Samet, A., Zanni-Merk, C., de Bertrand de Beuvron, F., Reich, C.: Combining evidential clustering and ontology reasoning for failure prediction in predictive maintenance. In: Proceedings of the 12th International Conference on Agents and Artificial Intelligence, ICAART 2020, Valletta, Malta (to appear, 2020)
3. Cram, D., Mathern, B., Mille, A.: A complete chronicle discovery approach: application to activity analysis. Exp. Syst., 321–346, May 2011
4. Dempster, A.: Upper and lower probabilities induced by multivalued mapping. AMS **38** (1967)
5. Dousson, C., Duong, T.V.: Discovering chronicles with numerical time constraints from alarm logs for monitoring dynamic systems. In: Proceedings of the International Joint Conference on Artificial Intelligence, San Francisco, CA, USA, July 31–August 06, pp. 620–626 (1999)
6. Dousson, C., Gaborit, P., Ghallab, M.: Situation recognition: Representation and algorithms. In: Proceedings of the International Joint Conference on Artificial Intelligence, Chambéry, France, August 28–September 3, pp. 166–174 (1993)
7. Fan, Y., Ye, Y., Chen, L.: Malicious sequential pattern mining for automatic malware detection. Exp. Syst. Appl. **52**, 16–25 (2016)
8. McCann, M., Li, Y., Maguire, L., Johnston, A.: Causality challenge: Benchmarking relevant signal components for effective monitoring and process control. In: Proceedings of the International Conference on Causality: Objectives and Assessment, Whistler, Canada, pp. 277–288 (2008)
9. Sellami, C., Miranda, C., Samet, A., Bach Tobji, M.A., de Beuvron, F.: On mining frequent chronicles for machine failure prediction. J. Intell. Manuf., 1 (2019)
10. Stone, M.: Cross-validatory choice and assessment of statistical predictions. J. Roy. Stat. Soc. Ser. B Methodol., 111–147 (1974)
11. Tax, N., Verenich, I., La Rosa, M., Dumas, M.: Predictive business process monitoring with LSTM neural networks. In: Dubois, E., Pohl, K. (eds.) CAiSE 2017. LNCS, vol. 10253, pp. 477–492. Springer, Cham (2017). https://doi.org/10.1007/978-3-319-59536-8_30

Rule-Based Classification
for Evidential Data

Nassim Bahri[1]($\boxtimes$), Mohamed Anis Bach Tobji[1,2],
and Boutheina Ben Yaghlane[3]

[1] Université de Tunis, ISG, LARODEC, Tunis, Tunisia
bahri.nassim@gmail.com
[2] Univ. of Manouba, ESEN, Manouba, Tunisia
anis.bach@esen.tn
[3] Université de Carthage, IHEC, LARODEC, Tunis, Tunisia
boutheina.yaghlane@ihec.rnu.tn

Abstract. In this paper, we tackle the problem of multi-rules based classification for evidential data, i.e., data where imperfection is modeled through the Evidence theory. In this setting, a new algorithm called EviRC is introduced. This method uses different pruning techniques to omit irrelevant rules and defines a new matching criteria between the rules and the instance to classify. The selected rules are then combined using the powerful combination rules of the Evidence theory. Extensive experiments were conducted on several data sets in order to evaluate the proposed method. The experiments produce interesting results in term of classification quality.

Keywords: Association rules · Classification · Dempster-Shafer theory · Associative classifier · Association Rule Mining

1 Introduction

Associative classifiers consists of a kind of classification techniques that rely on association rules to classify a new instance [22]. They select high-quality rules to build an accurate classifier. Initially introduced in [22], various related works were proposed such [20, 23] and [13]. However most of them deal with certain databases and does not consider imperfection inside the data. To handle imperfection in data, several models were introduced. The most known are the probabilistic databases [2], the possibilistic databases [10] and the evidential databases [9, 11, 19]. The latter model is considered flexible since it manages a wide variety of imperfection and is relevant in various domains including pattern mining [6, 18, 26], classification [16], skyline analysis [1, 12], database modeling and querying [7, 11], and clustering [4]. Various works used the evidential data model in real-world applications. In [28], a hard rating data set is transformed into soft rating one thanks to the evidence theory. In [5], the authors create an evidential educational database including students' feedbacks about their

© Springer Nature Switzerland AG 2020
J. Davis and K. Tabia (Eds.): SUM 2020, LNAI 12322, pp. 234–241, 2020.
https://doi.org/10.1007/978-3-030-58449-8_17

academic program. Finally, in [25], Samet et al. introduced a new opinion mining approach to evaluate the reliability level of a biomedical evidential database. This work studied the field of multi-rule based classification for evidential data. It considers associative classifier on databases without class label ambiguities.

In this paper, we present a new method denoted EviRC that stands for **Evidential Rule**-based **C**lassifier. This latter extracts valid classification rules from an evidential database (i.e., rules satisfying the minimum support and confidence thresholds). Then, two pruning techniques are performed to remove irrelevant rules. We introduce also a new measure to evaluate the relevance of the rule compared to the instance. Finally, we present the results of several experiments performed to evaluate the proposed approach.

2   Background Material

In this section, we briefly review the basic concepts related to the evidence theory, the evidential databases, and the evidential association rules.

2.1   Evidence Theory and Evidential Databases

The Dempster-Shafer theory (DST) also known as "Evidence theory" was introduced by Dempster in [14,15] and developed by Shafer in [27]. It is known to be a powerful framework for representing and reasoning under uncertainty. Its basic concepts are defined as follows: Let $\Theta = \{\theta_1, \theta_2, ..., \theta_n\}$ be a finite set of mutually exclusive and exhaustive propositions. Θ is called the *Frame of Discernment (FoD)* and represents the problem domain. A *Basic Belief Assignment function (bba)* is used to express the degree of belief committed to an element from the set 2^Θ such that:

$$m : 2^\Theta \to [0,1]; \quad m(\emptyset) = 0; \quad \sum_{A \subset \Theta} m(A) = 1 \qquad (1)$$

A proposition $A \in 2^\Theta$ is called a *focal element* only if $m(A) > 0$. $|A|$ refers to the cardinality of the element A. Note that an element A can be a singleton ($|A| = 1$) or a composite ($|A| > 1$). From the mass function two other measures are derived: (i) the belief function denoted *bel*. It quantifies the degree of truth given to a specified proposition A such that:

$$bel_\Theta : 2^\Theta \to [0,1]; \quad bel_\Theta(A) = \sum_{B \subseteq A} m(B) \qquad (2)$$

(ii) the plausibility function denoted *pl*. It quantifies the amount of belief that could be given to a subset A of Θ. This function is defined as follows:

$$pl_\Theta : 2^\Theta \to [0,1]; \quad pl_\Theta(A) = \sum_{B \cap A \neq \emptyset} m(B) = 1 - bel_\Theta(\bar{A}) \qquad (3)$$

An evidential database denoted $\mathcal{EDB}$ stores and manages imperfect data [19]. Imperfection in such model is handled using the evidence theory. An $\mathcal{EDB}$ is defined on n attributes and d tuples. Each attribute i (with $1 \leq i \leq n$) has a domain Θ_i. The value of the attribute i in the jth line is represented through a normalized *bba* denoted m_{ij} fulfilling the property expressed in Eq. (1).

2.2   Evidential Association Rules

The Association Rule Mining (ARM) consists on discovering hidden correlations between items in large databases [3]. An item in the evidential database corresponds to a focal element. It can be defined formally as $x_i \in 2^{\Theta_i}$ (the domain of attribute i). An itemset X is a conjunction of items that don't share the same domain. Thus, it is defined as: $X \in \prod_{i=1}^{n} 2^{\Theta_i}$, with n is the number of attributes constituting the evidential database. We define the inclusion operator which is a relation between two distinct itemsets [18]. It's defined as follows: Let X and Y be two evidential itemsets: $X \subseteq Y \iff \forall x_i \in X, x_i \in y_i$. Where x_i and y_i denoted respectively the ith item of X and Y. An evidential association rule R is an expression in the form: $X \to Y$ satisfying $X \cap Y = \emptyset$, where X and Y are two itemsets. We call X the antecedent of the rule and Y the consequent. A valid association rule, is a rule that meet the user defined thresholds. The most commonly used measures are the support and the confidence as stated in [3].

Definition 1. *Let X be an evidential itemset.*

$$m_j(X) = \prod_{x_i \in X} m_{ij}(x_i) \tag{4}$$

where m_j refers to the mass of the itemset X in the transaction j. Thus, the mass of X in the evidential database is computed as follows:

$$m_{\mathcal{EDB}}(X) = \frac{1}{d} \sum_{j=1}^{d} m_j(X) \tag{5}$$

where d is the size of the evidential database. Then, the support function of X is naturally defined as follows:

$$Sup_{\mathcal{EDB}}(X) = Bel_{\mathcal{EDB}}(X) \tag{6}$$

For the context of evidential ARM, the authors in [18] define the confidence measure of a rule R as follows:

$$Confidence(R) = Bel(Y|X) \tag{7}$$

where $bel(Y|X)$ represents the conditional belief. In [17], Fagin et al. propose a definition of the conditional belief. This definition is used in the existing associative classifiers based on the DST, namely [18] and [26] to compute the confidence of a rule as follows:

$$Bel(Y|X) = \frac{Bel(X \cap Y)}{Bel(X \cap Y) + Pl(X \cap \overline{Y})} \tag{8}$$

3   Problem and Proposed Approach: EviRC

The associative classification problem is formally defined as follows: Let $\mathcal{R}$ be the set of classification rules. A rule $r_i \in \mathcal{R}$ can be expressed in the following form [29]: $r_i : (a_i) \rightarrow y_i$. The left side of the rule is called the antecedent (considered as a precondition). It contains a set of attributes' values, whereas the consequent part of the rule (y_i) describes the class label. Indeed, the set of classification rules can contain redundant and conflicting rules [13].

Definition 2. *Given two rules R1: $X \rightarrow C1$ and R2: $X \rightarrow C2$, R1 and R2 are called conflicting rules. Given R3: $X \rightarrow C$, R4: $Y \rightarrow C$, with $X \subset Y$ and $confidence(R3) \geq confidence(R4)$. R3 and R4 are called redundant rules.*

The associative classification problem consists in classifying a new data instance using the set of extracted rules while considering the conflictual and the redundant rules. In this context, we introduce our method which consists of three major steps detailed as follows:

1. Rule mining Phase: Firstly, we mine association rules from an evidential database. Then, we keep only those satisfying the given support and confidence thresholds. Finally, we preserve the set of rules (called classification rules) in the form $P \rightarrow c_i$, where P is the rule's precondition and c_i is a class label. We denote by R_S the set of rules whose consequent parts are a class label.

2. Classifier Building: Once the classification rules are filtered, a post-mining step is needed to select a subset of the most pertinent rules. For this aim, two pruning measures were used. The first one is the PER that stands for Pessimistic Error Rate. It was initially introduced in the C4.5 method [24] to remove non-interesting rules. Later, it becomes widely used in the context of rule-based classification. The second used pruning criteria is the lift, also called the interest factor. It aims to analyse a statistical independence between the rules' variables in order to identify the misleading ones [29]. In our proposed method, only the rules whose variables are positively correlated are retained. Then, the next step consists of removing the redundant rules as done in the algorithm CMAR [20] because of their non informative nature. To do this, we rank the rules according to their confidence, support, and specificity. Then, only general and high-confidence rule are retained.

3. Rule Selection and Combination: From the retained rules, we extract a subset of rules matching the given instance in order to predict its class label. Given an instance X, characterized by a set of BBAs such that [26]:

$$X = \{m_i | m_i \in X, x_i^j \in \Theta_i, \sum_{x_i^j \in m_i} M(x_i^j) = 1\} \tag{9}$$

with x_i^j is the jth focal element of the BBA m_i and $M(\bullet)$ is a mass function. We denote by ρ_X the subset of classification rules covering the given instance. It contains rules whose antecedent parts have a non-null intersection with X. ρ_X is formally defined as:

$$\rho_X = \{R \in R_S, \forall m_i \in X, \exists x_i^j \in \Theta_i | x_i^j \cap R_a \neq \emptyset\} \tag{10}$$

where R_a is the antecedent of the rule. Next, for each retrieved rule, we compute a discount factor denoted df. It measures the relevance of the rule's precondition with the instance to classify and calculated as follows: for each element from the rule's premise $\{r_{ai} \in R_a, r_{ai} \in \Theta_i\}$, we compute a "distance" with the appropriate feature of the instance under classification.

$$d_i(r_{ai}, m_i) = \sum_{x_i^j \in m_i} \frac{|r_{ai} \cap x_i^i|}{|r_{ai} \cup x_i^i|} \times M(x_i^j) \tag{11}$$

where $i \in \overline{1..N}$ and N is the number of features. Thus, the discount factor is calculated as:

$$df(R, X) = \prod_{r_{ai} \in R_a} d_i(r_{ai}, m_i) \tag{12}$$

Next, each rule is transformed into a BBA according to the FoD of the class attribute:

$$\begin{cases} m_R^{\Theta_C}(\{R_c\}) = df \times Confidence(R); \\ m_R^{\Theta_C}(\{\Theta_C \setminus R_c\}) = df \times (1 - Confidence(R)); \\ m_R^{\Theta_C}(\{\Theta_C\}) = 1 - (m_R^{\Theta_C}(\{R_c\}) + m_R^{\Theta_C}(\{\Theta_C \setminus R_c\})). \end{cases} \tag{13}$$

where R_c is the conclusion part of the rule. Finally, the concluded BBAs are fused following the Dempster rule of combination [14]:

$$m_{\rho X} = \oplus_{i=1}^n m_{R_i}^{\Theta_C} \tag{14}$$

where $i \in \overline{1..|\rho_X|}$. In the last stage of this process, the class label is decided based on the pignistic probability computed from the final BBA (obtained from Eq. 14).

$$BetP(A) = \sum_{B \subseteq \Theta} m(B) \times \frac{|A \cap B|}{|B|} \quad \forall A \subseteq \Theta \tag{15}$$

4   Experiments and Results

To assess the performance of our proposed method, we conducted several experiments on different data sets. We developed the EviRC algorithm [8] under Python and we compared it with some existing methods.

Data Sets: To lead the experiments, we used four well-known data sets from the UCI benchmarks repository [21]: Balance Scale, Iris, Titanic and Vertebral. These data sets are perfect. To evaluate the performance of our proposed method, and due to the lack of real evidential data sets, we infect them with imperfection. For each feature, we introduce partial ignorance at a degree denoted $u\%$. This means that the feature value will be transformed into a normalized BBA where the mass of the original value is $(100 - u)\%$ and where the remaining $u\%$ is assigned to the frame of discernment.

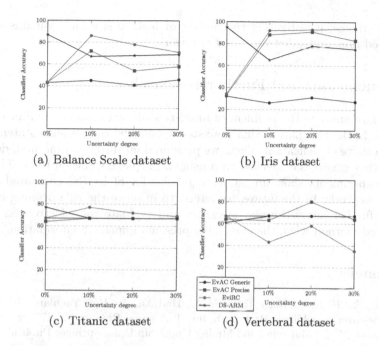

Fig. 1. The accuracy of classification methods on several data sets

Experimental Setup and Results: To perform the different experiments, we deal with a commonly used technique known as the B-Fold Cross-Validation. In this setup, we fix the value of B to 5. To evaluate the performance of our proposed method in comparison to the existing ones namely EvAC-Generic [26], EvAC-Precise [26], and DS-ARM [18], we conducted several experiments on the selected data sets (Data sets are available online on: https://bit.ly/3ezGSTP). In each experimentation, the support and confidence thresholds are respectively fixed to 20% and 50%. Then, the uncertainty degree (u) is varied from 0% to 30%. Figure 1 depicts the accuracy of several algorithms when varying the uncertainty degree on different data sets. From the Fig. 1, we observe that EviRC outperforms the other classification methods in 80% of the data sets when data are imperfect. These results confirm the importance of the pruning phase to build an accurate classifier. Indeed, considering all extracted rules in both EvAC (Generic and Precise) and DS-ARM introduces noise to the built classifier which affects the classification accuracy. Also, the rule selection strategy employed in EviRC contributes in improving its accuracy in comparison to the others methods. From Fig. 1, we note that DS-ARM is worse than EvAC and/or EviRC. This is explained by the fact that DS-ARM doesn't remove redundant rules when building the associative classifier. However, Fig. 1(d) revealed a weird behaviour of our proposed method in comparison to EvAC and DS-ARM. This may be related to the number of features describing the data set or even the features' domains. Hence, investigating the impact of feature size and domains will be considered in an extended version of this work.

5   Conclusions and Further Research

This paper considers the problem of multi-rules based classification for evidential data. In the first part of this analysis, we outlined the problem statement of the associative classification. Then, we presented our proposed method, denoted EviRC, that classifies evidential data using multiple association rules. The conducted experiments show an interesting behavior of EviRC compared to the existing methods. In the future, we intend to improve the rule pruning by considering further interestingness measures and evaluate the effectiveness of our introduced method when the class label presents imperfect information.

References

1. Abidi, A., Elmi, S., Bach Tobji, M.A., HadjAli, A., Ben Yaghlane, B.: Skyline queries over possibilistic RDF data. Int. J. Approx. Reason. **93**, 277–289 (2018)
2. Aggarwal, C.C.: Managing and Mining Uncertain Data. Springer Publishing Company, Incorporated (2009)
3. Agrawal, R., Srikant, R.: Fast algorithms for mining association rules. Proc. VLDB **1215**, 487–499 (1994)
4. Antoine, V., Quost, B., Marie-Hélène, M., Denoeux, T.: CECM: Constrained evidential C-means algorithm. Computat. Stat. Data Anal. **56**(4), 894–914 (2012)
5. Bach Tobji, M.A., Ben Yaghlane, B.: Extraction des itemsets fréquents à partir de données évidentielles: application à une base de données éducationnelles. vol. RNTI-E-21, pp. 211–232 (2011)
6. Bach Tobji, M.A., Ben Yaghlane, B., Mellouli, K.: A new algorithm for mining frequent itemsets from evidential databases. Proc. IPMU **8**, 1535–1542 (2008)
7. Bahri, N., Bach Tobji, M.A.: On indexing evidential data. Int. J. Approx. Reason. **106**, 63–87 (2019)
8. Bahri, N., Bach Tobji, M.A., Ben Yaghlane, B.: Evidential rule-based classifier (2020). https://gitlab.com/NassimBahri/evirc/
9. Bell, D.A., Guan, J.W., Lee, S.K.: Generalized union and project operations for pooling uncertain and imprecise information. Data Knowl. Eng. **18**(2), 89–117 (1996)
10. Bosc, P., Pivert, O.: About projection-selection-join queries addressed to possibilistic relational databases. IEEE Trans. Fuzzy Syst. **13**(1), 124–139 (2005)
11. Bousnina, F.E., Bach Tobji, M.A., Chebbah, M., Ben Yaghlane, B.: Modeling evidential databases as possible worlds. Int. J. Intell. Syst. **33**(6), 1146–1164 (2018)
12. Bousnina, F.E., Elmi, S., Chebbah, M., Bach Tobji, M.A., HadjAli, A., Ben Yaghlane, B.: Skyline operator over *tripadvisor* reviews within the belief functions framework. In: Jallouli, R., Zaïane, O.R., Bach Tobji, M.A., Srarfi Tabbane, R., Nijholt, A. (eds.) ICDEc 2017. LNBIP, vol. 290, pp. 186–197. Springer, Cham (2017). https://doi.org/10.1007/978-3-319-62737-3_16
13. Chen, G., Liu, H., Yu, L., Wei, Q., Zhang, X.: A new approach to classification based on association rule mining. Decis. Support Syst. **42**(2), 674–689 (2006)

14. Dempster, A.P.: Upper and lower probabilities induced by a multivalued mapping. Ann. Math. Stat. **38**(2), 325–339 (1967)
15. Dempster, A.P.: A generalization of Bayesian inference. J. Roy. Stat. Soc. Ser. B Methodolog. **30**(2), 205–247 (1968)
16. Denoeux, T.: A k-nearest neighbor classification rule based on dempster-shafer theory. IEEE Trans. Syst. Man Cybern. **25**(5), 804–813 (1995)
17. Fagin, R., Halpern, J.Y.: A new approach to updating beliefs. In: Proceedings of UAI, pp. 347–374 (1991)
18. Hewawasam, K.R., Premaratne, K., Shyu, M.L.: Rule mining and classification in a situation assessment application: a belief-theoretic approach for handling data imperfections. IEEE Trans. Syst. Man Cybern B Cybern. **37**(6), 1446–1459 (2007)
19. Lee, S.K.: An extended relational database model for uncertain and imprecise information. In: Proceedings of VLDB, pp. 211–220 (1992)
20. Li, W., Han, J., Pei, J.: CMAR: accurate and efficient classification based on multiple class-association rules. In: Proceedings 2001 IEEE International Conference on Data Mining, pp. 369–376. IEEE (2001)
21. Lichman, M.: UCI machine learning repository (2013). http://archive.ics.uci.edu/ml
22. Liu, B., Hsu, W., Ma, Y.: Integrating classification and association rule mining. In: KDD 1998, pp. 80–86 (1998)
23. Liu, B., Ma, Y., Wong, C.-K.: Classification using association rules: weaknesses and enhancements. In: Grossman, R.L., Kamath, C., Kegelmeyer, P., Kumar, V., Namburu, R.R. (eds.) Data Mining for Scientific and Engineering Applications. MC, vol. 2, pp. 591–605. Springer, Boston, MA (2001). https://doi.org/10.1007/978-1-4615-1733-7_30
24. Quinlan, J.R.: C4.5: Programs for Machine Learning, 1 edn. Elsevier (1992)
25. Samet, A., Guyet, T., Negrevergne, B., Dao, T.-T., Hoang, T.N., Tho, M.C.H.B.: Expert opinion extraction from a biomedical database. In: Antonucci, A., Cholvy, L., Papini, O. (eds.) ECSQARU 2017. LNCS (LNAI), vol. 10369, pp. 135–145. Springer, Cham (2017). https://doi.org/10.1007/978-3-319-61581-3_13
26. Samet, A., Lefèvre, E., Ben Yahia, S.: Evidential data mining: precise support and confidence. J. Intell. Inform. Syst. **47**(1), 135–163 (2016). https://doi.org/10.1007/s10844-016-0396-5
27. Shafer, G.: A Mathematical Theory of Evidence, vol. 1. Princeton University Press, Princeton (1976)
28. Wickramarathne, T.L., Premaratne, K., Kubat, M., Jayaweera, D.: Cofids: a belief-theoretic approach for automated collaborative filtering. IEEE Trans. Knowl. Data Eng. **23**(2), 175–189 (2011)
29. Tan, P.N., Steinbach, M., Karpatne, A., Kumar, V.: Introduction to Data Mining, 2 edn. Pearson (2018)

Undecided Voters as Set-Valued Information – Towards Forecasts Under Epistemic Imprecision

Dominik Kreiss$^{(\boxtimes)}$ and Thomas Augustin

Department of Statistics, LMU Munich, Ludwigstr. 33, Munich, Germany
{dominik.kreiss,thomas.augustin}@stat-uni.muenchen.de

Abstract. Increasing numbers of undecided individuals in pre-election polls throughout western democracies impose a severe challenge for election forecasting. While conventionally these voters are neglected relying on presumably unjustified assumptions, we sketch more nuanced approaches incorporating the potential valuable information in a set-valued manner. Hereby, each undecided voter is represented by the set of parties he or she is incapable to choose from. This set, containing one true, but unknown element, enables modelling under so-called *epistemic imprecision*. Depending on further assumptions, (imprecise) transition probabilities between the options can be estimated in order to achieve election forecasting. Starting with Dempster's upper and lower probabilities as the most cautious approach, two further ideas are introduced, providing initial methodology. Furthermore, extensions including Bayesian modeling are sketched. The theory is applied using data from the *German Longitudinal Election Study* for forecasting concerning the most recent German federal election of 2017. The results are promising, laying the groundwork for further research.

Keywords: Epistemic modeling · Election forecasting · Coarse data · Partial identification · Survey methodology

1 Introduction

If we think of an election in a multiparty system as a choice of individuals $i \in \{1, \cdots, N\}$ between the options $\{1, \cdots, j\} = S$, a decided individual in a pre-election poll is capable to single out one element of S as his or her choice, while an undecided is not. The position of the undecided can therefore be accurately represented by a nonempty subset $\ell \subset S$ containing all parties the individual is pondering between, hence all options that cannot be excluded.

One advantage of this set-valued information is the rather practical character, as most individuals are capable to state this subset ℓ precisely [8, p. 256 f], providing the opportunity to obtain this information by a pre-election survey. The idea of set-valued response in election choice was recently introduced by [8] in a political science framework, arguing that stepwise exclusion of options is the

© Springer Nature Switzerland AG 2020
J. Davis and K. Tabia (Eds.): SUM 2020, LNAI 12322, pp. 242–250, 2020.
https://doi.org/10.1007/978-3-030-58449-8_18

natural process of human choice [8, p. 256]. Furthermore, in her work about set-valued data, Plass [9, p. 2–3] argues that providing set-valued response categories might reduce nonresponse substantially. In conventional analysis, the undecided are overall neglected [11, p. 265], not only relying on disputable assumptions about the left out individuals but also missing out on valuable information about their position. Moreover, concerning the question which combination of parties will constitute the government, coalitions can be represented more directly by set-valued information. Despite these reasons, set-valued data is regrettably not yet included in most surveys but first approaches already exist as can be found in [8,9,11].

The subset ℓ, further on called *consideration set* following [8], determining the undecided individual's position, can be seen as a disjunctive random set, containing one ill-known true value. (e.g. [3]) Thus, to predict the undecided's choice on election day, we can develop models under *epistemic imprecision*, following [3, ch. 2], using the coarse information together with assumptions and further sources of information. A wide range of approaches are possible, reaching from Dempster's so to say agnostic bounds [4] up to point-valued estimation, relying on strong assumptions. We develop and apply three approaches weighting the justifiability of assumptions with the precision of the results and introduce methodology for overall election outcome forecasting using transition probabilities. We hereby break first ground introducing epistemic methodology to election forecasting.[1]

This paper is structured as follows: First, we briefly recall the underlying epistemic theory in Sect. 2.1 before introducing the general problem in Sect. 2.2 and three modeling approaches in Sect. 2.3. In Sect. 3, we apply the developed approaches to the most recent German federal election. The concluding remarks reflect on the approaches and future possibilities.

2 Methods

2.1 The Epistemic View of Set-Valued Information

Given the accurate, set-valued representation $\ell \in \mathscr{P}(S) = 2^S$ of an undecided individual with $\mathscr{P}(S)$ as the power set of the parties to choose from, there exists one true, yet unknown element $l \in \ell$ representing the undecided's choice on election day. The consideration sets ℓ result from individuals excluding their neglectable options, leading to a subset, which by definition contains the true element l. Hence, ℓ is a set consisting of distinguishable and finite elements containing incomplete information about the true value of interest l. This is the so-called epistemic view of set-valued information, following [3]. While we are looking for the random variable $Y(\omega)$ mapping from an underlying space Ω to S, we are only provided with incomplete information in the sense that $\forall \omega \in \Omega$

---

[1] In that sense we contribute to a solution of a "chicken-egg dilemma" (Fink), resulting from the lack of surveys including the set-valued question as well as missing methodology, providing applicable approaches for such data.

only $Y(\omega) \in \ell = \mathcal{Y}(\omega)$ is observable, where $\mathcal{Y}$ is a multi-valued mapping $\Omega \to 2^S$ representing the set of mappings $\{Y : \Omega \to S, Y(\omega) \in \mathcal{Y}(\omega) \; \forall \omega\}$. [1, p. 1504] We thus build an epistemic model of the random variable $Y(\omega)$, while for the undecided all that is known is $Y(\omega) \in \ell$.

The realization l can therefore be seen as a realization of an *ill-known random variable* incompletely described by a coarse version in the form of the set ℓ. Due to the lack of information about the true value l, prediction approaches have to incorporate further information and assumptions in order to obtain more concise or even point-valued results. By [3, p. 1503] this is described as representing both reality as well as knowledge of reality, explicitly accounting for the limited precision. Thus, one has to ponder between imprecise results and the justifiability of assumptions leading to more precise statements.[2]

2.2   From Set-Valued Information to Forecasting

Each individual from the sample is determined by both its consideration set $\ell \in \mathcal{P}(S)$ and its co-variables $X = x$ in some space $\mathcal{X}$, assessing their personal characteristics. The individual's consideration set from the pre-election survey is written as an event $\{\mathcal{Y} = \ell\}$ with $\ell \in \mathcal{P}(S)$ and his or her possibly unknown choice on election day $\{Y = l\}$ with $l \in S$. Given the consideration sets of participant $i \in \{1, \cdots, n\}$ in the pre-election poll, we want to obtain the expected frequency of each element of S within the population, with latent probability distribution $P(Y = l)$ for all $l \in S$, which is a multinomial distribution over the state space with $|S| - 1$ parameters. The observations Y_i are assumed to be identically and independently distributed and $P(Y = l)$ can be written in respect to the consideration sets and co-variables as

$$P(Y = l) = \sum_{(\ell,x) \in (2^S \times \mathcal{X})} P(Y = l, \mathcal{Y} = \ell, X = x) = \tag{1}$$

$$\sum_{(\ell,x) \in (2^S \times \mathcal{X})} \underbrace{P(Y = l | \mathcal{Y} = \ell, X = x)}_{Transition\ Probabilities} \cdot \underbrace{P(\mathcal{Y} = \ell | X = x)}_{Consideration\ Sets} \cdot \underbrace{P(X = x)}_{Co-Variables} \tag{2}$$

The probability distribution can therefore be factorized into three parts. First, the from now on so-called *transition probabilities,* determining the probability to vote for a specific party given the consideration set and co-variables. Second, the probability of the consideration sets given the co-variables and third, the one for the co-variables. While the second and third part can be directly estimated from the data of the pre-election survey alone, the first requires further assumptions and/or sources of information, as the eventual choice l from the options ℓ is not observable amongst the undecided. For the decided individuals, the transition probabilities are naturally one, while for the undecided either point- or interval-valued estimation is necessary. The transition probabilities can be seen as a further (imprecise) multinomial distribution over the individual's consideration set.

---

[2] See also Manski's Law of Decreasing Credibility [7, p. 1].

There are different directive questions concerning the estimation process of the transition probabilities resulting in several modeling approaches. First, one has to ponder whether results are obliged to be point-valued or not. Second, if the pre-election poll remains the only source of data and third, which assumptions are made in order to determine estimation. In the following section, three approaches relying on different constellations of these issues are discussed. Hereby, basic methodology to brake first ground is introduced and an outlook to improve these ideas is provided.

2.3   Approaches to Estimate Transition Probabilities

Starting with the idea of Dempster [4] as the **first approach**, only to use information available in the data alone, not relying on further assumptions nor information, the transition probabilities reflect the entire ambiguity of the individuals. Thus, as no information is available about which element of ℓ constitutes the true one, for every ℓ consisting of more than one element the transition probabilities take the whole range between 0 and 1.[3] Combination with the decided individuals and weighting according to Eq. (2) leads to interval-valued forecasting which tends to be wide. Hence, these so-called Dempster's bounds reach from worst- to best case scenario for each party, while the range of the interval reflects the ambiguity concerning the respective party. Even if the results might not be providing sufficient information depending on the question at hand, all information of the dataset that can be used, not relying on any assumption, is used in the process. The hereby estimated bounds can be seen as the extreme case, resulting from the most cautious way of modeling, leading to rather imprecise results.

As the other extreme, depending on the question at hand and preference, results are required to be point-valued, forcing overall stronger assumptions. Hereby, the parameters of the transition probabilities have to be estimated in a point-valued way to ensure overall point-valued forecasting.

As there is no information about the undecideds' choice provided, for the **second approach** we fall back on the decided individuals. Using the decided, the probability distribution $P(Y_i = l|X_i = x_i, I_d = 1)$ can be estimated from the data, with I_d as the indicator function for being decided. To enable point-valued estimation we then assume that, given the co-variables, the undecided choose identical to the decided. The consideration set hereby becomes the restriction of possible outcomes, while the tendency towards a party of the consideration set is predicted using the decided and co-variables as underlying data. Those predictions of affinity towards the parties of the undecided have to be scaled to comply with the multinomial distribution, excluding all options not in ℓ. Therefore, for all $l \in \ell$ the predicted affinity towards one party is divided by the sum of all the ones in the consideration set resulting in

$$\hat{P}(Y = l|\mathcal{Y} = \ell, X = x) = \frac{\hat{P}(Y = l|X = x, I_d = 1)}{\sum_{a \in \ell} \hat{P}(Y = a|X = x, I_d = 1)} \qquad (3)$$

---

[3] For more details and examples see for instance [11, p. 261] or [4, p. 325 ff].

leading to point-valued identification of every parameter.[4] There are several ways to estimate the conditional distributions for each individual necessary for Eq. (3), while we choose the most common approach of linear logit models (e.g. [5, p. 238 ff.]) Even though it is not impossible that the undecided, given the co-variables, behave in average identical to the decided while only excluding options outside the consideration set, some structural differences are likely to be ignored. Nevertheless, it can be argued that the drawbacks from neglecting the undecided overall outweighs this strong assumption.

The **third approach** includes information from the previous election, using data to estimate the transition probabilities of the former election $P(Y^+ = k | \mathcal{Y}^+ = k, X^+ = x^+)$, available within the post-election poll, with $+$ denoting the previous election. Incorporating data from different surveys is controversial, as both the political landscape and the selection of participants might differ severely. In order to obtain point-valued estimates with this information alone, it has to be assumed that the transition probabilities, given the co-variables, are constant between the elections. As this assumption is likely to be violated, there are reasons to rather incorporate the information in another (possibly hierarchical) way together with other sources of information. Nevertheless, point-valued forecasting can be achieved at the cost of these drawbacks.

These three approaches take first steps towards election forecasting including the undecided, while for **further research** each of them can be further developed and improved. Prior information to facilitate the estimation process in the form $p(Y = l | \mathcal{Y} = \ell)$ could be incorporated in the analysis, as well as set-valued prior information could be used to achieve more plausible interval-valued results. One could assume, building on the third approach, that given specific expert knowledge, the transition probabilities are constant between the elections. Also complex hierarchical Bayesian methodology using the sources of information from the decided individuals, the undecideds' choice of the former period and (set-valued) expert knowledge is possible. A natural way to make such point-valued approaches more robust would be to rely on appropriate neighbourhood models. Another instance where including expert knowledge could be important, is to either weaken assumptions or to deal with the missing not at random structures within the nonresponse of the survey. The three original approaches are computationally rather simple, but even the more complex methods suggested should still be scalable, as typical electoral polls rarely exceed 2000 participants.

3 Application

3.1 The Data from GLES

We applied the ideas developed above for the most recent German federal election of 2017 using the state of the art pre- and post-election surveys provided

---

[4] Note although intuitively this is a kind of random coarsening assumption, it differs from the usual CAR conditions.

for scientific use by the *GLES*.[5] Set-valued response is regrettably not directly included in this survey, but the assessment of the parties by the individuals as well as their statement about the certainty of their choice are, enabling construction of consideration set as described by [11, p. 261]. To facilitate a proof of concept of our methodology, we only focus on the most common case of indifference between exactly two parties as well as we only use the two binary co-variables *sex* and *residence in east or west Germany*.[6] Moreover, we examine the so-called second vote[7] for the six main parties anticipated to reach at least one seat in the parliament, not including non-voters and small parties. Furthermore, structures of the nonresponse in the dataset are not explicitly adjusted for.

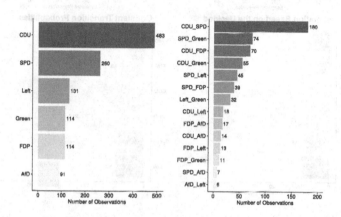

Fig. 1. Overview of occurrences of different groups amongst the participants questioned for the 2017 federal election by the GLES.

From the overall 1774 individuals used, 581 are undecided between exactly two parties, constituting about a third of the sample, while from the overall survey 11.87% were undecided between more than two options. Figure 1 illustrates the number of observations within the undecided and decided voters concerning the specific groups.

3.2   Results of the Different Approaches

We apply all three approaches discussed in Sect. 2.3 calculating overall forecasts according to Eq. (2), reliant on the same underlying dataset. The results are illustrated in Fig. 2 providing an overview of differences and similarities as well as general tendencies within the approaches.

---

[5] German Longitudinal Election Study: Pre- and post- election cross-section available under https://www.gesis.org/wahlen/gles/daten; last visited: 13.07.20.
[6] We are aware that two variables do not capture the entire structural properties of the individual in our proof-of-concept model, as should be by the co-variables in an ideal scenario to improve estimation for approaches 2 and 3.
[7] Vote for the party, which is usually used for forecasting.

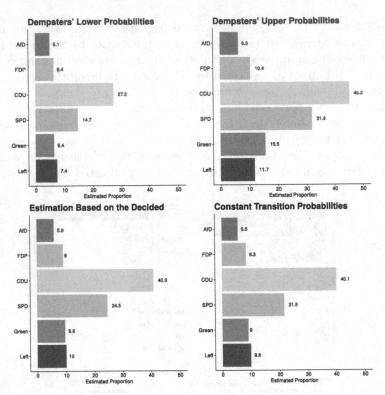

Fig. 2. Results from epistemic election forecasting of the three approaches based on the same underlying observations.

For the interval-valued Dempster bounds, upper and lower probabilities are illustrated with two separated plots. Hereby, the entire ambiguity is reflected between the upper and lower bounds, thus enclosing the other two approaches and showing the strongest deviation from conventional approaches. The second approach (estimation based on the decided) and the third (assuming constant transition probabilities between this and the last election) seem roughly similar here. The party CDU has by far the highest estimates throughout all approaches, but varies the most between upper and lower bounds. In contrast, the AfD has the lowest turnout with diminishing differences between the approaches. As the non-response structures are not adjusted for, the consideration sets are constructed and the variable selection merely served as a proof of concept, the results should be treated with caution concerning their political implications and validity.

Overall, the methodology has proven to be straightforwardly applicable producing plausible, but not yet sufficient results for final election outcome forecasting. Adjustments, necessary due to the missing not at random structures through weighting or expert knowledge and incorporation of further sources of information should yield substantially improved results.

4   Concluding Remarks

In this paper we introduced ideas in order to include the otherwise wasted information of the undecided from pre-election polls, by using their consideration sets. Several approaches are possible, weighting the precision of the results with the justifiability of the assumption, resulting in point- or interval-valued forecasting. We introduced and applied three approaches constituting possible directions, with the most cautious Dempster bounds and two point-valued ones based on different strong assumptions. Reliant on constructed consideration sets and simplifications, our forecasts are not yet perfected, but the potential is considerable. The approaches can be further developed and improved, as already sketched in Sect. 2.3, by making use of supplementary sources of information like expert knowledge or previous elections, for example, in an hierarchical Bayesian manner with imprecise probabilities. One further natural question would be the relationship with other approaches dealing with imprecise data using likelihood or loss minimisation like [2, 6, 10]. In contrast to conventional methodology, the approaches discussed here explicitly address and incorporate the ambiguity of the individuals by making use of their consideration sets, introducing new ideas to election forecasting in times of increasing relevance of undecided voters.

Acknowledgement. We are very thankful to the four anonymous reviewers for their helpful remarks.

References

1. Couso, I., Dubois, D.: Statistical reasoning with set-valued information: ontic vs. epistemic views. Int. J. Approx. Reason. **55**, 1502–1518 (2014)
2. Couso, I., Dubois, D.: A general framework for maximizing likelihood under incomplete data. Int. J. Approx. Reason. **93**, 238–260 (2018)
3. Couso, I., Dubois, D., Sánchez, L.: Random Sets and Random Fuzzy Sets as Ill-Perceived Random Variables. SAST. Springer, Cham (2014). https://doi.org/10.1007/978-3-319-08611-8
4. Dempster, A.: Upper and lower probabilities induced by a multivalued mapping. Ann. Math. Stat. **38**, 325–339 (1967)
5. Fahrmeir, L., Kneib, T., Lang, S., Marx, B.: Regression Models. Springer, Cham (2013). https://doi.org/10.1007/978-3-642-34333-9
6. Hüllermeier, E., Destercke, S., Couso, I.: Learning from imprecise data: adjustments of optimistic and pessimistic variants. In: Ben Amor, N., Quost, B., Theobald, M. (eds.) SUM 2019. LNCS (LNAI), vol. 11940, pp. 266–279. Springer, Cham (2019). https://doi.org/10.1007/978-3-030-35514-2_20
7. Manski, C.: Partial Identification of Probability Distributions. Springer, Cham (2003). https://doi.org/10.1007/b97478
8. Oscarsson, H., Rosema, M.: Consideration set models of electoral choice: theory, method, and application. Electoral. Stud. **57**, 256–262 (2019)
9. Plass, J.: Statistical modeling of categorical data under ontic and epistemic imprecision. PhD thesis, LMU Munich (2018). https://edoc.ub.uni-muenchen.de/22298/ (Accessed 13 July 2020)

10. Plass, J., Cattaneo, M., Augustin, T., Schollmeyer, G., Heumann, C.: Reliable inference in categorical regression analysis for non-randomly coarsened observations. Int. Stat. Rev. **87**(3), 580–603 (2019)
11. Plass, J., Fink, P., Schöning, N., Augustin, T.: Statistical modelling in surveys without neglecting 'The undecided'. In: Augustin, T., Doria, S., Miranda, E., Quaeghebeur, E. (eds.) ISIPTA 2015, pp. 257–266. SIPTA (2015)

Multi-dimensional Stable Matching Problems in Abstract Argumentation

Francesco Santini[✉]

Dipartimento di Matematica e Informatica, Università degli Studi di Perugia,
Perugia, Italy
francesco.santini@unipg.it

Abstract. We show how different multi-dimensional extensions of the *stable matching* (or *marriage*) problem can be represented by Abstract Argumentation frameworks: the set of stable extensions exactly corresponds to the set of solutions in the original problem. We show how to allow incomplete preference lists and ties in the same problems, and consequently frameworks. All the proposed problems are NP-Complete or NP-Hard: efficient stable-semantics solvers can help their solutions, and in general cross-fertilisation can benefit both the fields.

1 Introduction

In his pioneering work on *Abstract Argumentation* [5], P.M. Dung set a wide scenario by connecting stable models in *Logic* and *Game Theory* to *Abstract Argumentation Frameworks* (*AAF*s), which are essentially directed graphs where arguments are represented as nodes, and the attack relation is represented by edges. From such a simple abstraction, it is possible to capture important properties in many related fields.

The *Stable Matching* (or *marriage*, *SM*) problem [9,10] is exactly one of such areas. It is the problem of finding a *stable matching* between two equally sized sets of elements given an ordering of preferences for each element. For example, given the sets of men and women, a match is stable when there does not exist any *man-woman* marriage by which both *man* and *woman* would be individually better off than they are with the person to which they are currently matched. Stable matchings correspond to *stable extensions* if an *AAF* is correctly generated from the given *SM* problem [5].

In 1976, D. Knuth proposed twelve open questions on the *SM* problem [11], one of which required to generalise it from two to three parties, thus obtaining a 3-dimensional *SM*. The proposed entities to be matched were split into *women*, *men*, and *dogs*. Since Knuth did not specify any precise definition of "preference" or "blocking triples", one can conceive a number of ways to define such a problem.

In this paper we extend the SM-to-AAF encoding in [5] with the purpose to deal with different definitions of n-dimensional *SM* (*nDSM*), where $n \geq 2$. We also consider incomplete lists of preference and ties. The cross-fertilisation can be fruitful in several ways: Argumentation-based solvers, which are constantly

© Springer Nature Switzerland AG 2020
J. Davis and K. Tabia (Eds.): SUM 2020, LNAI 12322, pp. 251–260, 2020.
https://doi.org/10.1007/978-3-030-58449-8_19

increasing their performance [1],[1] could be tested and compared on frameworks generated from *SM problems*, and complexity results about *nDSM* may suggest the complexity of Argumentation-related problems on different graph topologies (see Sect. 5).

2   Background

In this section we report the necessary background information about *AAF*s and *SM*s.

Definition 1 (Abstract Frameworks [5]**).** *An Abstract Argumentation Framework (AAF) is a pair* $\langle A_{rgs}, R \rangle$ *of a set* A_{rgs} *of arguments and a binary relation* R *on* A_{rgs}, *called attack relation.* $\forall a_i, a_j \in A_{rgs}$, $a_i R a_j$ *(or* $R(a_i, a_j)$*) means that* a_i *attacks* a_j *(R is asymmetric).*

A *semantics* is the formal definition of a method (either declarative or procedural) ruling the argument evaluation process. In the *extension*-based approach, a semantics definition specifies how to derive a set of extensions from an *AAF*, where an extension B of an *AAF* $\langle A_{rgs}, R \rangle$ is simply a subset of A_{rgs}. In Definition 2 we define conflict-free sets:

Definition 2 (Conflict-free sets [5]**).** *A set* $B \subseteq A_{rgs}$ *is conflict-free iff no two arguments a and b in B exist such that a attacks b.*

This is what we need to define the only semantics on which we focus in this paper:

Definition 3 (Stable semantics [5]**).** *A conflict-free set* $B \subseteq A_{rgs}$ *is a stable extension iff for each argument which is not in B, there exists an argument in B that attacks it.*

In [5] P.M. Dung shows how his theory based on Abstract Argumentation can be used to investigate the logical structure of the classical *SM* problem solutions. An instance of the classical *SM* problem [9] comprises a set $|M| = n$ of n men and a set $|W| = n$ of women, where each of these individuals ranks all the members of the opposite sex in a strict linear order (without ties). All men and women must be matched together in a couple such that no element x of couple c_i prefers an element y of different couple c_j that also prefers x: this statement represents the overall stability condition. If such an (x, y) pair exists in the match, then it is defined as *blocking*; a matching MT is stable if no blocking pair exists.

To assemble a framework representing an *SM* problem, the set A_{rgs} has cardinality $|M| \cdot |W| = n^2$, and each argument is labelled by using an element from M, and one from W. With $w_i \succ_k w_j$ we mean that man m_k prefers woman w_i to w_j (the same holds for women). In the following we will also use $p(m_i, w_j)$ to denote the rank of w_j in the preference list of m_i: e.g., $p(m_i, w_j) = 1$ means w_j is the most preferred woman for m_i.

---

[1] See also the International Competition on Computational Models of Argumentation (ICCMA) page: http://argumentationcompetition.org/index.html.

Definition 4 (SMs and AAFs [5]**).** *Given an SM, a corresponding AAF is* $A_{rgs} = \{(m \times w) \mid m \in M, w \in W\}$*, and* $R \subseteq A_{rgs} \times A_{rgs}$ *such that* $R((m_k, w_l), (m_i, w_j))$ *iff*

- *$i = k$ and $w_l \succ_i w_j$, or*
- *$j = l$ and $m_k \succ_j m_i$.*

As a result, the set of stable extensions and the set of stable marriages correspond.

Theorem 1 (Stable extensions/marriages [5]**).** *A set* $B \subseteq A_{rgs}$ *represents a solution to the SM problem iff* B *is a stable extension of the corresponding AAF.*

In Fig. 1 we report in tabular form the list of preferences for both men and women. Figure 2 encodes the *SM* problem in Fig. 1 as proposed in Definition 4. The only stable extension is $\{(m_1, w_2), (m_2, w_1)\}$, which consequently represents the only stable marriage.

	w_1	w_2
m_1	1	2
m_2	1	2

	m_1	m_2
w_1	2	1
w_2	1	2

Fig. 1. An *SM* problem given in tabular form, with men preferences on the left and women preferences on the right, e.g., $p(m_1, w_1) = 1$ and $p(w_2, m_1) = 2$.

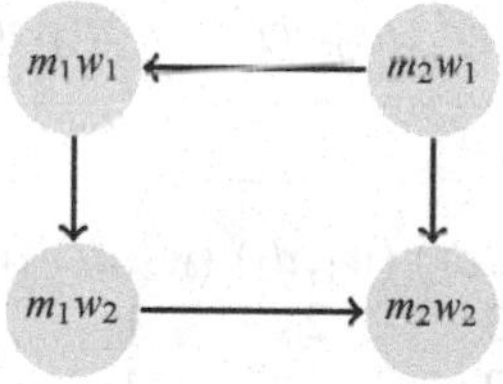

Fig. 2. The *AAF* encoding of the *SM* problem in Fig. 1. The only stable extension is $\{(m_1, w_2), (m_2, w_1)\}$.

3 Encoding Multi-dimensional Matching Problems

In this section we propose three different extensions of the SM problem, first to dimension three, and then to dimension n. We show how to obtain an *AAF* whose stable extensions represent the solutions of such problems.

3.1 3-Gender Stable Matching Problem

An instance of the *3-gender Stable Matching Problem* (*3GSM*) involves three finite sets X, Y, and Z. These sets have equal cardinality k, and their intersection is the empty, i.e. $X \cap Y, Y \cap Z, X \cap Z = \emptyset$. A match in *3GSM* is a complete match of the three sets, that is a subset of $X \times Y \times Z$ with cardinality k such that each element of X, Y, and Z appears exactly once. For each element x of X, we define its preference, denoted by $\succ_x$, to be a linear order on the elements of $Y \times Z$. The meaning of $(y_1, z_1) \succ_x (y_2, z_2)$ is that x prefers to be matched to (y_1, z_1) than to (y_2, z_2). We can analogously define $\succ_y$ and $\succ_z$, respectively over the elements in $X \times Z$ and $X \times Y$. It is known that some instances of *3GSM* do not have stable matchings, but in general this problem may have more than one solution; moreover, deciding if *3GSM* has at least a matching is NP-complete [12].

A matching MT is unstable if there exists a triple $t \in X \times Y \times Z$ which is not in MT, and each component of t prefers the pair of remaining components to the pair it is matched with in MT. Formally, a stable match is a match MT, such that, for all $(x, y, z) \notin MT$ and for the triples $(x, y_1, z_1), (x_2, y, z_2), (x_3, y_2, z) \in MT$, it holds that $(y_1, z_1) \succ_x (y, z)$, or $(x_2, z_2) \succ_y (x, z)$, or $(x_3, y_3) \succ_z (x, y)$.

To encode such a problem into a corresponding $\langle A_{rgs}, R \rangle$, we propose the classical example with men/women/dogs, that is $X = M = \{m_1, m_2\}$, $Y = W = \{w_1, w_2\}$, $Z = D = \{d_1, d_2\}$, and the preference lists in Fig. 3.

Definition 5 (3GSM and AAF). *Given a 3GSM problem, a corresponding AAF is defined by* $A_{rgs} = \{(m, w, d) \mid m \in M, w \in W, d \in D\}$ *and* $R \subseteq A_{rgs} \times A_{rgs}$, *such that* $R((m_g, w_h, d_i), (m_j, w_k, d_l))$ *iff*

- $g = j$ *and* $(w_h, d_i) \succ_g (w_k, d_l)$, *or*
- $h = k$ *and* $(m_g, d_i) \succ_h (m_j, d_l)$, *or*
- $i = l$ *and* $(m_g, w_h) \succ_i (m_j, w_k)$.

	(w_1, d_1)	(w_1, d_2)	(w_2, d_1)	(w_2, d_2)		(m_1, d_1)	(m_1, d_2)	(m_2, d_1)	(m_2, d_2)
m_1	3	2	1	4	w_1	1	2	3	4
m_2	2	1	4	3	w_2	2	4	3	1

	(m_1, w_1)	(m_1, w_2)	(m_2, w_1)	(m_2, w_2)
d_1	3	2	1	4
d_2	2	1	4	3

Fig. 3. A *3GSM* problem given in tabular form, with preferences of men/women/dogs.

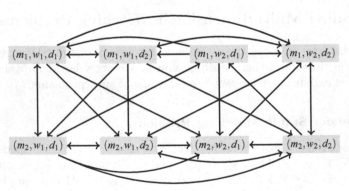

Fig. 4. The AAF encoding of the *3GSM* problem in Fig. 3.

The stable extensions are $\{(m_1, w_2, d_1), (m_2, w_1, d_2)\}$, $\{(m_1, w_2, d_2), (m_2, w_1, d_1)\}$ and $\{(m_1, w_1, d_1), (m_2, w_2, d_2)\}$, and they have been computed using ConArg [3,4].[2] Theorem 2 formally links the solutions of both the problems.

Theorem 2. *A set $B \subseteq A_{rgs}$ represents a solution of a 3GSM problem iff B is a stable extension of the corresponding AAF as defined in Definition 5.*

Proof. ($\Rightarrow$) Let B be a solution of the *3GSM*. Since attacks in Definition 5 are only posed between arguments with the same $m/w/d$, B is consequently conflict-free. Let us now suppose $(m_g, w_h, d_i) \notin B$. This means there exists (m_g, w_k, d_l) such that $(w_k, d_l) \succ_h (w_h, d_i)$, or (m_j, w_h, d_l) such that $(m_j, d_l) \succ_h (n_g, d_i)$, or (m_j, w_k, d_i) such that $(m_j, w_k) \succ_i (m_g, w_h)$. Therefore, according to Definition 5 (m_g, w_h, d_i) is attacked by at least one of such arguments: all arguments outside B are attacked by B means B is stable.

($\Leftarrow$) Let B be a stable extension. We also suppose that B does not represent a total function of M to $W \times D$ (or from W to $M \times D$ or from D to $M \times W$). Consequently, we derive that there exists $(m_g, w_h, d_i) \in A_{rgs} \setminus B$ such that m_g is not matched with any (w_h, d_i): this means that (m_g, w_h, d_i) is not attacked by B, which is a contradiction because B is stable. Hence B corresponds to a total function: all men, women, and dogs in *3GSM* are matched. In addition, by defining the attacks as in Definition 5, the stability of B enforces the fact there does not exist a triple $t = (m_g, w_h, d_i) \in M \times W \times D$ such that is $t \notin B$ and each component of t prefers the pair that it is matched with in t to the pair that it is matched with in B.

From the properties of the related *3GSM* problem it is easy to prove that:

Proposition 1. *An AAF assembled as proposed in Definition 5 may have no stable extension or several of them.*

The problem in this subsection can be extended to deal with *n GSM* problems.

Proposition 2. *Given n genders, Definition 5 can be straightforwardly extended to let an AAF represent an nGSM, with $n > 3$. The set of stable extensions corresponds to the solution of such an nGSM as well.*

Proof (Sketch of). It is possible to extend Definition 5 in this way: the set of arguments can be modelled by $A_{rgs} = \{(x_1, x_2, \ldots, x_n) \mid x_1 \in X_1, x_2 \in X_2, \ldots, x_n \in X_n\}$, and the conditions on the attack relationship are n, one for each gender. The proof of Theorem 2 can be accordingly extended by considering n components for a tuple t.

3.2   3-Person Stable Assignment Problem

The *3-person stable assignment* problem (*3PSA*) is a three-dimensional generalisation of the *stable roommate problem* [8], which partitions $2n$ persons into

---

[2] ConArg Web interface: http://www.dmi.unipg.it/conarg/.

n pairs of stable roommates: differently from stable marriages, any two elements of a non-bipartioned set can be matched. Also this problem has been already represented as an AAF by P.M. Dung in [5], with the purpose to show that a stable extension may not exist in some frameworks. A 3PSA instance of size n involves a set S of cardinality $3k$, where k is an integer. The preference of $s \in S$, denoted $\succ_s$, is a linear order on the set of unordered pairs $\{\{s_1, s_2\} \mid s_1 \neq s_2 \text{ and } s_1, s_2 \in S \setminus \{s\}\}$.

A *stable assignment* MT in *3PSA* is a partition of S into k disjoint three-element subsets, such that for all $\{s_1, s_2, s_3\} \notin MT$ and for the subsets $\{s_1, s_{11}, s_{12}\}, \{s_2, s_{21}s_{22}\}, \{s_3, s_{31}, s_{32}\} \in MT$, either $\{s_{11}, s_{12}\} \succ_{s_1} \{s_2, s_3\}$, $\{s_{21}, s_{22}\} \succ_{s_2} \{s_1, s_3\}$, or $\{s_{31}, s_{32}\} \succ_{s_3} \{s_1, s_2\}$. The NP-completeness of *3PSA* follows from that of *3GSM* because the former is a generalisation of the latter [12]; for the same reason, even in this case a solution may not exist.

Definition 6 (3PSA and AAF). *Given a 3PSA problem over a set S of cardinality $3k$, a corresponding AAF is defined by $A_{rgs} = \{\{q, r, s\} \mid q, r, s \in S\}$, and $R \subseteq A_{rgs} \times A_{rgs}$ such that $R(\{q_f, r_h, s_i\}, \{q_j, r_k, s_l\})$ iff*

- *$f = j$ and $\{r_h, s_i\} \succ_f \{r_k, s_l\}$, or*
- *$h = k$ and $\{q_f, s_i\} \succ_h \{q_j, s_l\}$, or*
- *$i = l$ and $\{q_f, r_h\} \succ_i \{q_j, r_k\}$.*

Similarly to *3GSM*, even for the frameworks based on *3PSA* we can derive the same conclusions.

Theorem 3. *A set $B \subseteq A_{rgs}$ represents a solution of a 3PSA problem iff B is a stable extension of the corresponding AAF as defined in Definition 6.*

Proposition 3. *Given a set S of cardinality nk, Definition 6 can be straightforwardly extended to let an AAF represent an $nPSA$, with $n > 3$. The set of stable extensions corresponds to the solution of such an $nPSA$ as well.*

3.3 Cyclic Preferences

In an instance of the *three-dimensional stable matching problem with cyclic preferences* (*c3DSM*), we have three disjoint sets of entities X, Y and Z such that $|X| = |Y| = |Z| = n$ (we can consider them as genders, for example). Each agent $x \in X$ has a total order of the agents in Y according to the preference of x. Similarly, each agent $y \in Y$ has a total order of the agents in Z, and each agent $z \in Z$ has a total order of the agents in X: this is the reason why preferences are cyclic. A triple $(x, y, z) \notin MT$, is blocking if $(x, y_1, z_1), (x_2, y, z_2), (x_3, y_3, z) \in MT$ and $y \succ_x y_1$, $z \succ_y z_2$, and $x \succ_z x_3$.

A match M is a solution of *c3DSM* if there exists no blocking triple for MT. We know from [7] and [13] that if $n = 3, 4, 5$ this problem admits a stable matching, and from [13] that if $n = 3$, it admits two solutions. Moreover, the complexity of this problem is still an open question [7,13]. As for the other problems in this section, we can define an *AAF* which represents a c3DSM problem.

Definition 7 (c3DSM and AAF). *Given a c3DSM problem, a corresponding AAF is defined by $A_{rgs} = \{(x, y, z) \mid x \in X, y \in Y, z \in Z\}$ and $R \subseteq A_{rgs} \times A_{rgs}$, such that $R((x_g, y_h, z_i), (x_j, y_k, z_l))$ iff*

- $g = j$ and $y_h \succ_g y_k$, or
- $h = k$ and $z_i \succ_h z_l$, or
- $i = l$ and $x_g \succ_i x_j$.

Even for this extension, similar formal results can be obtained as in previous subsections.

Theorem 4. *A set $B \subseteq A_{rgs}$ represents a solution of the c3DSM problem iff B is a stable extension of the corresponding AAF as defined in Definition 7.*

Proposition 4. *Given n genders, Definition 7 can be straightforwardly extended to let an AAF represent an cnDSM, with $n > 3$. The set of stable extensions corresponds to the solution of such an cnDSM as well.*

4 Incomplete Preference Lists and Ties

The original *SM* problem has been extended in the literature in several ways [9] in order to more naturally model the preferences of the elements in the considered set.

A different variant of the SM problem allows incomplete preference lists: an *SM* problem has incomplete lists (*SMI*) if an individual can exclude a partner whom she/he does not want to be matched with [9]. In practice some preferences are just omitted. Therefore, function p is partial: there exists some (m_i, w_j) for which p is not defined, i.e., $p(m_i, w_j) \uparrow$ and/or $p(w_j, m_i) \uparrow$.[3] A further extension is represented by preference lists that allow ties, i.e., in which it is possible to express the same preference for more than one possible partner: the problem is usually named as "SM with ties" (*SMT*) [9].

We define *3GSMIT* as the *3GSM* problem including incomplete preference lists and ties. In the same way as in Sect. 3 we can model this problem with an *AAF*.

	(w_1, d_1)	(w_1, d_2)	(w_2, d_1)	(w_2, d_2)		(m_1, d_1)	(m_1, d_2)	(m_2, d_1)	(m_2, d_2)
m_1	2	1	–	3	w_1	1	2	3	3
m_2	2	1	4	3	w_2	2	4	3	1

	(m_1, w_1)	(m_1, w_2)	(m_2, w_1)	(m_2, w_2)
d_1	3	2	1	3
d_2	2	1	–	3

Fig. 5. A *3GSMIT* problem given in tabular form, with men/women/dogs preferences.

---

[3] Conversely, $p(m_i, w_j) \downarrow$ and/or $p(w_j, p_i) \downarrow$ if p is defined on that couple.

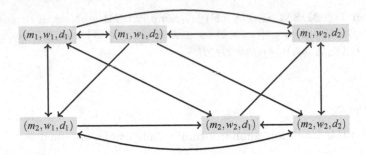

Fig. 6. The *AAF* from the *3GSM* problem with incomplete preferences and ties in Fig. 5.

Definition 8 (3GSMIT and AAF). *Given a 3GSMIT problem, a corresponding AAF is* $A_{rgs} = \{(m, w, d) \mid m \in M, w \in W, d \in D, p(m, w) \downarrow \wedge p(w, m) \downarrow$
$\wedge p(m, d) \downarrow \wedge p(d, m) \downarrow, \wedge p(w, d) \downarrow \wedge p(d, w) \downarrow\}$ *and* $R \subseteq A_{rgs} \times A_{rgs}$, *s.t.*
$R((m_g, w_h, d_i), (m_j, w_k, d_l))$ *iff*

- $g = j$ *and* $(w_h, d_i) \succeq_g (w_k, d_l)$, *or*
- $h = k$ *and* $(m_g, d_i) \succeq_h (m_j, d_l)$, *or*
- $i = l$ *and* $(m_g, w_h) \succeq_i (m_j, d_k)$.[4]

In Fig. 5 we report a modified version of the *3GSM* problem in Sect. 3.1: "−" means a preference is not defined, and now we also have some ties. The set of stable extensions in the encoded framework in Fig. 6 is different w.r.t. the result for Fig. 4: besides $\{(m_1, w_2, d_2), (m_2, w_1, d_1)\}$ and $\{(m_1, w_1, d_1), (m_2, w_2, d_2)\}$ which are maintained even for the *AAF* in Fig. 6, we have $\{(m_1, w_1, d_2), (m_2, w_2, d_1)\}$ instead of $\{(m_1, w_2, d_1), (m_2, w_1, d_2)\}$ (both these arguments are not present in Fig. 6 anymore).

Moreover, if from the same example we remove e.g. the preference towards (m_2, d_2) from the list of w_2, argument (m_2, w_2, d_2) disappears from Fig. 6. In this case the stable extensions are $\{(m_1, w_1, d_2), (m_2, w_2, d_1)\}$ and $\{(m_1, w_1, d_1)\}$. The second stable extension is not a perfect match: e.g. m_2 is not matched to any couple. This requires us to redefine the correspondence between stable extensions and solutions of *3GSMIT*.

Theorem 5. *Being* $\langle A_{rgs}, R \rangle$ *a framework representing a 3GSMIT problem with* $|M| = |W| = |D| = k$, B *is a stable solution of the problem iff there exists a stable extension* $B \subseteq A_{rgs}$ *such that* $|B| = k$ *and all* $m_i \in M, w_j \in W, d_k \in D$ *appear as labels of arguments in* B.

5   Conclusion

We have shown how different multi-dimensional extensions of the *SM* problem can be modelled by AAFs whose stable extensions exactly cover the set of solutions in the corresponding matching problem. The 2-sex encoding was originally

---

[4] Note that the stability conditions now use $\succeq$ instead of $\succ$ because of ties.

proposed by P.M. Dung in [5]. These problems find real-world applications in many fields, for instance the assignment of graduating medical students to their first hospital appointments. $nGSM$, $nPSA$, and $cnDSM$ are NP-complete problems or the complexity is still open. Solving them with an Argumentation-based solver can be a convenient option [3,4].

In the future, studying the complexity of these problems can help in understanding the complexity of solving the stable semantics in n-partite frameworks [6], and vice-versa. Moreover, we plan to use *Weighted Abstract Argumentation* [2] to study and model strategies in *Cooperative Games*, from a more general approach than just focusing on marriages. We think of using weights on arguments/attacks to study (strictly or weakly) dominant strategies. Finally, we plan to generate a synthetic benchmark of AAFs obtained from random n-dimensional matching problems.

References

1. Bistarelli, S., Rossi, F., Santini, F.: Not only size, but also shape counts: abstract argumentation solvers are benchmark-sensitive. J. Log. Comput. **28**(1), 85–117 (2018)
2. Bistarelli, S., Rossi, F., Santini, F.: A novel weighted defence and its relaxation in abstract argumentation. Int. J. Approx. Reason. **92**, 66–86 (2018)
3. Bistarelli, S., Santini, F.: ConArg: a constraint-based computational framework for argumentation systems. In: IEEE 23rd International Conference on Tools with Artificial Intelligence, ICTAI, pp. 605–612. IEEE Computer Society (2011)
4. Bistarelli, S., Santini, F.: Modeling and solving AFs with a constraint-based tool: ConArg. In: Modgil, S., Oren, N., Toni, F. (eds.) TAFA 2011. LNCS (LNAI), vol. 7132, pp. 99–116. Springer, Heidelberg (2012). https://doi.org/10.1007/978-3-642-29184-5_7
5. Dung, P.M.: On the acceptability of arguments and its fundamental role in non-monotonic reasoning, logic programming and n-person games. Artif. Intell. **77**(2), 321–357 (1995)
6. Dvořák, W., Dunne, P.E.: Computational problems in formal argumentation and their complexity. IFCoLog J. Logics Appl. (FLAP) **4**(8), 2557–2622 (2017)
7. Eriksson, K., Sjöstrand, J., Strimling, P.: Three-dimensional stable matching with cyclic preferences. Math. Soc. Sci. **52**(1), 77–87 (2006)
8. Irving, R.W.: An efficient algorithm for the "stable roommates" problem. J. Algorithms **6**(4), 577–595 (1985)
9. Iwama, K., Miyazaki, S.: A survey of the stable marriage problem and its variants. In: International Conference on Informatics Education and Research for Knowledge-Circulating Society, ICKS 2008, pp. 131–136. IEEE Computer Society (2008)
10. Knuth, D.E.: Stable Marriage and its Relation to Other Combinatorial Problems: An Introduction to the Mathematical Analysis of Algorithms. American Mathematical Society, Providence (1997)
11. Knuth, D.E.: Mariages stables et leurs relations avec d'autres problemes combinatoires: introduction a l'analysis mathematique des algorithmes-. Les Presses de l'Université de Montréal (1976)

12. Ng, C., Hirschberg, D.S.: Three-dimensional stable matching problems. SIAM J. Discrete Math. **4**(2), 245–252 (1991)
13. Pashkovich, K., Poirrier, L.: Three-dimensional stable matching with cyclic preferences. CoRR abs/1807.05638 (2018). http://arxiv.org/abs/1807.05638

Modal Interpretation of Formal Concept Analysis for Incomplete Representations

Mohammed Sadou[1] , Yassine Djouadi[1(✉)], and Allel Hadj-Ali[2]

[1] RIIMA Laboratory, Department of Computer Science,
USTHB University, Algiers, Algeria
{m.sadou,y.djouadi}@univ-alger.dz
[2] LIAS/ENSMA, Poitiers, France
allel.hadjali@ensma.fr

Abstract. Formal Concept Analysis (FCA) theory relies on relational representation called formal context. Dealing with incomplete representations is a challenging issue in FCA. In this spirit, a Kripke structure has been proposed as semantics of three-valued Kleene's logic for the exclusive definition of validity and possibility of attribute implications. Existing approaches consider possible intents as a set of attributes possibly satisfied by a given set of objects (a possible extent is dually considered). It appears that such a consideration is counter intuitive and quite misleading. Indeed, considering a possible intent as a whole granule of knowledge, in which each attribute is possibly satisfied by "all" objects, is a binding measure, in the sense that we cannot have possible intermediate granules of knowledge (i.e. intents or extents).

In this paper, we propose as a first stage, a new Kripke structure as semantics of modal logic. This structure is based on a completely different consideration, namely: "an object is a possible world". As a second stage, a distance δ that measures the implication strength $o \rightarrow Y$ (dually $p \rightarrow X$) between an object o and a set of properties Y (dually between a property p and a set of objects X) is proposed. This distance which is recursively defined upon the accessibility relation, allows to bring an ordered set of possible worlds.

Keywords: Modal logic · FCA · Incomplete representations

1 Introduction

Formal concept analysis (FCA for short), introduced by Wille [17], provides a theoretical framework that aims to extract interesting clusters (granules) of knowledge called formal concepts and dependencies. These formal concepts are obtained from a relational representation of data, called formal context. In classical setting, a formal context is a binary relationship between a set of objects and a set of properties, usually represented by an object-property table. A table entry may contain cross mark or blank mark, depending on whether an object has the property in question or not. FCA considers Galois derivation operator

© Springer Nature Switzerland AG 2020
J. Davis and K. Tabia (Eds.): SUM 2020, LNAI 12322, pp. 261–269, 2020.
https://doi.org/10.1007/978-3-030-58449-8_20

for extracting and organizing formal concepts on lattice called concept lattice. A formal concept is then a pair of objects and properties that are in mutual correspondence. The set of objects is the extent of the concept and the set of properties is the intent. Formal Concept Analysis has gained recognition in many areas due to its potential of knowledge discovery [15]. For several real life applications criteria, the classical definition of FCA has been extended to many forms. Fuzzy set theory has led to an extension of the original FCA setting by allowing a fuzzy value for the proposition "object o has a property p". Fuzzy formal concepts result from such representation [2]. Recently in [8,9], Dubois and Prade have proposed a possibility-theoretic reading of FCA. Besides the operator of sufficiency, typically used in FCA, three other operators have been proposed namely possibility, necessity and dual sufficiency [1,6,7].

In [5] authors treat in an exhaustive way, the several possible meanings of the proposition objects "object o has a property p" as it may be uncertain, gradual, interval-valued, unknown (missing) or even not applicable. Authors also notice the suitable theoretical framework for each case.

Related to total ignorance in Formal Concept Analysis [3], a three-valued Kleene's logic is investigated to obtain as much of attribute implicational knowledge about a given context. A Kripke structure [14], based on completion, represents semantics of this logic. In this approach [3], a possible intent is considered as a set of properties possibly satisfied by a given set of objects (a possible extent is dually considered).

It appears that such a consideration is counter intuitive and quite misleading. Indeed, considering a possible intent as a whole granule of knowledge, in which each attribute is possibly satisfied by "all" objects, is a binding measure, in the sense that we cannot have possible intermediate granules of knowledge (i.e. intents or extents).

The aim of this paper is to take benefits from the theoretical framework supplied by modal logic (Possible Worlds Semantics) and its underlying notion of accessibility relation. For this purpose, as first stage, we propose a new Kripke structure for Formal Concept Analysis. This structure is based on a completely different consideration, namely: "an object is a possible world". As a second stage, we propose to evaluate to which extent an object o satisfies a set of properties Y, by evaluating the implication $o \rightarrow Y$. The accessibility relation is used to this purpose by means of the notion of transitions between possible worlds.

The paper is organized as follows. Section 2 gives a survey on FCA and existing approaches related to incomplete formal contexts. Section 3 presents briefly Logical Concept Analysis [10,11]. The proposed modal structure is presented in Sect. 4. Whereas the next section gives an interpretation of such a structure w.r.t. incomplete formal contexts. Endly, immediate perspectives are outlined.

2   Formal Concept Analysis

2.1   Classical Setting

Formal concept analysis mainly relies on a formal context. In the classical setting, a formal context is a triple $\mathcal{K} = (\mathcal{O}, \mathcal{P}, \mathcal{I})$ where $\mathcal{I}$ is a binary relationship between a set $\mathcal{O}$ of objects and a set $\mathcal{P}$ of properties (also called attributes in the sequel). Usually, a formal context is represented under the form of a table which formalizes this relationship, the rows correspond to objects and the columns to properties. The notation $o\mathcal{I}p$ means that object $o \in \mathcal{O}$ satisfies the property $p \in \mathcal{P}$. In [17], a Galois powerset derivation operator $(.)^{\triangle}$ has been considered for a given set of objects X (similarly for a given set of properties Y). This powerset operator is defined s.t. $X^{\triangle} = \{p \in \mathcal{P} | \forall o \in \mathcal{O}\ (o \in X \Rightarrow o\mathcal{I}p)\}$. Similarly for a set of properties Y, we define the set $Y^{\triangle}$ of objects that satisfy all properties in Y s.t. $Y^{\triangle} = \{o \in \mathcal{O} | \forall p \in \mathcal{P}\ (p \in Y \Rightarrow o\mathcal{I}p)\}$.

A formal concept in $\mathcal{K}$ is a pair $\langle X, Y \rangle$ such that $X^{\triangle} = Y$ and $Y^{\triangle} = X$, where X (resp. Y) is called the extent (resp. the intent). The set of all formal concepts in $\mathcal{K}$ equipped with a partial order $\leq$ defined by: $\langle X_1, Y_1 \rangle \leq \langle X_2, Y_2 \rangle$ iff $X_1 \subseteq X_2$ (or equivalently $Y_2 \subseteq Y_1$) forms a complete lattice, called a concept lattice.

2.2   Incomplete Representations

It is widely agreed that in many areas, the relationship between a given set of objects and the set of their corresponding properties may be incomplete. For this purpose, incomplete representations of formal contexts have been addressed by Obiedkov [16], Burmeister et al. [3]. The authors have proposed to introduce a third value, denoted "?", in the formal context, resulting in an incomplete formal context $\mathcal{K}^? = (\mathcal{O}, \mathcal{P}, \{+, -, ?\}, \mathcal{I})$ (also called three-valued formal context). Let $o \in \mathcal{O}$ and $p \in \mathcal{P}$, the interpretation of the relationship $\mathcal{I}$ is given as follows: $(o, p, +) \in \mathcal{I}$ means that the object o has the property p; $(o, p, -) \in \mathcal{I}$ means that the object o does not have the property p and $(o, p, ?) \in \mathcal{I}$ means that it is unknown, whether the object o has the property p or not.

In the proposal of Burmeister and Holzer [3], an incomplete formal context is defined as the family of all complete Boolean formal contexts, called completions, obtained by modifying unknown entries $(o, p, ?)$ into known ones $((o, p, +)$ **or** $(o, p, -))$. For a set $X \subseteq \mathcal{O}$ of objects (dually for a set Y of properties), Burmeister and Holzer [3] have also proposed two powerset derivation operators $(.)^{\square}$ and $(.)^{\lozenge}$, defined as follows:

$X^{\square} = \{p \in \mathcal{P} \mid (o, p, +) \in \mathcal{I}$ for all $o \in X\}$, is the set of all properties in $\mathcal{P}$ applying to all objects in X.

$X^{\lozenge} = \{p \in \mathcal{P} \mid ((o, p, +) \in \mathcal{I}$ or $(o, p, ?) \in \mathcal{I})$ for all $o \in X\}$ is the set of all properties in $\mathcal{P}$ possibly applying to all objects in X.

Applying $(.)^{\square}$ (resp. $(.)^{\lozenge}$) powerset operator to a given set of objects X generates a certain intent $X^{\square}$ (resp. a possible intent $X^{\lozenge}$). Certain extents (resp. possible extents) are dually defined in [3].

3   Logical Concept Analysis

Logical models were studied to provide a rich and uniform representation of information and its semantics. In this spirit, logical models have been considered for FCA upon the hypothesis that the objects and the properties can be interpreted by logical formulae. Ferré et al. have proposed a Logical Concept Analysis framework where the powerset of the set of properties is considered as an arbitrary logic, to which are associated a deduction relation [10,11]. It becomes that an object satisfies a set of properties if we can infer the formula representing the properties from the formula representing the object. The authors have reformulated a formal context $(\mathcal{O}, \mathcal{P}, \mathcal{I})$ by $(\mathcal{O}, 2^{\mathcal{P}}, i)$, where $2^{\mathcal{P}}$ is a powerset of $\mathcal{P}$ and i is a mapping from $\mathcal{O}$ to $2^{\mathcal{P}}$ defined by $i(o) = \{p \in \mathcal{P} \mid o\mathcal{I}p\}$. The auhors have also proposed to consider two alternative powerset derivation operators σ and τ, defined as follows for a set X (resp. Y) of objects (resp. of properties):

$$\sigma(X) = \bigcap_{o \in \mathcal{O}} i(o) \text{ and } \tau(Y) = \{o \in \mathcal{O} \mid i(o) \supseteq \mathcal{O}\}.$$

Although Logical Concept Analysis approaches provide a rich representation framework as well as a way for reasoning on logical formulae (descriptions of objects), it may be argued that all existing approaches do not allow to consider incomplete representations. For this purpose, we propose to consider Kripke structure in order to deal with incomplete formal contexts. Indeed, let us assimilate an object (in the sense of FCA) with a possible word (in the sense of modal logic). Let also o_d be a partial description of an object o, this might be a set of sentences or a single property. Thus, possible words semantics brings a way in order to evaluate to which extent an object o satisfies a set of properties Y, by evaluating the implication $o_d \rightarrow Y$. If $(o_d \rightarrow Y)$ is true then Y is true in o. If $o_d \rightarrow Y$ is not true then we go to the nearest object o' that has a closer description to the object o, if A is true in the object o' then $o_d \rightarrow Y$ is true, otherwise Y is recursively evaluated in o', and so on. Such an inference model needs a suited modal structure. The next section illustrates our proposition dedicated to this purpose.

4   Possible World Semantics Proposal

4.1   Model

Modal logic [12] adds to the language of classical logic two "modalities" operators, namely possibility ($\lozenge$) and necessity ($\square$). These operators are also described trough the idea of possible worlds. A possible truth is one that can be ultimately false and a necessity truth is one that can only be true [4]. The most common semantics of modal logics introduces the notion of worlds and imposes further structures called accessibility relations [13]. It becomes that a model $\mathcal{M}$ of modal logic (possible worlds semantics) is a triple $\mathcal{M} = (W, R, V)$ of a non empty set of possible worlds W, a binary accessibility relation R between worlds, and a valuation function assigning to a given proposition variable $p \in P$, a set of worlds where p is true. If a world w_1 is accessible from the world w, then w_1 is called a "possible world" of w, that is when the proposition p is true in one of the possible worlds of a world w, then it is said that p is possibly true in the world w, denoted by $w \models \Diamond p$; if p is true in every possible worlds of the world w, then p is necessary true in w, denoted by $w \models \Box p$.

Based on this principle, the proposed model, in which the set of propositional variables P corresponds to the set of properties $\mathcal{P}$, is given trough the Definition 1. Whereas the Definition 2. introduces the "possible satisfaction" of a subset $Y \subseteq P$ of propositional variables by an object o. It is worth noticing that such a set of propositional variables Y is considered with a conjunctive semantics.

Definition 1. (Model) *For a formal context* $(2^{\mathcal{O}}, \mathcal{P}, f)$, *the corresponding model* $\mathcal{M} = (W, R, V)$ *is given as:* $W = \mathcal{O}$; $V = f$ *(s.t. f is a mapping from $\mathcal{P}$ to* $2^{\mathcal{O}}$ *defined by* $f(p) = \{o \in \mathcal{O} \mid o \mathcal{I} p\}$*); and the world o_1 is R accessible to the world o_0, denoted $o_0 R o_1$, if the world o_0 can be transformed into o_1 by adding a propositional variable in o_0.*

Definition 2. *A set of propositional variables $Y \subseteq P$ is possible in the world o_0, noted $o_0 \models \Diamond Y$, iff there exists o_1 such that $o_0 R o_1$ and $o_1 \models Y$.*

Let us denote by "R-step" a single transition (single step) among the accessibility relation. It may be remarked that applying once $\Diamond$ operator, corresponds to such a transition. Thus, $k \times R$-step (cf. Definition 3.) between the world o_0 and the world o_k represents the transformation of world o_0 to the world o_k by adding, in each step, a property p_i.

Definition 3. *A set of propositional variables Y is reached from o_0 by $k \times R$-step $(k > 1)$, we write $o_0 \models \Diamond^k Y$, iff there exists o_1 such that $o_0 R o_1$ and $o_1 \models \Diamond^{k-1} Y$.*

Let (o_0, o_{i-1}, o_i, o_k) be a set of possible worlds. We propose to consider a distance between two given worlds as the succession of transformations between these two worlds. We give hereafter the definition of the proposed distance. It is worth noticing that the distance between two successive worlds can be calculated by different ways. For instance, we assume that $d(o_i, o_{i+1}) = 1$ if $o_i R o_{i+1}$.

Definition 4. *The distance between the world o_0 and the world o_k, denoted $d(o_0, o_k)$, is the sum of the elementary distances between o_{i-1} to o_i such that $o_{i-1} R o_i$. Formally,* $d(o_0, o_k) = \sum_{i=0}^{k} d(o_{i-1}, o_i)$ *with $o_{i-1} R o_i$.*

The degree to which extent a possible world (object) o satisfies a set of propositional variables denoted $\delta(o_0 \rightarrow Y)$, may be evaluated by the degree of the implication $o_0 \rightarrow Y$. We propose to calculate $\delta(o_0 \rightarrow Y)$ as follows (μ is an arbitrary constant s.t. $\mu > |\mathcal{P}|$):

$$\delta(o_0 \rightarrow Y) = \begin{cases} 0 & \text{if } o_0 \models Y \\ \sum_{i=0}^{k} d(o_{i-1}, o_i) & \text{if } o_{i-1} R o_i \text{ and } o_k \models Y \\ \mu & \text{if } \not\exists o_i, o_{i+1}, o_i R o_{i+1} \text{ and } o_{i+1} \models Y \end{cases}$$

4.2   Possible Formal Concepts

The modal structure $\mathcal{M} = (W, R, V)$, above proposed, induces new powerset derivation operators, namely $(.)^{\Diamond_p} : 2^P \to 2^W$ and $(.)^{\Diamond_o} : 2^W \to 2^P$, respectively defined for $Y \subseteq P$ and $X \subseteq W$ as: $(Y)^{\Diamond_p} = \{o \in W | o \models \Diamond Y\}$ and $(X)^{\Diamond_o} = \{p \in P \mid \sum_i \delta(o_i \to p) \leq 1, o_i \in X\}$. On this basis, we consider in our proposal, "possible formal concepts" as given in the following definition, whereas the Proportions 1 and 2. give some useful algebraic properties

Definition 5. *A possible formal concept is a pair (X, Y) such that $X^{\Diamond_o} = Y$, and $Y^{\Diamond_p} = X$.*

Proposition 1. *Let X_1, X_2 two sets of possible worlds and Y_1, Y_2 two sets of propositional variables, the following properties are satisfied:*

(1): $X_1 \subseteq X_2 \Rightarrow X_2^{\Diamond_o} \subseteq X_1^{\Diamond_o}$
(2): $Y_1 \subseteq Y_2 \Rightarrow Y_2^{\Diamond_p} \subseteq Y_1^{\Diamond_p}$

Proofs are omitted in this paper due to the lack of space

Proposition 2. *Let $\mathcal{B}_p$ denotes the set of all possible formal concepts. The set $(\mathcal{B}_p \cup \langle W, \emptyset_p \rangle \cup \langle \emptyset_O, P \rangle)$ equipped with a partial order (denoted $\leqslant$) defined as: $(X_1, Y_1) \leqslant (X_2, Y_2)$ iff $X_1 \subseteq X_2$ (or, equivalently $Y_2 \subseteq Y_1$) forms a possible concepts lattice.*

5   Incomplete Representations

In the previous section, a generic modal structure (model) has been proposed for FCA. This section details how the proposed model may be used for handling incomplete formal contexts $\mathcal{K}^? = (\mathcal{O}, \mathcal{P}, \{+, -, ?\}, \mathcal{I})$. Indeed, in addition to what is certainly known, the accessibility relation allows to represent properties possibly satisfied by an object. Formally, a subset of properties Y is "possibly satisfied" in the world o (denoted $o \models \Diamond Y$) iff we can change one unknown entry "?" in o_0 into a known one "+" and then $o_0 \models Y$. The following proposition gives an useful way for approximation of possible formal concepts (i.e. intermediate formal concept).

Proposition 3. *These two properties are verified:*

(1) $X^\square \subseteq X^{\Diamond_o} \subseteq X^\Diamond$
(2) $Y^\square \subseteq Y^{\Diamond_p} \subseteq Y^\Diamond$

Recall also that both $(.)^\square$ and $(.)^\Diamond$ operators have been proposed in [3]. Thus, the above proposition gives a comparative study of our approach w.r.t the one proposed in [3].

Example 1. Let us consider the incomplete formal context $\mathcal{K}^?$ presented in Table 1 for which $V(p_1) = \{o_1, o_3\}$, $V(p_2) = \{o_1, o_2\}$, and $V(p_3) = \{o_1\}$. The Fig. 1 gives the modal representation of $\mathcal{K}^?$. The set $Y = \{p_2, p_3\}$ is satisfied in the object o_1 ($o_1 \models Y$). The set Y is also possibly satisfied in the object o_2 ($o_2 \models \Diamond Y$), and it is not possibly satisfied in $o3$. On the other hand, the set Y is possibly possibly satisfied in o_3 ($o_3 \models \Diamond^2 Y$). Applying the modal operators to $\{o_1, o_2\}$ (resp. to $\{p_2, p_3\}$) results on: $\{o_1, o_2\}^{\Diamond_o} = \{p_2, p_3\}$, (resp. $\{p_2, p_3\}^{\Diamond_p} = \{o_1, o_2\}$). It becomes that the pair $\langle \{o_1, o_2\}, \{p_2, p_3\} \rangle$ forms a possible formal concept of the incomplete formal context $\mathcal{K}^?$.

Table 1. Incomplete formal context $\mathcal{K}^?$

$\mathcal{K}^?$	p_1	p_2	p_3
o_1	+	+	+
o_2	−	+	?
o_3	+	?	?

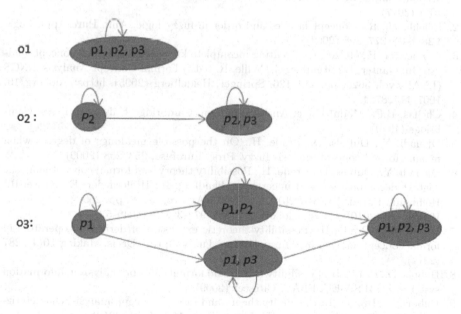

Fig. 1. Possible worlds representation of $\mathcal{K}^?$

6   Conclusion

In this paper we have proposed a theoretical framework based on possible words semantics, in order to deal with incomplete representations, namely incomplete formal contexts. The proposed modal structure relies essentially on an accessibility relation that allows 1 to k transitions between possible worlds by adding to each transition a property, achieving then possible formal concepts. Powerset operators dedicated to this purpose are also given and some of their properties are established. We have also proposed to evaluate to which extent an object o satisfies a set of properties Y by introducing the measure $\delta(o \rightarrow Y)$. As immediate perspectives:

- We intend to investigate similar properties between $\Diamond_o$(dually $\Diamond_p$) operator and possibility operator $(.)^{\Pi}$ already considered by Dubois and prade [8].
- Noting that the closure property cannot be established as such, we expect to introduce constraints on the accessibility relation in order to get this important property.
- A scalable application to information retrieval is in progress.

References

1. Ait-Yakoub, Z., Djouadi, Y., Dubois, D., Prade, H.: Asymmetric composition of possibilistic operators in formal concept analysis: application to the extraction of attribute implications from incomplete contexts. Int. J. Intell. Syst. **32**(12), 1285–1311 (2017)
2. Bělohlávek, R.: Concept lattices and order in fuzzy logic. Ann. Pure Appl. Logic **128**(1–3), 277–298 (2004)
3. Burmeister, P., Holzer, R.: Treating incomplete knowledge in formal concept analysis. In: Ganter, B., Stumme, G., Wille, R. (eds.) Formal Concept Analysis. LNCS (LNAI), vol. 3626, pp. 114–126. Springer, Heidelberg (2005). https://doi.org/10.1007/11528784_6
4. Chellas, B.F.: Modal Logic: An Introduction. Cambridge University Press, Cambridge (1980)
5. Djouadi, Y., Dubois, D., Prade, H.: On the possible meanings of degrees when making formal concept analysis fuzzy. Proc. Eurofuse, 253–258 (2009)
6. Djouadi, Y., Dubois, D., Prade, H.: Possibility theory and formal concept analysis: context decomposition and uncertainty handling. In: Hüllermeier, E., Kruse, R., Hoffmann, F. (eds.) IPMU 2010. LNCS (LNAI), vol. 6178, pp. 260–269. Springer, Heidelberg (2010). https://doi.org/10.1007/978-3-642-14049-5_27
7. Djouadi, Y., Prade, H.: Possibility-theoretic extension of derivation operators in formal concept analysis over fuzzy lattices. Fuzzy Optim. Decis. Making **10**(4), 287 (2011)
8. Dubois, D., Prade, H.: Possibility theory and formal concept analysis in information systems. In: IFSA-EUSFLAT. Citeseer (2009)
9. Dubois, D., Prade, H.: Possibility theory and formal concept analysis: characterizing independent sub-contexts. Fuzzy Sets Syst. **196**, 4–16 (2012)
10. Ferré, S., Ridoux, O.: A logical generalizationof formal concept analysis. In: Ganter, B., Mineau, G.W. (eds.) ICCS-ConceptStruct 2000. LNCS (LNAI), vol. 1867, pp. 371–384. Springer, Heidelberg (2000). https://doi.org/10.1007/10722280_26
11. Ferré, S., Ridoux, O., Sigonneau, B.: Arbitrary relations in formal concept analysis and logical information systems. In: Dau, F., Mugnier, M.-L., Stumme, G. (eds.) ICCS-ConceptStruct 2005. LNCS (LNAI), vol. 3596, pp. 166–180. Springer, Heidelberg (2005). https://doi.org/10.1007/11524564_11

12. Hughes, G.E., Cresswell, M.J., Cresswell, M.M.: A New Introduction to Modal Logic. Psychology Press, London (1996)
13. Kracht, M., Kutz, O.: Logically possible worlds and counterpart semantics for modal logic. In: Philosophy of Logic, pp. 943–995. Elsevier (2007)
14. Kripke, S.A.: Semantical analysis of modal logic I normal modal propositional calculi. Math. Logic Quart. **9**(5–6), 67–96 (1963)
15. Kumar, C.: Knowledge discovery in data using formal concept analysis and random projections. Int. J. Appl. Math. Comput. Sci. **21**(4), 745–756 (2011)
16. Obiedkov, S.: Modal logic for evaluating formulas in incomplete contexts. In: Priss, U., Corbett, D., Angelova, G. (eds.) ICCS-ConceptStruct 2002. LNCS (LNAI), vol. 2393, pp. 314–325. Springer, Heidelberg (2002). https://doi.org/10.1007/3-540-45483-7_24
17. Wille, R.: Restructuring lattice theory: an approach based on hierarchies of concepts. In: Ferré, S., Rudolph, S. (eds.) ICFCA 2009. LNCS (LNAI), vol. 5548, pp. 314–339. Springer, Heidelberg (2009). https://doi.org/10.1007/978-3-642-01815-2_23

A Symbolic Approach for Counterfactual Explanations

Ryma Boumazouza[ID], Fahima Cheikh-Alili[ID], Bertrand Mazure[ID],
and Karim Tabia[(✉)][ID]

CRIL, Univ. Artois and CNRS, 62300 Lens, France
{boumazouza,cheikh,mazure,tabia}@cril.univ-artois.fr

Abstract. We propose a novel symbolic approach to provide counter-factual explanations for a classifier predictions. Contrary to most explanation approaches where the goal is to understand which and to what extent parts of the data helped to give a prediction, counterfactual explanations indicate which features must be changed in the data in order to change this classifier prediction. Our approach is symbolic in the sense that it is based on encoding the decision function of a classifier in an equivalent CNF formula. In this approach, counterfactual explanations are seen as the Minimal Correction Subsets (MCS), a well-known concept in knowledge base reparation. Hence, this approach takes advantage of the strengths of already existing and proven solutions for the generation of MCS. Our preliminary experimental studies on Bayesian classifiers show the potential of this approach on several datasets.

Keywords: eXplainable AI · MCS · Counterfactual explanation

1 Introduction

Recently, a symbolic approach for explaining classifiers has been proposed in [5]. This approach first compiles a classifier into an **equivalent** and **tractable** symbolic representation then enumerates some forms of explanations such as prime implicants. It has many nice features in terms of tractability, explanation enumeration and formal analysis of classifiers. This paper proposes a novel approach that is designed to equip such symbolic approaches [5] with a module for counterfactual explainability. Intuitively, we view the process of computing counterfactual explanations as the one of computing Minimal Correction Subsets (MCS generation) where the knowledge base stands for the classifier and the data instance in hand. As we will show later, our symbolic approach for counterfactual generation has many nice features added to the fact of lying on well-known concepts and efficient existing techniques for MCS generation.

The inputs to our approach are a classifier's decision function f compiled into an equivalent symbolic representation in the form of an Ordered Decision Diagram (ODD) and a data instance x. Our contribution is to model the problem of

J. Davis and K. Tabia (Eds.): SUM 2020, LNAI 12322, pp. 270–277, 2020.
https://doi.org/10.1007/978-3-030-58449-8_21

counterfactual generation as the one of MCS generation. We will show the properties of this encoding and highlight the links between MCS and counterfactual explanations. Our experiments show that using existing MCS generation tools, one can efficiently compute counterfactual explanations as far as a classifier can be compiled into an ODD which is the case of Bayesian network classifiers [5], decision trees and some neural nets [4].

2 From a Symbolic Representation of a Classifier to an Equivalent CNF Encoding

Our approach for counterfactual explanation proceeds in two steps : The first one is encoding a symbolic representation (given in the form of an ODD) into an equivalent CNF representation. The second step consists in computing MCSs meant as counterfactual explanations given the CNF representation of the classifier and any data instance. In this section, we describe the first step of our approach. For the sake of simplicity, the presentation is limited to binary classifiers with binary features but the approach still applies to non binary classifiers as stressed in [5].

Definition 1 *(Binary Classifier). A binary classifier is defined by two sets of variables: A feature space $X = \{X_1,...,X_n\}$ where $|X| = n$, and a binary class variable denoted Y. Both the features and the class variable take values in $\{0,1\}$.*

Definition 2 *(Decision Function of a Classifier). A decision function of a classifier (X,Y) is a function $f : X \rightarrow Y$ mapping each instantiation x of X to $y = f(x)$.*

Definition 3 *(Ordered Decision Diagram ODD). An Ordered Decision Diagram (ODD) is a rooted, directed acyclic graph, defined over an ordered set of discrete variables, and encoding a decision function. Each node is labeled with a variable X_i , $i = 1, \ldots, n$ and has an outgoing edge for each value x_i of the variable X_i, except for the sink nodes, which represent the terminal nodes.*

An Ordered Binary Decision Diagram OBDD is an ODD where all the variables are binary. If there is an edge from a node labeled X_i to a node labeled X_j, then $i < j$ (more on tractable representations such as ODDs can be found in [6]).

Example 1. Figure 1a shows a naive Bayes classifier for deciding whether a student will be admitted to a university (class variable: Admit (A)). The features of an applicant are: work-experience (WE), first-time-applicant (FA), entrance-exam (E) and gpa (GPA). In Fig. 1b, we provide the OBDD representing the classifier decision function f with the variable ordering (WE, FA, E, GPA). Here, the sinks correspond to the values of the class variable (A).

Let us now focus on our target representation. A CNF (Clausal Normal Form) formula is a conjunction of clauses. A clause is a formula composed of a disjunction of literals. A literal is either a Boolean variable or its negation. A quantifier-free formula is built from atomic formulae using conjunction $\wedge$, disjunction $\vee$,

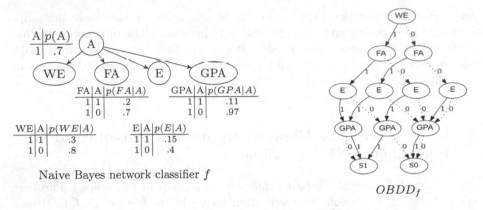

Naive Bayes network classifier f

$OBDD_f$

Fig. 1. A Naive Bayes network classifier and its corresponding OBDD.

and negation $\neg$. An interpretation μ assigns values from $\{0, 1\}$ to every Boolean variable. Let Σ be a CNF formula, μ satisfies Σ iff μ satisfies all clauses of Σ.

There are several methods to encode a decision diagram as a CNF formula. For instance in [2], the authors proposed a method called "Single-Cut-Node" to store a BDD (Binary Decision Diagram) as a CNF where they model BDD nodes as multiplexers. A second method called "The No-Cut method" creates clauses starting from f corresponding to the "off-set" and a last method called "The Auxiliary-Variable-Cut" which combines the two other methods. For the sake of simplicity and clarity, we choose the simplest method which does not involve adding new variables during the encoding process since we want to restrict our explanations to the input variables of the classifier. We implement a simple way to encode the symbolic representation of a classifier as a CNF formula based on the "The No-Cut" method [2]. In our case, since we are dealing with binary Boolean functions (binary features and class variable), our tractable representation of the decision function f is an OBDD. We use along with this paper positive/true/1 and negative/false/0 interchangeably. Let us first define an "off-set" of a Boolean function and a CNF formula.

Definition 4 (Off-Set of a Boolean function). *The Off-set of a Boolean function f, denoted as f^0, is $f^0 = \{v \in \bigcup_{i=1}^{n}\{0,1\}^i, |f(v) = 0\}$ If $f^0 = \{0,1\}^n$, then f is unsatisfiable. Otherwise, f is satisfiable.*

Intuitively, f^0 is the set of counter-models of f. This concept contains the counter-models we need to enumerate in order to construct our CNF's clauses. The OBDD is used to enumerate all the paths from the root to the 0-sink node (the off-set), where each element of f^0 corresponds to a path within it.

Definition 5 (CNF encoding of an OBDD). *Let f be the decision function encoded by an ordered binary decision diagram $OBDD_f$. Let also Off-set$(OBDD_f)$ be the off-set of $OBDD_f$. We define the obtained CNF formula from $OBDD_f$ as $\Sigma_f = \wedge \neg e_i$ where $e_i \in f^0$ and $i = 1..|f^0|$.*

Let α be the associated formula of f. The intuition is that $\neg \alpha \equiv \vee \, e_i$ where $e_i \in f^0$. Then f comes down to negating $\vee \, e_i$ allowing to obtain directly f in the form of a CNF. Following Definition 5, we have:

- Every variable of the feature space $X = \{X_1,...,X_n\}$ of the classifier will correspond to a Boolean variable in the CNF Σ_f.
- The class variable Y is captured by the truth value of the CNF (Σ_f).
- Modeling a prediction made by the classifier for a given data instance x comes down to the truth value of: (CNF $\Sigma_f \wedge \Sigma_x$) where Σ_x stands for the data instance x encoded as a CNF by a set of unit[1] clauses.

Our encoding guarantees the logical equivalence between the $OBDD_f$ and the obtained CNF Σ_f. The following proposition formally states this result.

Proposition 1. *Let f be a binary decision function and $OBDD_f$ its compiled representation. Let also Σ_f be the CNF representation of the decision function f obtained following Definition 5. Then an interpretation μ is model (resp. countermodel) of Σ_f iff it is mapped to 1 (resp. to 0) by f.*

Lemma 1. *Given a binary classifier, a data instance x and the predicted class $f(x) = y$, $(\Sigma_f \wedge \Sigma_x)$ is SAT (resp. UNSAT) iff $f(x) = 1$ (resp. $f(x) = 0$).*

We stress that both of the compilation of the classifier to a symbolic representation and the encoding into a CNF formula, is done only once in our approach and can be re-used to explain as many instances as wanted.

3 Generating Counterfactual Explanations

Intuitively, a counterfactual explanation for an instance of interest x and a classifier f is the minimum changes to x needed to alter the output of f.

Definition 6 *(Counterfactual Explanation). Let x be a complete data instance and $f(x)$ its prediction by the decision function f. A counterfactual explanation $\acute{x}$ of x is such that:*

- *$\acute{x} \subseteq x$ ($\acute{x}$ is a subset or a part of x)*
- *$f(x[\acute{x}]) = 1 - f(x)$ (prediction inversion)*
- *There is no $\hat{x} \subset \acute{x}$ such that $f(x[\hat{x}]) = f(x[\acute{x}])$ (minimality)*

In Definition 6, the term $x[\acute{x}]$ denotes the data instance x where variables included in $\acute{x}$ are reversed. In our setting, a counterfactual $\acute{x}$ is defined as a part of the data instance x such that $\acute{x}$ is minimal and $\acute{x}$ allows to flip the prediction $f(x)$. The explanation comes as follow: an output y is returned because variables from the features space X had values $(x_1, x_2...)$. If instead, X had values $(x'_1, x'_2, ...)$ while all other variables had remained constant, the output y' would have been returned. Counterfactual explanations are expected to explain both

---

[1] A unit clause involves only one Boolean variable represented by a literal.

the outcome of a prediction and how that would change if *things had been differ-ent*. Our main idea is to model the counterfactual explanation task as a Partial Max-SAT problem. Recall that our CNF encoding of an OBDD representation of a classifier's decision function f ensures that a negative (resp. positive) predic-tion leads to an unsatisfiable (resp. satisfiable) CNF Boolean formula. Namely, $f(x) = 1$ iff $(\Sigma_f \wedge \Sigma_x)$ is satisfiable. In the case where $f(x) = 0$, $\Sigma_f \wedge \Sigma_x$ is unsat-isfiable and it is possible to identify the subsets of Σ_x allowing to restore the consistency of $\Sigma_f \wedge \Sigma_x$ (recall that Σ_f is satisfiable unless the classifier f always predicts 0 regardless of the instance x). This is a well-known problem dealt with in many areas such as knowledge base reparation, consistency restoration, etc. We will see later how to provide explanations when the outcome is positive, namely $f(x) = 1$, using the same mechanisms. It is important to note that our CNF is composed of two parts: Σ_f and Σ_x, where Σ_f encodes the classifier and it is satisfiable and Σ_x encode the data instances x and represented as a set of unit clauses. In the case of negative predictions, since $\Sigma_f \wedge \Sigma_x$ is unsatisfi-able (inconsistent), we can compute a sort of reparation set that is composed of the subsets of data instance x that cause the unsatisfiability of $\Sigma_f \wedge \Sigma_x$. This is known as the minimal correction subset (MCS).

Definition 7 (MSS). *A maximal satisfiable subset (in short, MSS) Φ of a CNF Σ is a subset (of clauses) $\Phi \subseteq \Sigma$ that is satisfiable and such that $\forall \ \alpha \in \Sigma \setminus \Phi$, $\Phi \cup \{\alpha\}$ is unsatisfiable.*

Definition 8 (MCS (Co-MSS)). *A minimal correction subset (in short MCS, also called Co-MSS) Ψ of a CNF Σ is a set of formulas $\Psi \subseteq \Sigma$ whose complement in Σ, i.e., $\Sigma \setminus \Psi$, is an MSS of Σ.*

In our case, given a data instance x and a function f, and their respective CNFs Σ_x and Σ_f, an MCS Φ ensures the minimality property and tells what clauses to remove from $\Sigma_f \wedge \Sigma_x$ to restore its consistency. Note that although the number of MCSs can be exponential in the worst case, it remains low in many benchmarks.

3.1   Counterfactuals for Negative Predictions Through MCSs

Our main idea for explaining a negative prediction for an instance x is to compute its MCSs. An MCS identifies the subset of clauses to be repaired to restore the satisfiability of the CNF formula. In order for an MCS to correspond to a counterfactual explanation, it should contain only the unit clauses belonging to Σ_x indicating what features need to be removed or flipped such that the whole CNF (namely, $\Sigma_f \wedge \Sigma_x$) becomes again satisfiable. This leads to splitting the CNF into two subsets: hard constraints and soft ones.

Definition 9 (Partial Max-SAT). *Given a CNF formula Σ where some clauses are hard and some are soft, Partial Max-SAT is the problem of find-ing a truth assignment that satisfies all hard constraints and the maximum of soft ones.*

In order to solve Partial Max-SAT, we will consider the general setting, where a formula is composed of two disjoint sets of clauses $\Sigma = \Sigma_H \cup \Sigma_S$ [1], where Σ_H denotes the hard clauses and Σ_S denotes the soft ones. The set of hard clauses is Σ_f while soft clauses is Σ_x representing the data instance x to explain. The CNF encoding of the classifier Σ_f as a set of hard clauses is presented in the previous section. The CNF encoding of the data instance Σ_x as soft clauses is done as follows. Let Σ_x be the *soft clauses*, defined as follows:

- Each clause $\alpha \in \Sigma_x$ is composed of exactly one literal $(\forall \alpha \in \Sigma_x, |\alpha| = 1)$
- Each literal representing a Boolean variable of Σ_x corresponds to a Boolean variable $\{X_i \in X / i \in [1, n]\}$ of the feature space of the decision function f.

Following our approach, an MCS for $\Sigma_f \wedge \Sigma_x$ comes down to a subset of soft clauses, namely a part of x that is enough to remove in order to restore the consistency, hence to flip the prediction $f(x) = 0$. Proposition 2 states that each MCS computed for $\Sigma_f \wedge \Sigma_x$ represents a counterfactual explanation $\acute{x} \subseteq x$ for the prediction $f(x) = 0$ and vice versa.

Proposition 2. *Let f be the decision function, $OBDD_f$ its compiled symbolic representation, x be a data instance predicted negatively $(f(x) = 0)$ and $\Sigma_f \wedge \Sigma_x$ an unsatisfiable CNF. Let $CF(x, f)$ be the set of counterfactuals of x wrt. f. Let $MCS(\Sigma_{f,x})$ the set of MCSs of $\Sigma_f \wedge \Sigma_x$. Then:*

$$\forall \acute{x} \subseteq x, \acute{x} \in CF_f(x, f(x)) \iff \acute{x} \in MCS(\Sigma_{f,x}) \tag{1}$$

The MCS enumeration is done over the *soft clauses*, which practically should reduce the time needed to enumerate all the MCS since we will have less clauses to consider. As for positively predicted instances, we can simply work on the negation of $OBDD_f$ representation of the decision function f namely, we will rely on the CNF $\Sigma_{\neg f} \wedge \Sigma_x$ to compute the counterfactuals in a similar way.

4 Experiments

This section provides our experiments to evaluate our approach of generating counterfactuals. Given a data instance of interest x and the decision function f, we encode both f and x as a CNF formula. We start by considering typical binary classification problems and focus on *Binary Naive Bayes Classifiers (BNC)*. To test our approach, we compiled synthetic naive Bayes classifiers to OBDDs using the approach proposed in [6], then we encode these OBDDs into CNF formulas before getting into the generation of the MCSs. We mention that for each network size, we used an average of 4 to 10 networks to run our experiments.

4.1 Compiling Bayes Classifiers into OBDDs

Table 1 summarizes the compilation experiments we ran on BNCs with different sizes. For each category of classifiers having the same number of features, we compute the average size of their corresponding OBDD (number of nodes). As expected, we notice a large increase in the OBDD size as the number of features of the classifier grows up. Moreover, it seems that the OBDD size also strongly depends on the classifier's parameters in addition to the number of features.

Table 1. Average size of the OBDD representations.

Nb_Features	5	10	16	20	22	25
OBDD$_{size}$	9	42	370	1020	2546	8626

4.2 Dumping OBDDs as CNF Formulas

Next step will be to encode the obtained OBDD into CNF Boolean formulas. We aim in the following to compare the size of both OBDDs and CNFs of classifiers with different sizes. As observed experimentally in Table 2, the time and size (number of clauses) of the generated CNF are strongly correlated to the size of its OBDD. While the compilation time scales linearly with the number of nodes of the OBDD, the size of the CNF can be much smaller, depending on the classifier's parameters (threshold of the used BNC and variable order used for OBDD). We remind that this encoding is done only once for a given classifier and then can subsequently be used for explaining any number of instances.

4.3 MCSs Generation

Once we got the CNF ($\Sigma_f \wedge \Sigma_x$), we can get to the generation of MCSs. In our experiments, we use the boosting algorithm for the MCSs generation proposed in [3] and implemented in *EnumELSRMRCache*[2]. Recall that our input is a CNF composed of hard clauses (encoding the classifier) and soft ones (encoding the data instance to explain). The data instances were randomly generated. The aim here is to compare the number of counterfactual explanations with the size of OBDD and CNF representations. Table 2 summarizes the results of the average number of counterfactual explanations generated given a data instance and a classifier. As expected, the number of explanations increases with the CNF size in general, but remains strongly related to: (1) the classifier and OBDD parameters (variable ordering), and (2) to the data instance itself. As shown in Table 2, the average run-time does not seem to depend on the number of MCSs

Table 2. Average size of OBDD/CNF, runtime (ms), and number of the counterfactuals (MCSs).

#Vars	5	10	16	20	22	25
OBDD$_{size}$	9	42	370	1020	2546	8626
CNF$_{size}$	3	64	2598	27122	123878	684847
Encoding_Runtime (ms)	1.4	32.3	2725	241806	430471	327888500
#MCS	3	23	101	305	272	364
Runtime (*ms*)	1.9	2.3	17.8	1762	9299.4	109148.5

---

[2] Available at http://www.cril.univ-artois.fr/enumcs/.

generated but more on the number of features of the classifier, which is expected since the time-consuming part of generating the MCSs is related to the size of the representations in terms of number of clauses and their size.

To sum up the results, it can be said that as long as we can get a symbolic tractable representation of a classifier, our approach can provide counterfactual explanations. The number of different MCSs of the CNF Boolean formulas remains low in our case since our approach computes the MCSs over the soft clauses only, which experimentally significantly reduces the time of MCSs enumeration.

5   Concluding Remarks

The approach proposed in this paper allows to equip the symbolic approach proposed in [6] with a module for counterfactual explanations. Our approach is simple and takes advantage of well-defined concepts and proven tools for MCSs. Moreover, our approach is specifically designed to provide actionable explanations.

Acknowledgment. This work is done as part of a PhD thesis benefiting from the support of the Région Hauts-de-France.

References

1. Biere, A., Heule, M., van Maaren, H.: Handbook of Satisfiability, vol. 185. IOS Press (2009)
2. Cabodi, G., Nocco, S., Quer, S.: Improving SAT-based bounded model checking by means of BDD-based approximate traversals. In: 2003 Design, Automation and Test in Europe Conference and Exhibition, pp. 898–903. IEEE (2003)
3. Grégoire, É., Izza, Y., Lagniez, J.M.: Boosting MCSes enumeration. In: IJCAI, pp. 1309–1315 (2018)
4. Shi, W., Shih, A., Darwiche, A., Choi, A.: On tractable representations of binary neural networks. arXiv preprint arXiv:2004.02082 (2020)
5. Shih, A., Choi, A., Darwiche, A.: A symbolic approach to explaining Bayesian network classifiers. arXiv preprint arXiv:1805.03364 (2018)
6. Shih, A., Choi, A., Darwiche, A.: Compiling Bayesian network classifiers into decision graphs. In: Proceedings of the AAAI Conference on Artificial Intelligence, vol. 33, pp. 7966–7974 (2019)

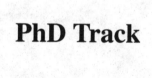

PhD Track

Modelling Multivariate Ranking Functions with Min-Sum Networks

Xiaoting Shao[1]([⊠]), Zhongjie Yu[1], Arseny Skryagin[1], Tjitze Rienstra[2], Matthias Thimm[2], and Kristian Kersting[1]

[1] TU Darmstadt, Darmstadt, Germany
{xiaoting.shao,yu,arseny.skryagin,kersting}@cs.tu-darmstadt.de
[2] Universitat Koblenz, Koblenz, Germany
{rienstra,thimm}@uni-koblenz.de

Abstract. Spohnian ranking functions are a qualitative abstraction of probability functions, and they have been applied to knowledge representation and reasoning that involve uncertainty. However, how to represent a ranking function which has a size that is exponential in the number of variables still remains insufficiently explored. In this work we introduce *min-sum networks* (MSNs) for a compact representation of ranking functions for multiple variables. This representation allows for exact inference with linear cost in the size of the number of nodes.

Keywords: Spohnian ranking functions · Graphical models

1 Introduction

Spohnian ranking functions are a qualitative order-of-magnitude abstraction of probability functions. These can be used to measure uncertainty using ranks [15] represented by natural numbers or ∞, which can be understood as a degree of surprise: 0 for not surprising, 1 for surprising, 2 for very surprising, and so on, and ∞ for impossible.

These functions have been applied to problems of knowledge representation and reasoning that involve uncertainty but where probabilities are unknown or irrelevant, such as belief revision and non-monotonic inference [4,7,12]. One of the fundamental issues when using ranking functions in practice is the representation of a ranking function, which has a size that is exponential in the number of variables. The same problem arises in probabilistic modeling, where it is solved by using probabilistic graphical models (PGMs) as compact representations of probability distributions [10]. Because ranks behave much like probabilities if + is replaced with *min* and × with +, it is sometimes possible to adapt PGMs to represent and reason about ranking functions. For example, the ranking-based counterpart of a Bayesian network is called a *ranking network* or *OCF network* [2,7,9,15].

In this paper we introduce *min-sum networks* (MSNs) for compact representation of ranking functions. They are an adaptation of *sum-product networks*

© Springer Nature Switzerland AG 2020
J. Davis and K. Tabia (Eds.): SUM 2020, LNAI 12322, pp. 281–288, 2020.
https://doi.org/10.1007/978-3-030-58449-8_22

(SPNs) [13]. An SPN is a rooted directed acyclic graph with a recursively defined structure: a node is either a *sum-node* with weighted edges pointing to its children; a *product-node* with non-weighted edges pointing to its children; or a leaf node representing a univariate distribution. Compared to many PGM models, SPNs support exact inference with linear cost in the size of the number of nodes. This advantage, combined with the ability to handle missing data makes SPNs to a very attractive choice for modeling any data set. Indeed, several SPN learning techniques have shown comparable or better performance than other state-of-the-art models in tasks such as image classification and natural language processing [3,5].

The tractability of SPN's inference carries over to MSNs. More precisely, the rank of an event or proposition according to the ranking function represented by an MSN can be computed with cost linear in the size of the number of nodes in the MSN. One issue, however, is that the MSN needs to be constructed first. To address this, we propose a method to learn an MSN based on a set of observations in such a way that more probable events are less surprising (lower ranked) than less probable events.

The overview of this paper is as follows. We present the necessary basics concerning ranking theory in Sect. 2. In Sect. 3 we define min-sum networks, while Sect. 4 deals with the problem of learning min-sum networks. We conclude in Sect. 5.

2   Ranking Functions

Ranking functions are a qualitative abstraction of probability functions where events receive *ranks* [15]. A rank is a non-negative integer or ∞ and can be understood as a degree of surprise: 0 for not surprising, 1 for surprising, 2 for very surprising, and so on, and ∞ for impossible. Formally, a ranking a *ranking function* (also known as an *ordinal conditional function* or *kappa function*) is defined as follows.

Definition 1. *A ranking function over a set Ω is a function $\kappa : \Omega \to \mathbb{N}_0^\infty$ such that $\kappa(w) = 0$ for at least one $w \in \Omega$. A ranking function κ is extended to a function over propositions or events (i. e., subsets of Ω) by defining $\kappa(X) = \infty$ if $X = \emptyset$, and $\kappa(X) = min(\{\kappa(w) \mid w \in X\})$, otherwise. The rank of A conditional on B is denoted $\kappa(A \mid B)$ and is defined by $\kappa(A \mid B) = \kappa(A \cap B) - \kappa(B)$.*

A ranking function κ induces beliefs using the principle that A is believed if and only if the complement $\overline{A} = \Omega \setminus X$ is surprising (i.e. $\kappa(\overline{A}) > 0$). Similarly, A is believed *conditional on B* if and only if $\kappa(\overline{A} \mid B) > 0$.

3   Min-Sum Networks

Here we provide the definition of MSN, which is a ranking-based variation on SPNs [13]. We first need to introduce some notation and terminology.

Random variables will be denoted by uppercase letters (e.g. X_i). We restrict our attention to Boolean random variables. We use $\mathbf{X}$ to denote a collection $\{X_1, \ldots, X_n\}$ of random variables and denote by $val(\mathbf{X})$ the set of *realisations* of $\mathbf{X}$, i.e., $val(\mathbf{X}) = \{T, F\}^n$ (T for true and F for false). Elements of $val(\mathbf{X})$ are denoted by lowercase boldface letters (e.g. $\mathbf{x}$). A realisation of some subset of $\{X_1, \ldots, X_n\}$ will be called *evidence*. Given a random variable X_i we use x_i and $\bar{x}_i$ to denote indicator variables. We say that x_i (resp. $\bar{x}_i$) is *consistent* with evidence $\mathbf{e}$ iff $\mathbf{e}$ does not assign false (resp. true) to X_i.

A min-sum network (MSN) over variables $\mathbf{X} = \{X_1, \ldots, X_n\}$ is an acyclic directed graph N whose leaves are the variables $x_1, \ldots, x_n, \bar{x}_1, \ldots, \bar{x}_n$. Given a node i in N we denote by N_i the subgraph rooted at node i and we denote by $Ch(i)$ the set of children of node i. The internal nodes of N are either *min-nodes* or *sum-nodes*. Each edge from a *min-node* i to another node j has an associated *weight* w_{ij} which is a non-negative integer or ∞, satisfying $min(\{w_{ij}\}) = 0$. The rank of evidence $\mathbf{e}$ according to N is defined recursively: $N(\mathbf{x}) = \sum_{j \in Ch(i)} N_j(\mathbf{x})$ if the root of N is a *sum-node*; and $N(\mathbf{x}) = min(\{w_{ij} + N_j(\mathbf{x}) | j \in Ch(i)\})$, if the root of N is a *min-node*. Further, we define a leaf L_{X_i} to consist of one *min* node, two weights w_{x_i} and $w_{\bar{x}_i}$ and two indicators x_i, $\bar{x}_i$.

3.1 Validity, Consistency and Completeness

Similar to the SPNs, we introduce the three key properties, which allows us to link MSNs with the ranking theory and ensure the error-free inference: validity, completeness, and consistency. The link is that the values of $N(\mathbf{x})$ for all $\mathbf{x} \in val(\mathbf{X})$ define a ranking function by

$$\Phi_N(\mathbf{e}) = min(\{N(\mathbf{x}) \mid \mathbf{x} \in val(\mathbf{X}), \mathbf{e} \text{ is consistent with } \mathbf{x}\}).$$

We define all three properties in the following.

Definition 2. *An MSN N over variables $\mathbf{X}$ is valid, iff $N(\mathbf{e}) = \Phi_N(\mathbf{e})$ for all evidence $\mathbf{e}$ of $\mathbf{X}$.*

By contrast, the definitions of completeness and consistency are the same as those of SPNs:

Definition 3. *An MSN N is complete, iff all children of the same min-node have the same scope.*

Definition 4. *An MSN N is consistent, iff there is no sum-node and variable X such that x appears in one child of this node and $\bar{x}$ in another child.*

We combine the three properties and show in the following theorem that:

Theorem 1. *An MSN is valid if it is complete and consistent.*

Since there are only two mutually exclusive configurations $\{x_i = 1, \bar{x}_i = 0\}$ and $\{x_i = 0, \bar{x}_i = 1\}$, we define the delta functions

$$\delta_{x_i} := \begin{cases} 1, & x_i = 1 \\ \infty, & x_i = 0 \end{cases} \quad \text{and} \quad \delta_{\bar{x}_i} := \begin{cases} 1, & \bar{x}_i = 1 \\ \infty, & \bar{x}_i = 0 \end{cases}.$$

Table 1. Rankings in the Wet Grass example.

R	T	T	T	T	T	T	T	T	F	F	F	F	F	F	F	F
N	T	T	T	T	F	F	F	F	T	T	T	T	F	F	F	F
H	T	T	F	F	T	T	F	F	T	T	F	F	T	T	F	F
S	T	F	T	F	T	F	T	F	T	F	T	F	T	F	T	F
Rank	4	1	∞	∞	∞	∞	∞	∞	5	∞	8	2	3	∞	6	0

So, we can express the value of the leaf L_i for any arbitrary variable X_i as

$$
\begin{aligned}
L_i &= \min\left(\delta_{x_i} \cdot w_{x_i}, \delta_{\bar{x}_i} \cdot w_{\bar{x}_i}\right) \\
&= \delta_{x_i} \cdot w_{x_i} \cdot \mathbb{1}_{\{\delta_{\bar{x}_i} \cdot w_{\bar{x}_i} > \delta_{x_i} \cdot w_{x_i}\}} + \delta_{\bar{x}_i} \cdot w_{\bar{x}_i} \cdot \mathbb{1}_{\{\delta_{\bar{x}_i} \cdot w_{\bar{x}_i} > \delta_{\bar{x}_i} \cdot w_{\bar{x}_i}\}}.
\end{aligned}
\tag{1}
$$

Consequently, the leaf L_i can take the two values for the two configurations w_{x_i} or $w_{\bar{x}_i}$. Using the same idea

$$
\begin{aligned}
&\min\left(\{w_{ij} + N_j(\mathbf{x}) | j \in Ch(i)\}\right) \\
&= \sum_{j \in Ch(i)} (w_{ij} + N_j(\mathbf{x})) \cdot \mathbb{1}_{\{\sum_{k \in Ch(i) \setminus \{j\}} (w_{ik} + N_k(\mathbf{x})) > w_{ij} + N_j(\mathbf{x})\}}.
\end{aligned}
\tag{2}
$$

With (1) and (2) we write $N(\mathbf{x})$ as a series which will be referred to as an *expansion* of the MSN. Therefore, an MSN is valid if its expansion has the same value as $\Phi_N(\mathbf{e})$ for all evidence $\mathbf{e}$: each configuration has exactly one partial sum (condition 1), each partial sum is convergent for exactly one configuration (condition 2). From condition 2 we conclude that $N(\mathbf{x}) = w_{\mathbf{x}} < \infty$ and consequently $\Phi_N(\mathbf{e}) = \sum_{x \in \mathbf{e}} N(x) = \sum_{x \in \mathbf{e}} w_x = \sum_{k \in n(\mathbf{e})} w_k$, where $n(\mathbf{e})$ is the number of the configurations complying with condition 2. From condition 1, we conclude $n(\mathbf{e}) << |val(\mathbf{X})| = 2^n$, therefore $\Phi_N(\mathbf{e}) = \sum_{k \in n(\mathbf{e}): w_k < \infty} w_k = N(\mathbf{e}) < \infty$ and MSN is valid.

Now we prove by induction from the leaves to the root that, if the MSN is complete and consistent, then its expansion is its network series. The rest of the proof follows analogously to that one in [13], emphasising the necessity of *completeness* for *min-node* and *consistency* for *sum-node*.

3.2   Min-Sum Network Example: Wet Grass

The Wet Grass [1] example is a well-known example of probabilistic graphical models. It consists of a collection of four boolean random variables $\mathbf{X} = \{R, N, H, S\}$, where R stands for "it has been raining"; S for "Holmes' sprinkler was on"; N for "Holmes' neighbor's grass is wet"; and H for "Holmes' grass is wet". In this paper, we turn Wet Grass into a ranking example. Table 1 lists the ranks of all possible configurations of the four random variables. For instance, it is not surprising if it has not been raining, the sprinkler was off, and both lawns are not wet, i.e., $\kappa(\mathbf{x}) = 0$ for $\mathbf{x} = \{R = F, N = F, H = F, S = F\}$. However, it is impossible if it has not been raining, the sprinkler was off, but both lawns are wet, i.e., $\kappa(\mathbf{x}) = \infty$ for $\mathbf{x} = \{R = F, N = T, H = T, S = F\}$.

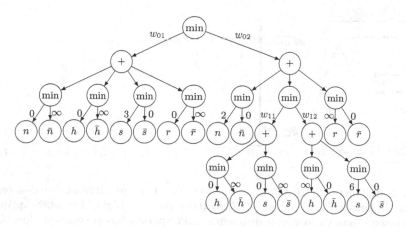

Fig. 1. An example of a valid MSN. The weights of the *min-nodes* are $w_{01} = 1$, $w_{02} = 0$, $w_{11} = 3$, $w_{12} = 0$.

The ranking function of the random variables in Table 1 can be modeled with a manually designed valid MSN, shown in Fig. 1. To query the rank of input evidence, a bottom-up pass needs to be operated. Denote $L_X^{w_i}$ the leaf of random variable X from the *sum-node* with weight w_i. For example, with $\mathbf{x} = \{R = F, N = T, H = F, S = T\}$, the rank of $\mathbf{x}$ according to the MSN is

$$N(\mathbf{x}) = \min(\{(L_N^{w_{01}} + L_H^{w_{01}} + L_S^{w_{01}} + L_R^{w_{01}}) + w_{01},$$
$$L_N^{w_{02}} + \min\{(L_H^{w_{11}} + L_S^{w_{11}}) + w_{11}, (L_H^{w_{12}} + L_S^{w_{12}}) + w_{12}\} + L_R^{w_{02}} + w_{02}\})$$
$$= \min(\{(\infty + 0 + \infty + 3) + 1, 2 + \min\{(\infty + 0) + 3, (0 + 6) + 0\} + 0\})$$
$$= \min(\{\infty, 2 + 6 + 0\}) = 8,$$

which means one gets very surprising if it has not been raining and the sprinkler was on, while Mr. Holmes' lawn is dry and his neighbor's is wet. Given the same evidence, the corresponding rank from Table 1 always matches the rank calculated by the valid MSN from Fig. 1, i.e. $N(\mathbf{x}) = \kappa(\mathbf{x})$. Thus, the MSN in Fig. 1 models exactly the ranks in Table 1.

4 Learning MSNs

A common way to estimate the parameters θ of a statistical model is to compute the Maximum Likelihood Estimation (MLE) in such a way that under the assumed statistical model the observational data is most probable. Denote $D \triangleq \{\mathbf{x}_1, \mathbf{x}_2, \ldots, \mathbf{x}_m\}$ as the observational data, MLE is then defined as $\hat{\theta} \triangleq \arg\max_\theta \log p(D|\theta)$ [11]. In analogy, we can learn the parameters θ of an MSN by minimizing the rank on the observational data, that is,

$$\hat{\theta} \triangleq \arg\min_\theta N(D|\theta). \tag{3}$$

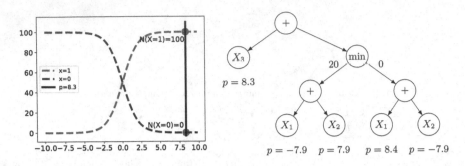

Fig. 2. Left: The ranking function for one variable. The vertical black line denotes the optimal parameter for X_3 in the MSN on the right side. Right: The MSN optimized using samples from the synthetic distribution. The parameters are marked blue. (Color figure online)

We assume the training examples D are independent and identically distributed. Then we can rewrite the rank as $N(D|\theta) = \sum_{i \in m} N(\mathbf{x}_i|\theta)$. Intuitively, minimizing $N(D|\theta)$ yields θ that makes the observed data most probable under the assumed MSN. The parameters θ of an MSN consist of weights $\mathbf{w}$ at *min-nodes* and the univariate distributional parameter p in the leaf nodes. We define the ranking function on the leaves as the following form

$$(2 * (\text{sigmoid}(p) - 0.5) * (X - 0.5) + 0.5) * C, \tag{4}$$

where $C = 100$. At inference time, we round up the output to get natural numbers. Figure 2 (left) shows a plot of this ranking function.

To find the θ that minimize the objective function $N(D|\theta)$, gradient descent is commonly used if the target function is differentiable. We implemented MSNs in python and Tensorflow so that differentiation and gradients can be computed automatically by Tensorflow. We now demonstrate a concrete example using this learning method. First we construct a synthetic distribution with three Bernoulli variables X_1, X_2 and X_3 of which X_1 and X_2 are dependent on each other and X_3 is independent of any. The joint probabilistic distribution can be factorized as $P(X_1, X_2, X_3) = P(X_1) * P(X_2|X_1) * P(X_3|X_1, X_2) = P(X_1) * P(X_2|X_1) * P(X_3)$ using the chain rule and independence information. 300 samples are generated from this distribution for learning the parameters of the MSN. The sample counts for each configuration of the three variables are shown in Table 2. We take an MSN with randomly initialized parameters and use gradient-based method to optimize the objective function (3) on the 300 samples. The optimization yields the network in Fig. 2 (right) with its parameters marked blue.

The ranks of all the possible configurations computed by this network are listed in Table 2. The ranks of all the observed configurations are sorted as $N(0, 1, 0) < N(1, 0, 0) < N(0, 1, 1) = N(0, 0, 0) < N(1, 0, 1) < N(0, 0, 1)$, which correspond exactly to reversely sorted empirical probability of all the observed configurations $P(0, 1, 0) > P(1, 0, 0) > P(0, 1, 1) > P(0, 0, 0) > P(1, 0, 1) >$

Table 2. The sample counts and the ranks computed by the MSN of every possible configuration of a synthetic distribution.

X_1, X_2, X_3	0,0,0	0,0,1	0,1,0	0,1,1	1,0,0	1,0,1	**1,1,0**	**1,1,1**
Rank	100	200	0	100	20	120	**100**	**200**
Count	28	8	111	38	92	23	**0**	**0**

$P(0, 0, 1)$. Besides, we expect the ranks for the unseen configurations to be as high as possible because they are very unlikely to happen. Here, we have two unseen configurations $X_1 = 1, X_2 = 1, X_3 = 0$ and $X_1 = 1, X_2 = 1, X_3 = 1$ whose ranks are respectively 100 and 200. That means, $X_1 = 1, X_2 = 1, X_3 = 1$ is least likely to happen which is correct. But $X_1 = 1, X_2 = 1, X_3 = 0$ is more likely to happen than, for example, $X_1 = 1, X_2 = 0, X_3 = 1$, which is the only rank that does not match the empirical probability. Besides, optimization may get stuck at a local optimum which possibly again leads to ranking computations that are not consistent with empirical probabilities. We leave this challenge for future research. By definition, the *min-nodes* encode mixtures of their children, and the *sum-nodes* assume independence between their children [14]. Take the MSN in Fig. 2 (right) as an example, X_3 is independent of X_1 and X_2, that means we can get the marginal ranking function of X_3 by simply removing the other independent branch. This yields an univariate ranking function with $p = 8.3$. The marginal sample counts for $X_1 = 0$ and $X_1 = 1$ are respectively 231 (111+92+28) and 69 (38+23+8), which means the rank of $X_1 = 1$ should be zero and the rank of $X_1 = 0$ should be larger than zero. Recall Fig. 2 (left), the vertical black line denotes the optimal parameter $p = 8.3$ for X_3 and this parameter yields $N(X_1 = 1) = 100$ and $N(X_1 = 1) = 0$, which matches the empirical probability well.

5   Conclusion and Future Work

Based on the notion of SPNs, we have introduced MSNs for compact representation and tractable inference with ranking functions. Ranking functions are used in models of belief revision and non-monotonic inference [4,7,12], and we believe that these applications may benefit from using min-sum networks for representing ranking functions. One obstacle, however, is that min-sum networks must first be constructed, and for this purpose, we proposed a method to learn a min-sum network based on a set of observations. There is a number of directions for future work. One is to improve our learning method. In particular, Giang and Shenoy [6] have studied desirable properties for transformations from probability functions to ranking functions. An interesting question is whether a min-sum network can be learned from an (empirical) probability distribution in such a way that these desirable properties are satisfied. Another interesting question is whether min-sum networks can be constructed or learned on the basis of qualitative information. For instance, the non-monotonic inference system called *System Z* involves determining the unique "most normal" ranking function that satisfies a given knowledge base containing default rules [12]. If this ranking function can be constructed directly as a min-sum network then this network could be used to answer certain types of queries with a cost that is linear with respect to the size of the network. Finally, our approach is based on adapting SPNs for representing ranking functions. It may also be possible to adapt SPNs in a similar way to represent other representations of uncertainty, such as possibility measures, belief functions and plausibility measures [8].

Acknowledgement. KK, MT, TR and XS acknowledge funding of the German Science Foundation (DFG) via the project "CAML", KE 1686/3-1. ZY and KK acknowledge the funding of the German Federal Ministry of Education and Research (BMBF) project "MADESI", 01IS18043B. AS and KK acknowledge the support of the BMBF and the Hessian Ministry of Science and the Arts (HMWK) within the National Research Center for Applied Cybersecurity ATHENE.

References

1. Barber, D.: Bayesian Reasoning and Machine Learning. Cambridge University Press, Cambridge (2012)
2. Benferhat, S., Tabia, K.: Belief change in OCF-based networks in presence of sequences of observations and interventions: application to alert correlation. In: Pacific Rim International Conference on Artificial Intelligence (2010)
3. Cheng, W.C., Kok, S., Pham, H.V., Chieu, H.L., Chai, K.M.A.: Language modeling with sum-product networks. In: Fifteenth Annual Conference of the International Speech Communication Association (2014)
4. Darwiche, A., Pearl, J.: On the logic of iterated belief revision. Artif. Intell. (1997)
5. Gens, R., Domingos, P.: Discriminative learning of sum-product networks. In: Advances in Neural Information Processing Systems (2012)
6. Giang, P.H., Shenoy, P.P.: On transformations between probability and Spohnian disbelief functions. In: Proceedings of Uncertainty in Artificial Intelligence (1999)
7. Goldszmidt, M., Pearl, J.: Qualitative probabilities for default reasoning, belief revision, and causal modeling. Artif. Intell. (1996)
8. Halpern, J.Y.: Reasoning about Uncertainty. MIT Press, Cambridge (2017)
9. Kern-Isberner, G., Eichhorn, C.: Intensional combination of rankings for OCF-networks. In: The Twenty-Sixth International FLAIRS Conference (2013)
10. Koller, D., Friedman, N.: Probabilistic Graphical Models: Principles and Techniques. MIT Press, Cambridge (2009)
11. Murphy, K.P.: Machine Learning: A Probabilistic Perspective. MIT Press, Cambridge (2012)
12. Pearl, J.: A natural ordering of defaults with tractable applications to default reasoning. In: Proceedings of Theoretical Aspects of Reasoning about Knowledge (1990)
13. Poon, H., Domingos, P.: Sum-product networks: a new deep architecture. In: International Conference on Computer Vision Workshops (2011)
14. Spohn, W.: A survey of ranking theory. In: Arló-Costa, H., Hendricks, V.F., van Benthem, J. (eds.) Readings in Formal Epistemology. SGTP, vol. 1, pp. 303–347. Springer, Cham (2016). https://doi.org/10.1007/978-3-319-20451-2_17
15. Spohn, W.: The Laws of Belief: Ranking Theory and Its Philosophical Applications. Oxford University Press, Oxford (2012)

An Algorithm for the Contension Inconsistency Measure Using Reductions to Answer Set Programming

Isabelle Kuhlmann[(✉)] and Matthias Thimm

University of Koblenz-Landau, Koblenz, Germany
{iskuhlmann,thimm}@uni-koblenz.de

Abstract. We present an algorithm for determining inconsistency degrees wrt. the contension inconsistency measure [7] which utilizes three-valued logic to determine the minimal number of atoms that are assigned truth value B (paradoxical/both true and false). Our algorithm is based on an answer set programming encoding for checking for upper bounds and a binary search algorithm on top of that. We experimentally show that the new algorithm significantly outperforms the state of the art.

Keywords: Inconsistency Measurement · Answer set programming · Contension inconsistency measure

1 Introduction

Dealing with inconsistent information is an important aspect in rational accounts to formal reasoning. In applications such as decision-support systems, a knowledge base is usually compiled by merging the formalised knowledge of many different experts. It is unavoidable that different experts contradict each other and that the merged knowledge base becomes inconsistent. One way of dealing with inconsistent information is to abandon classical inference and define new ways of reasoning. Some examples of such formalisms are, e. g., paraconsistent logics [2], default logic [12], answer set programming [3], and computational models of argumentation [1]. Moreover, the field of belief revision [9] deals with the particular case of inconsistencies in dynamic settings.

The field of *Inconsistency Measurement*—see the seminal work [6] and the recent book [8]—provides an *analytical* perspective on the issue of inconsistency. An *inconsistency measure* is a function that maps a knowledge base to a non-negative real number, the interpretation of that number being that larger values indicate a larger inconsistency within the knowledge base. The field of inconsistency measurement has proposed a series of different approaches to measure inconsistency, focusing on aspects such as minimal inconsistent subsets [4,10], or non-classical semantics [7,13], see [14] for an overview.

In this paper, we are considering algorithmic problems involving the contension inconsistency measure from [7]. This measure uses Priest's three-valued

© Springer Nature Switzerland AG 2020
J. Davis and K. Tabia (Eds.): SUM 2020, LNAI 12322, pp. 289–296, 2020.
https://doi.org/10.1007/978-3-030-58449-8_23

logic [11] to determine the minimal number of atoms in the language that are "conflicting" in the knowledge base under consideration (we will provide formal details in Sect. 2). In [15] it has been shown that the problem of deciding whether a certain value is an upper bound for the inconsistency degree wrt. the contension measure in NP-complete. Although this is an intractable problem, it still belongs to the rather "easier" problems in inconsistency measurement, as the corresponding decision problems wrt. many other measures are even higher in the polynomial hierarchy [15]. In this paper, we are presenting an algorithm for determining the inconsistency degree wrt. the contension measure of an arbitrary knowledge base using answer set programming (ASP) [3]. The latter is a declarative problem solving formalism that is suitable for addressing NP-hard problems (and beyond). More specifically, the contributions of this work are the following:

1. We present an ASP encoding of the problem whether a certain number is an upper bound for the inconsistency degree and integrate it into a binary search algorithm for determining the actual value (Sect. 3).
2. We report on some preliminary experiments that show that this algorithm significantly outperforms the state of the art (Sect. 4).

In addition, Sect. 2 covers the necessary preliminaries and Sect. 5 concludes the work by giving an overview of the contributions and possible future work.

2   Preliminaries

Let At be some fixed set of propositions and let $\mathcal{L}(\mathsf{At})$ be the corresponding propositional language constructed using the usual connectives $\wedge$ (*conjunction*), $\vee$ (*disjunction*), and $\neg$ (*negation*).

Definition 1. *A knowledge base* $\mathcal{K}$ *is a finite set of formulas* $\mathcal{K} \subseteq \mathcal{L}(\mathsf{At})$. *Let* $\mathbb{K}$ *be the set of all knowledge bases.*

If X is a formula or a set of formulas we write $\mathsf{At}(X)$ to denote the set of propositions appearing in X.

Semantics for a propositional language is given by *interpretations* where an *interpretation* ω on At is a function $\omega : \mathsf{At} \to \{\mathsf{true}, \mathsf{false}\}$. Let $\Omega(\mathsf{At})$ denote the set of all interpretations for At. An interpretation ω *satisfies* (or is a *model* of) a proposition $a \in \mathsf{At}$, denoted by $\omega \models a$, if and only if $\omega(a) = \mathsf{true}$. The satisfaction relation $\models$ is extended to formulas in the usual way.

For $\Phi \subseteq \mathcal{L}(\mathsf{At})$ we also define $\omega \models \Phi$ if and only if $\omega \models \phi$ for every $\phi \in \Phi$. A formula or set of formulas X_1 *entails* another formula or set of formulas X_2, denoted by $X_1 \models X_2$, if and only if $\omega \models X_1$ implies $\omega \models X_2$. If there is no ω with $\omega \models X$ we also write $X \models \bot$ and say that X is *inconsistent*.

2.1   The Contension Inconsistency Measure

Let $\mathbb{R}_{\geq 0}^{\infty}$ be the set of non-negative real values including infinity. The most general form of an inconsistency measure is as follows.

Definition 2. *An inconsistency measure* $\mathcal{I}$ *is a function* $\mathcal{I} : \mathbb{K} \to \mathbb{R}_{\geq 0}^{\infty}$ *that satisfies* $\mathcal{I}(\mathcal{K}) = 0$ *if and only if* $\mathcal{K}$ *is consistent, for all* $\mathcal{K} \in \mathbb{K}$.

The intuition we intend to be behind any concrete approach to inconsistency measure $\mathcal{I}$ is that a larger value $\mathcal{I}(\mathcal{K})$ for a knowledge base $\mathcal{K}$ indicates more severe inconsistency in $\mathcal{K}$ than lower values. Moreover, we reserve the minimal value (0) to indicate the complete absence of inconsistency.

With regard to an inconsistency measure $\mathcal{I}$, EXACT$_{\mathcal{I}}$ denotes the problem of deciding whether a given value $a \in \mathbb{R}_{\geq 0}^{\infty}$ is the inconsistency value $\mathcal{I}(\mathcal{K})$ of a knowledge base $\mathcal{K}$ [15]. UPPER$_{\mathcal{I}}$ and LOWER$_{\mathcal{I}}$ denote the problems whether a given value $a \in \mathbb{R}_{\geq 0}^{\infty}$ is an upper or a lower bound of $\mathcal{I}(\mathcal{K})$, respectively. For any function $\mathcal{I}$ according to Definition 2 that satisfies $\mathcal{I}(\mathcal{K}) = 0$ if and only if $\mathcal{K}$ is consistent, the decision problems UPPER$_{\mathcal{I}}$ and EXACT$_{\mathcal{I}}$ are NP-hard. LOWER$_{\mathcal{I}}$ is coNP-hard [15]. Moreover, VALUE$_{\mathcal{I}}$ is the natural function problem which returns the value of $\mathcal{I}(\mathcal{K})$ for a given knowledge base $\mathcal{K}$.

In [7], Grant and Hunter introduce an inconsistency measure based on semantics of Priest's three-valued logic [11]. In addition to *true* (T) and *false* (F), this logic includes a third value which indicates *paradoxical*, or *both true and false* (B). Table 1 shows the truth tables for this logic.

Table 1. Truth tables for Priest's propositional three-valued logic [11]

x	y	$x \wedge y$	$x \vee y$		x	$\neg x$
T	T	T	T			
T	B	B	T			
T	F	F	T		x	$\neg x$
B	T	B	T		T	F
B	B	B	B		B	B
B	F	F	B		F	T
F	T	F	T			
F	B	F	B			
F	F	F	F			

Let i be a three-valued interpretation, i.e., a function that assigns one of the three truth values to each atom in a knowledge base $\mathcal{K}$, denoted as $i :$ At$(\mathcal{K}) \mapsto \{T, F, B\}$. The domain of a certain interpretation i can be divided into two groups corresponding to their truth value [7]. More specifically, there is the group of atoms which are assigned a classical truth value (T or F), and there is the group of atoms which are assigned B. The former is defined as follows:

$$\text{Binarybase}(i) = \{\alpha \mid i(\alpha) = T \text{ or } i(\alpha) = F\}$$

Because the other group comprises those atoms which take part in conflicts, it is denoted

$$\text{Conflictbase}(i) = \{\alpha \mid i(\alpha) = B\}.$$

Further, a *model* is defined as an interpretation where each formula ϕ in $\mathcal{K}$ is assigned either T or B. Thus, the set of models is defined as follows:

$$\text{Models}(\mathcal{K}) = \{i \mid \forall \phi \in \mathcal{K}, i(\phi) = T \text{ or } i(\phi) = B\}$$

Example 1. Consider knowledge base $\mathcal{K}_1 = \{a \wedge b, \neg a \vee b, \neg b \wedge \neg c\}$. A model of $\mathcal{K}_1$ is the interpretation which assigns T to a, F to c, and B to b, denoted i_1. Consequently, Binarybase and Conflictbase wrt. i_1 are the following:

$$\text{Binarybase}(i_1) = \{a, c\} \qquad \text{Conflictbase}(i_1) = \{b\}$$

There exist also other models, for example the interpretation that assigns B to all $x \in \{a, b, c\}$ is a model of every knowledge base.

The previous definitions allows us to formulate the contension inconsistency measure $\mathcal{I}_c$ wrt. a knowledge base $\mathcal{K}$ as

$$\mathcal{I}_c(\mathcal{K}) = \min\{|\text{Conflictbase}(i)| \mid i \in \text{Models}(\mathcal{K})\}.$$

Consequently, $\mathcal{I}_c$ describes the minimum number of atoms in $\mathcal{K}$ that are assigned truth value B. Considering the exemplary knowledge base $\mathcal{K}_1$ presented in Example 1, we can easily see that it is inconsistent. More specifically, the first and third formula, $a \wedge b$ and $\neg b \wedge \neg c$, respectively, contradict each other in the sense of classical propositional logic. Since at least b (i.e., one atom) must be assigned the truth value B to make the knowledge base consistent in three-valued logic, the minimal size of Conflictbase is 1. Thus, $\mathcal{I}_c(\mathcal{K}_1) = 1$.

It has been shown [15] that $\text{UPPER}_{\mathcal{I}_c}$ is NP-complete, $\text{LOWER}_{\mathcal{I}_c}$ is coNP-complete, $\text{EXACT}_{\mathcal{I}_c}$ is DP-complete, and $\text{VALUE}_{\mathcal{I}_c}$ is $\text{FP}^{\text{NP}[\log]}$-complete. To the best of our knowledge, the currently only existing algorithm for computing $\mathcal{I}_c(\mathcal{K})$ is the one given in TweetyProject[1]. This algorithm follows a naive approach by searching for a solution in a brute force fashion. The given knowledge base is first converted to CNF and then checked for consistency. If the knowledge base is consistent, 0 is returned correspondingly. If it is not, for each proposition x, each clause containing x is removed and the resulting knowledge base is checked for consistency again. This is equivalent to setting x to B in three-valued logic. If one of the new knowledge bases is consistent, 1 is returned correspondingly. If, again, none of the knowledge bases turned out to be consistent, two propositions are set to B, i.e., all possible pairs of propositions are iteratively removed, then all triples, and so forth.

2.2   Answer Set Programming

Answer Set Programming (ASP) [3] is a declarative programming paradigm based on logic programming which is targeted at difficult search problems. ASP incorporates features of Reiter's default logic [12] and logic programming.

---

[1] http://tweetyproject.org/api/1.14/net/sf/tweety/logics/pl/analysis/ ContensionInconsistencyMeasure.html.

An *extended logic program* incorporates both negation-as-failure (not) and classical negation ($\neg$). Such a program comprises rules of the form

$$H \leftarrow A_1, \ldots, A_n, \text{not } B_1, \ldots, \text{not } B_m. \tag{1}$$

where H, as well as A_i, $i \in \{1, \ldots, n\}$ and B_j, $j \in \{1, \ldots, m\}$ are literals. In (1), $\{H\}$ is called the *head* of the rule, and $\{A_1, \ldots, A_n, B_1, \ldots, B_m\}$ is called the *body* of the rule. We refer to a set of literals X as *closed under* a positive program P, i.e., a program that contains no instance of not, if and only if for any rule $r \in P$, the head of r is contained in X whenever the body of r is a subset of X. The smallest of such sets wrt. a positive program P is denoted as $\mathsf{Cn}(P)$, and it is always uniquely defined. For an arbitrary program P, a set X is called an *answer set* of P if $X = \mathsf{Cn}(P^X)$ where $P^X = \{H \leftarrow A_1, \ldots, A_n \mid H \leftarrow A_1, \ldots, A_n, \text{not } B_1, \ldots, \text{not } B_m. \in P, \{B_1, \ldots, B_m\} \cap X = \emptyset\}$.

A rule with an empty body is referred to as a *fact*, a rule with an empty head is a *constraint*. It should be noted that the head of a rule does not necessarily consist of a single literal – some dialects of ASP allow for constructions such as a *choice rule*, a rule where the head comprises a set of literals of which basically any subset can be set to true. There is also the notion of *cardinality constraints* with lower and upper bounds. Such rules are of the form

$$l\{A_1, \ldots, A_n, \text{not } B_1, \ldots, \text{not } B_m\}u \tag{2}$$

The intuition behind this is that a cardinality constraint is only satisfied by an answer set if at least l and at most u of the literals $A_1, \ldots, A_n, B_1, \ldots, B_m$ are included in the answer set. Cardinality constraints can be used as body elements as well as heads of rules [5].

3 Measuring Contension Inconsistency Using ASP

In order to utilise ASP for measuring $\mathcal{I}_c$, i.e., to compute $\text{VALUE}_{\mathcal{I}_c}$, wrt. a knowledge base $\mathcal{K}$, we will encode the problem $\text{UPPER}_{\mathcal{I}_c}$ in ASP and then send calls to an ASP solver in an iterative manner. By using binary search on the search space of possible inconsistency values, only logarithmic many calls are required. More precisely, wrt. the contension inconsistency measure, the maximum inconsistency value corresponds to the number of atoms n. Thus, the starting point of the binary search is $n/2$.

As a first step in encoding $\text{UPPER}_{\mathcal{I}_c}$, we create three new propositional atoms $e_{x_T}, e_{x_B}, e_{x_F}$ for each atom x. Thus, the new atoms form a representation of the evaluation of the atom in three-valued logic. For the "guess" part of the ASP, at most u atoms $e_{x_B^i}$, $i \in \{1, \ldots, n\}$ are set to true. This can be modeled as a rule consisting of a cardinality constraint (as introduced in (2)): $0\{e_{x_B^1}, \ldots, e_{x_B^n}\}u$. where u is the upper bound we want to show.

For the "check" part of the ASP, we first need to model that for an atom x only one of its corresponding atoms e_{x_T}, e_{x_B}, e_{x_F} can be evaluated to true:

$$e_{x_T} \leftarrow \text{not } e_{x_B}, \text{not } e_{x_F}.,$$
$$e_{x_B} \leftarrow \text{not } e_{x_T}, \text{not } e_{x_F}.,$$
$$e_{x_F} \leftarrow \text{not } e_{x_B}, \text{not } e_{x_T}.$$

The formulas in $\mathcal{K}$ are comprised of the set of atoms At as well as the operators $\wedge$, $\vee$, and $\neg$. Hence, each operator must be encoded in ASP as well. More specifically, we construct rules that model the evaluation of the formulas $x \wedge y$, $x \vee y$, and $\neg x$ as follows (with new symbols $e_{...}$):

$$x \wedge y \mapsto \quad e_{x \wedge y_T} \leftarrow e_{x_T}, e_{y_T}., \quad e_{x \wedge y_F} \leftarrow e_{x_F}.,$$
$$e_{x \wedge y_F} \leftarrow e_{y_F}., \quad e_{x \wedge y_B} \leftarrow \text{not } e_{x \wedge y_F}, \text{not } e_{x \wedge y_T}.$$

$$x \vee y \mapsto \quad e_{x \vee y_F} \leftarrow e_{x_F}, e_{y_F}., \quad e_{x \vee y_T} \leftarrow e_{x_T}.,$$
$$e_{x \vee y_T} \leftarrow e_{y_T}., \quad e_{x \vee y_B} \leftarrow \text{not } e_{x \vee y_F}, \text{not } e_{x \vee y_T}.$$

$$\neg x \mapsto \quad e_{\neg x_B} \leftarrow e_{x_B}., \quad e_{\neg x_T} \leftarrow e_{x_F}.,$$
$$e_{\neg x_F} \leftarrow e_{x_T}.$$

More complex formulas can be reduced to these rules. Finally, we need to ensure that all formulas are evaluated either T or B. To achieve this, we add the integrity constraint $\leftarrow e_{\phi_F}$. for each formula ϕ.

Example 2. We continue Example 1. The ASP corresponding to $\mathcal{K}_1$ would contain the following rules:

1. Cardinality constraint: $0\{e_{a_B}, e_{b_B}, e_{c_B}\}2$. Here, we use 2 as the upper bound.
2. Ensure that each atom only gets one evaluation:

$$e_{a_T} \leftarrow \text{not } e_{a_B}, \text{not } e_{a_F}., \quad e_{a_B} \leftarrow \text{not } e_{a_T}, \text{not } e_{a_F}., \quad e_{a_F} \leftarrow \text{not } e_{a_B}, \text{not } e_{a_T}.,$$
$$e_{b_T} \leftarrow \text{not } e_{b_B}, \text{not } e_{b_F}., \quad e_{b_B} \leftarrow \text{not } e_{b_T}, \text{not } e_{b_F}., \quad e_{b_F} \leftarrow \text{not } e_{b_B}, \text{not } e_{b_T}.,$$
$$e_{c_T} \leftarrow \text{not } e_{c_B}, \text{not } e_{c_F}., \quad e_{c_B} \leftarrow \text{not } e_{c_T}, \text{not } e_{c_F}., \quad e_{c_F} \leftarrow \text{not } e_{c_B}, \text{not } e_{c_T}.$$

3. Encodings for all formulas:
   (a) $a \wedge b$:

$$e_{a \wedge b_T} \leftarrow e_{a_T}, e_{b_T}., \quad e_{a \wedge b_F} \leftarrow e_{a_F}., \quad e_{a \wedge b_F} \leftarrow e_{b_F}.,$$
$$e_{a \wedge b_B} \leftarrow \text{not } e_{a \wedge b_F}, \text{not } e_{a \wedge b_T}.$$

   (b) $\neg a \vee b$:

$$e_{\neg a \vee b_F} \leftarrow e_{\neg a_F}, e_{b_F}., \quad e_{\neg a \vee b_T} \leftarrow e_{\neg a_T}., \quad e_{\neg a \vee b_T} \leftarrow e_{b_T}.,$$
$$e_{\neg a \vee b_B} \leftarrow \text{not } e_{\neg a \vee b_F}, \text{not } e_{\neg a \vee b_T}.$$

   (c) $\neg b \wedge \neg c$:

$$e_{\neg b \wedge \neg c_T} \leftarrow e_{\neg b_T}, e_{\neg c_T}., \quad e_{\neg b \wedge \neg c_F} \leftarrow e_{\neg b_F}., \quad e_{\neg b \wedge \neg c_F} \leftarrow e_{\neg c_F}.,$$
$$e_{\neg b \wedge \neg c_B} \leftarrow \text{not } e_{\neg b \wedge \neg c_F}, \text{not } e_{\neg b \wedge \neg c_T}.$$

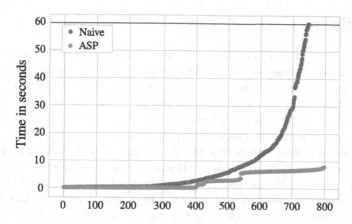

Fig. 1. Comparison of the (naive) state of the art implementation of the contension inconsistency measure and the (ASP) implementation of the algorithm proposed in this work. The red horizontal line visualises the timeout of 60 s. (Color figure online)

(d) Negations:

$$e_{\neg a_B} \leftarrow e_{a_B}., \ e_{\neg a_T} \leftarrow e_{a_F}., \ e_{\neg a_F} \leftarrow e_{a_T}.,$$
$$e_{\neg b_B} \leftarrow e_{b_B}., \ e_{\neg b_T} \leftarrow e_{b_F}., \ e_{\neg b_F} \leftarrow e_{b_T}.,$$
$$e_{\neg c_B} \leftarrow e_{c_B}., \ e_{\neg c_T} \leftarrow e_{c_F}., \ e_{\neg c_F} \leftarrow e_{c_T}.$$

4. Integrity constraints: $\leftarrow e_{a \wedge b_F}., \ \leftarrow e_{\neg a \vee b_F}., \ \leftarrow e_{\neg b \wedge \neg c_F}.$

4   Preliminary Experiments

The algorithm presented in the previous section is implemented in Java by use of the TweetyProject[2] library. The library already provides an implementation of the contension inconsistency measure that constitutes the state of the art.

In order to evaluate the proposed ASP algorithm, we compare its implementation with the naive one. We created a total of 800 random knowledge bases of different sizes and complexities. The knowledge bases are comprised of around 15-20 formulas which contain 0-10 connectors. To achieve this, we utilised a sampler (namely, the `SyntacticRandomSampler`[3]) provided by the TweetyProject. The generated knowledge bases are built on signatures that contain either 5, 10, or 15 propositional atoms. Then we applied both algorithms on each of these knowledge bases and measured the execution time. A timeout was set to 60 s. Figure 1 displays the measured execution time regarding each knowledge base, sorted from low to high wrt. both algorithms. Clearly, the ASP algorithm performs more efficiently. While applying the naive algorithm produced a timeout in 53 cases, the ASP implementation required only a maximum of 7.97 s to return the inconsistency value.

---

[2] http://tweetyproject.org/index.html.
[3] http://tweetyproject.org/api/1.14/net/sf/tweety/logics/pl/util/
SyntacticRandomSampler.html.

5 Conclusion

In this paper, we introduced an algorithm for calculating the contension inconsistency measure by means of reductions to answer set programming. By providing rules for encoding three-valued evaluations of propositional formulas in ASP rules, an inconsistency value can be retrieved using only logarithmic many calls to an answer set solver. In Sect. 4 we compared an implementation of a state of the art algorithm for calculating contension inconsistency with the proposed method. The evaluation shows that the ASP algorithm clearly outperforms the state of the art. This quite positive result leads to the conclusion that reductions to ASP are a reasonable method to approach problems in the field of inconsistency measurement. Consequently, it would be useful to explore the calculation of other inconsistency measures using reductions to ASP as well.

References

1. Atkinson, K., et al.: Toward artificial argumentation. AI Mag. **38**(3), 25–36 (2017)
2. Béziau, J.Y., Carnielli, W., Gabbay, D. (eds.): Handbook of Paraconsistency. College Publications, London (2007)
3. Brewka, G., Eiter, T., Truszczynski, M.: Answer set programming at a glance. Commun. ACM **54**(12), 92–103 (2011)
4. De Bona, G., Grant, J., Hunter, A., Konieczny, S.: Towards a unified framework for syntactic inconsistency measures. In: Proceedings of the AAAI 2018 (2018)
5. Gebser, M., Kaminski, R., Kaufmann, B., Schaub, T.: Answer set solving in practice. Synth. Lect. Artif. Intell. Mach. Learn. **6**(3), 1–238 (2012)
6. Grant, J.: Classifications for inconsistent theories. Notre Dame J. Formal Logic **19**(3), 435–444 (1978)
7. Grant, J., Hunter, A.: Measuring consistency gain and information loss in stepwise inconsistency resolution. In: Liu, W. (ed.) ECSQARU 2011. LNCS (LNAI), vol. 6717, pp. 362–373. Springer, Heidelberg (2011). https://doi.org/10.1007/978-3-642-22152-1_31
8. Grant, J., Martinez, M. (eds.): Measuring Inconsistency in Information. College Publications, London (2018)
9. Hansson, S.: A Textbook of Belief Dynamics. Kluwer Academic Publishers (2001)
10. Hunter, A., Konieczny, S.: Measuring inconsistency through minimal inconsistent sets. In: Proceedings of the KR 2008, pp. 358–366 (2008)
11. Priest, G.: Logic of paradox. J. Philos. Logic **8**, 219–241 (1979)
12. Reiter, R.: A logic for default reasoning. Artif. Intell. **13**(1–2), 81–132 (1980)
13. Thimm, M.: Measuring inconsistency with many-valued logics. Int. J. Approximate Reasoning **86**, 1–23 (2017)
14. Thimm, M.: On the evaluation of inconsistency measures. In: Measuring Inconsistency in Information. College Publications (2018)
15. Thimm, M., Wallner, J.: On the complexity of inconsistency measurement. Artif. Intell. **275**, 411–456 (2019)

Author Index

Printed in the United States
By Bookmasters